Foundations of Robotics

Damith Herath · David St-Onge
Editors

Foundations of Robotics

A Multidisciplinary Approach with Python
and ROS

Editors
Damith Herath
Collaborative Robotics Lab
Human Centred Technology Research
Centre
University of Canberra
Canberra, ACT, Australia

David St-Onge
Department of Mechanical Engineering
École de technologie supérieure
Montreal, QC, Canada

Kinova Inc., Quebec
This open-access book project has been fully funded by Kinova Robotics.
https://www.kinovarobotics.com/

ISBN 978-981-19-1982-4   ISBN 978-981-19-1983-1   (eBook)
https://doi.org/10.1007/978-981-19-1983-1

Cover art by Laurent Pinabel (CC-BY-NC-ND)

This Springer imprint is published by the registered company Springer Nature Singapore Pte Ltd.
The registered company address is: 152 Beach Road, #21-01/04 Gateway East, Singapore 189721, Singapore

Foreword by Ken Goldberg

Robots are Mirrors of Ourselves

> *Man is a robot with defects.*
> *Emile Cioran*

Robots will always be fascinating because they reflect our very human fears and hopes. Robots are a perennial subject for artists and writers, who often wish for an accessible introduction to understanding how they work. Robots are also of great interest to engineers, who often wish for an accessible introduction to understanding their context in history and culture. This book, edited by leading artists and engineers Damith Herath and David St-Onge, provides both.

The word "robot" emerged in 1920, shortly after the 1918 Pandemic. It was coined, by playwright Karel Capek, from a Czech word for hard work and central to his popular script about human-like machines rebelling against unfair working conditions. Although the word was new, the concept of human-like machines has a long history, dating back to Egyptian hydraulic machines, Pygmalian's sculpture in ancient Greece, the medieval Golem, the alchemists' automata, and Frankenstein.

In 1942, Isaac Asimov introduced three "Laws" of robotics. Osamu Tezuka's Astro Boy emerged in 1952, and artist Nam Jun Paik exhibited a series of sculptural robots in 1962. Countless robots have appeared in artworks, science fiction books, films, and television series.

Real robots are of great interest for application in industry, exploration, defence, healthcare, and service. Robotics research has a long history, dating back to Nikola Tesla's demonstration of a radio-controlled boat in 1898 and the emergence of "teler-obots" to handle radioactive materials during World War II. The IEEE Robotics and Automation Society held its first conference in 1984 (a significant year for several reasons), and there are now dozens of conferences and journals devoted to robotics research.

In 2022, during a huge resurgence of interest in AI and as the 2019 Pandemic begins to subside, robots continue to attract interest and maintain a strong hold on our collective imagination. Books, films, and newspapers promote sensational stories about human-like robots "stealing" jobs and making humans obsolete. Companies such as Tesla, Google, GM, and Toyota are actively working on autonomous driving. Flying drones are being used for cinematography, inspection, and surveillance. Robots that sort packages are being adopted to keep up with skyrocketing demand for e-commerce.

Although many artists and designers have worked with robots, almost all courses in robotics today are taught in engineering departments: in computer science or in engineering: electrical, mechanical, or industrial. As a result, current robotics textbooks are geared for engineers. They focus on mathematical models of coordinate frames, wrench mechanics, and control theory, assuming that readers have completed coursework in geometry, calculus, physics, and programming.

This book is different. It is written for students of all ages and backgrounds who want to learn about the broad fundamentals of robotics. This includes artists, designers, and writers who want to learn more about the technical workings of robots, and engineers who want to learn more about the cultural history of robots.

The book begins with a review of the rich history of robots. It then introduces chapters on teaching, designing, and programming, with details on the open access standard Robot Operating System (ROS) and a concise review of core mathematical concepts. The book then goes into the details of robot perception and actuation, with chapters on algorithms for robot control, motion planning, and manipulation. It also introduces active research topics such as bio-inspired robot design, human–robot interaction, ethics, and recent advances in robot learning.

This book provides the "foundation" for understanding how robots work.

It is the accessible introduction that artists and engineers have been waiting for.

February 2022

Ken Goldberg
William S. Floyd Jr. Distinguished
Chair in Engineering
UC Berkeley
Berkeley, CA, USA
http://goldberg.berkeley.edu

Foreword by Sue Keay

The world is changing. Robotics and robotics technology is becoming increasingly pervasive. We have robots in our homes, in the form of things like vacuum cleaner robots, and in many cases, we don't even notice their presence. For robots to become truly useful to humans, to understand us and operate in ways that make sense to us, and to be able to operate reliably and seamlessly in the cluttered, disorganised and unstructured world that we live in requires robot-builders that have a deep understanding of the complexity of not only technology but of the humans that use it and the complex environments that we inhabit. For this, traditional engineering-type learning is no longer sufficient. In the future, the pathway into a career in robotics is likely to be more complicated than via degrees in mechanical, electrical and mechatronics engineering, or computer science. A multidisciplinary approach, more human-centric design considerations, as well as pedagogy, safety, psychology, research design, and ethics is needed, all subjects of this text.

The authors have a very human-centred approach to robotics and a keen eye for how to incorporate arts, creativity, and the social sciences into this traditionally engineering-heavy field. They also combined this with a deep understanding of industry context, how to meaningfully apply robotics R&D to solve industry problems, and the importance of keeping human workers engaged in the process through the use of collaborative robotics. After leading the development of Australia's first national robotics roadmap, these are all themes that have come across very strongly both in case studies of the creation and use of robotics technology as well as in public consultations that I have been engaged in. The one burning issue for all robotics companies in Australia (and all around the world) is access to robotics talent, and that's where this book, Foundations of Robotics, plays an important role.

Foundation of Robotics provides the tools and building blocks necessary to train our next generation of robot technologists and equip them with a taste of the multidisciplinary considerations that are required to build modern robots. Importantly, the book also stresses the importance of diversity and culture, if we are to build robotic technologies that are truly representative of the communities within which they are used. In many cases, the robots will be used in industrial settings, and this is why the partnership with Kinova to develop this book is especially important to ensure that learning is related to industry best practices and that practical examples and exercises are given to students to consolidate their learnings.

I highly commend this book to you, whether you are a student of robotics, a teacher, an experienced researcher, a hobbyist, an enthusiast, or just an interested observer. Damith, David, and their team of contributing authors are leading the way in expanding the horizons of future roboticists and smoothing the path for more extensive deployment of robotics technologies, especially cobots, in the future.

February 2022

Dr. Sue Keay, FTSE
Chair, Board of Directors
Robotics Australia Group

Chair, Advisory Board
Australian Cobotics Centre

Preface

These are exciting times to be engaged in robotics!

Over the last couple of decades, we have had great fun building and programming some fascinating and interactive robots. Robotics is becoming pervasive, and robots are ever more in contact with ordinary humans away from research labs and manufacturing confines. However, as exciting as it may be, little has evolved in how robotics is taught at universities. Increasingly, this is becoming problematic as traditionally trained engineers are called to develop robots that could have an impact and interactions with the community at large. In our own practice, we have realised the evolving multidisciplinary nature of robotics. Recently, both of us have been developing new undergraduate programmes in robotics. While there are several exceptional textbooks that deal with various facets of robotics (books by such luminaries as Khatib, Siciliano, Thrun, Corke, Dudek comes to mind), we were at a loss in finding a comprehensive introductory textbook that touches on some critical elements of modern robotics that are usually omitted in traditional engineering programmes. Thus, the initial impetus came almost by necessity to develop a book that we can use in our courses that is true to our multidisciplinary backgrounds.

Traditionally, robotics is aligned with one of the following foundational disciplines, Mechanical, Electrical and Computer Engineering (and these have their roots in physics). Depending on the alignment of the department, the course you study will have a flavour that accentuates the particular alignment to the point that sometimes even the terminology will be different (e.g. a robot may be referred to as a cyber-physical system!). To complicate matters further, you will soon find out that roboticists possibly do not agree on a singular definition for what a robot is. What all these allude to is that robotics is still a young and emerging discipline (compared to its founding roots), and we must collectively develop and contribute to its body of knowledge in an inclusive and mindful way so as to embrace its ever-expanding disciplinary boundaries. The book you are holding is our contribution to the field. We believe that a foundational book in robotics should be broadly multidisciplinary yet grounded in the fundamentals essential to understanding the standard building blocks of robotics.

   In developing this book, we wanted to approach it not only from designing a robot from first principles firmly rooted in engineering but also from the point of view of the human element, present during the design process and throughout the robot's journey post-fabrication. We started by asking what should a modern foundational textbook in robotics look like, particularly tapping into our experience working in the worlds of robotic art and human–robot interaction research. The natural realisation was that this book requires collaboration at the highest level with colleagues from many disciplines. What you are about to read is a fresh new look at robotics based on our own interactions with students and colleagues tempered by a desire to present robotics in a more humanistic light.

   A second intention has been to make the material relevant to the industrial practice and accessible. We believe one of the unique aspects of the book is the industry expert interviews dotted throughout the book. They inspire and provide insider insights as to what goes into making real robots for real commercial applications. We hope you enjoy the little personal stories shared by various experts in the field. We are ever so grateful to be associated with Kinova in this aspect. The Kinova team provided helpful feedback throughout the book's development, providing insight into shaping the academic content of the book. The reader can be assured that the foundational concepts presented here will not be lost in the practical realities of working in the real world with real robots. We are also grateful to Kinova for funding[1] the project to publish the book as a Springer open access book. Considering the ever-increasing cost of student textbooks, we hope that free accessibility to this book provides many aspiring roboticists access to relevant academic material without hindrance. Modern robotics is also about entrepreneurship. We like to invite you to read the inspiring story behind Kinova's founding as narrated by Charles Deguire, the president and CEO of Kinova, embedded in the first chapter—we hope the book will ignite a spark of entrepreneurship in you!

The book is divided into three main parts.
We believe that robot design should be part of an ecosystem influenced by culture, contemporary thinking, and ancillary technologies of the day. Thus, the first part, *Contextual Design,* brings together an eclectic collection of ideas that will lay the contextual foundation on which the rest of the book is built. This part begins with a colourful historical perspective highlighting the mythological beginnings of robotics, its trends, and the importance of craft, arts, and creation in evolving modern robotics. We then explore the parallel pedagogical evolution in robotics. The second chapter highlights some of its missteps and approaches you can take to learn and teach robotics as a student or a teacher successfully. Next, the chapter on Design Thinking provides pointers to useful tools and ways of thinking in solving problems, robotic or otherwise. The final three chapters in this part provide introductory material on software, ROS—the Robot Operating System and mathematics, the ancillary *technologies* upon which modern robotics is being constructed.

---

[1] Although Kinova Robotics has generously funded this project, they have never interfered with the academic independence of the editors and the authors in developing the book.

The second part develops your understanding of the foundational technical domains: the *Embedded design*. We start with an introduction to sensors, actuators, and algorithms, the building blocks of a robot. The eighth and ninth chapters develop the key ideas relevant to mobile robots—robots that can move around in the world (think self-driving cars!). The tenth chapter is a deep dive into robot arms that enable them to manipulate the environment. Then we explore how to assemble a swarm of robots. Concepts and challenges in deploying multi-robot systems are discussed in detail. Finally, the part concludes with a chapter revolving around prototyping and discussing the embedded design process. Topics including 3D printing and computer-aided design are discussed in practical detail, giving you the confidence to understand how to combine theoretical knowledge with actual implementations of prototypes that allows you to build and test your robot designs.

While most industrial robots are still destined to be confined in isolated factory settings where human interaction is minimal, a paradigm shift is happening now in how we interact with robots. Increasingly, robots are being designed and deployed to be interactive and to be able to work with humans. The *Interaction Design* part explores the implications and some of the emerging new technical domains that underpin this (r)evolution. It is no longer enough to test your robots for their technical ability. They now need to be evaluated for their ability to work with or alongside humans. The first chapter in this part takes you through the emerging domain of human-robot interactions from a psychological perspective. It provides you with a thorough guide on developing user studies to test your hypotheses about robots interacting with humans with helpful case studies and statistical tools. Safety takes an elevated meaning in this new interactive world. The fourteenth chapter discusses the existing and emerging international safety standards related to various types of robots and robot deployments. It provides practical approaches and tools to deploy robots safely in interactive and collaborative settings.

The robots and techniques we discussed in Part II rely on clearly defined world models and constraints restricting their use to relatively simple environments or use cases. While these techniques have allowed us to deploy robots successfully in a wide variety of tasks, we are now starting to see their limitations. As you would imagine, the human world is highly complex. Such simplistic models are no longer adequate to deploy robots in natural human-centred interactive settings (think self-driving cars again!). The chapter on *Machine Learning* discusses some of the cutting-edge ideas being developed in robotics. These emerging ideas enable robots to operate in more complex worlds and to attempt complicated tasks (as humans do) successfully. As robots begin to interact with us in such complex ways, they can no longer be treated as mere tools. On the one side, they are increasingly becoming human-like, and on the other, they are increasingly permeating and challenging our way of life. As a robot designer, you now have a fundamental responsibility to think about the broader implications of your robot design. The final chapter on *Robot Ethics* is a systematic guide to help you navigate the robot design process with an ethical framework.

As detailed in the second chapter, no amount of theoretical work and instructions alone is sufficient to properly acquire the skills needed to design and deploy robots successfully. A hands-on, project-based approach is an essential pedagogical component in robotics. Therefore, the book includes two comprehensive projects that capture most of the theoretical elements covered in the book. In addition, we have included the necessary software and other resources needed to complete these projects on the companion website. We hope you make use of these resources to the fullest.

We have endeavoured to make each chapter relatively self-contained, so if you are after a specific topic, it is bound to be covered in its entirety within a single chapter. Each chapter has a section at the beginning that describes the key learning objectives and a summary at the end. This should enable you to identify a particular topic you are after quickly. The parts and the chapters are laid out in a way that you can also read them consecutively, building on from one to the other.

However you use it, we hope that you enjoy the book and be inspired by the truly interdisciplinary nature of the field.

Please visit the companion website of the book for teaching and learning resources, updates and errata at: https://foundations-of-robotics.org
Book's GitHub: https://github.com/Foundations-of-Robotics

Canberra, Australia   Damith Herath
Montreal, Canada   David St-Onge
January 2022

Acknowledgements

A book of this nature is simply the result of direct and indirect collaboration and support of many colleagues, students, family, and friends. So many have inspired and supported us along the journey to reach this point. Foremost, we want to thank all the contributing authors to the book. This is an enormous undertaking on their part, particularly during a pandemic. They have tirelessly worked around the clock to develop high-quality content within the brief time frame in which this project was set out. Our authors include early career researchers, graduate students, industry veterans, and senior academics from many disciplines. A genuinely multidisciplinary and multi-generational effort!

To maintain the academic integrity of the content and make sure the chapters are presented in the best possible way, we have enlisted the help of several colleagues both from academia and the industry to carefully review and provide feedback to the authors of the chapters. In particular, we thank Ilian Bonev, Matt Bower, Jacob Choi, Jenny L. Davis, Samira Ebrahimi-Kahou, Sabrina Jocelyn, Rami Khushaba, Sarath Kodagoda, Dominic Millar, Adel Sghaier, Bill Smart, and Elizabeth Williams for their contributions to the high-quality review of the chapters.

We want to acknowledge the efforts and extend our thanks to the student team which worked on putting the teaching labs together. The hexapod team includes Chris Lane, Bryce Cronin, Charles Raffaele, Dylan Morley, and Jed Hodson. The ROS mobile manipulator projects were built upon the efforts of Nerea Urrestilla Anguiozar, Rafael Gomes Braga, and Corentin Boucher. Without their tireless efforts, none of this would have been possible.

This book is possible and openly accessible, thanks to the trust and support of our industrial partner, Kinova Robotics. More specifically, we thank Jonathan Lussier and Jean Guilbault for jumping into the project early on and sharing their thoughts all through the production. In addition, Marc-André Brault and Maude Goulet managed the interviews within each chapter, an essential feature of the book. We also want to thank all the industrial experts featured in our interviews dotted throughout the book.

We appreciate the patience and support of the Springer Nature team for their tireless efforts to bring this book to life and into your hands.

And of course, we thank you, the readers of the book. Whether you are a robotics student or an academic adapting the text for your teaching, we hope this book inspires you to see robotics in a whole new light. We would love to hear your feedback, so please feel free to drop us an email or drop in for a cuppa if you are in our part of the world!

Finally, we wish to dedicate this book to our respective families, for they sustained us and endured the madness of being academics!

Contents

Editors and Contributors

About the Editors

Damith Herath Associate Professor, Collaborative Robotics Lab, University of Canberra, Australia.

Damith Herath is an Associate Professor in Robotics and Art at the University of Canberra. Damith is a multi-award winning entrepreneur and a roboticist with extensive experience leading multidisciplinary research teams on complex robotic integration, industrial and research projects for over two decades. He founded Australia's first collaborative robotics startup in 2011 and was named one of the most innovative young tech companies in Australia in 2014. Teams he led in 2015 and 2016 consecutively became finalists and, in 2016, a top-ten category winner in the coveted Amazon Robotics Challenge—an industry-focused competition amongst the robotics research elite. In addition, Damith has chaired several international workshops on Robots and Art and is the lead editor of the book *Robots and Art: Exploring an Unlikely Symbiosis*—the first significant work to feature leading roboticists and artists together in the field of Robotic Art. e-mail: Damith.Herath@Canberra.edu.au

David St-Onge Associate Professor, Department of Mechanical Engineering, ÉTS Montréal.

David St-Onge (Ph.D., Mech. Eng.) is an Associate Professor in the Mechanical Engineering Department at the École de technologie supérieure and director of the INIT Robots Lab (initrobots.ca). David's research focuses on human-swarm collaboration more specifically with respect to operators' cognitive load and motion-based interactions. He has over 10 years' experience in the field of interactive media (structure, automatization and sensing) as workshop production director and as R&D engineer. He is an active member of national clusters centered on human-robot interaction (REPARTI) and art-science collaborations (Hexagram). He participates in national training programs for highly qualified personnel for drone services (UTILI), as well as for the deployment of industrial cobots (CoRoM). He led the team effort to present the first large-scale symbiotic integration of robotic art at the IEEE International Conference on Robotics and Automation (ICRA 2019). e-mail: david.st-onge@etsmtl.ca

Contributors

Beltrame Giovanni Department of Computer and Software Engineering, Polytechnique Montréal, Montreal, Canada

Belzile Bruno Department of Mechanical Engineering, École de Technologie Supérieure, Montréal, Canada

Boucher Corentin Department of Mechanical Engineering, École de Technologie Supérieure, Montréal, Canada

Bouteiller Yann Department of Computer and Software Engineering, Polytechnique Montréal, Montreal, Canada

Cawthorne Dylan Unmanned Aerial Systems Center, University of Southern Denmark, Odense, Denmark

Grant Janie Discipline of Psychology, Faculty of Health, University of Canberra, Academic Fellow, Graduate Research, Canberra, Australia

Haskard Adam Bluerydge, Canberra, ACT, Australia

Herath Damith Collaborative Robotics Lab, University of Canberra, Canberra, Australia

Hinwood David University of Canberra, Bruce, Australia

Peng Fanke UniSA Creative, University of South Australia, Canberra, Australia

Petraki Eleni Faculty of Education, University of Canberra, Canberra, Australia

Pianca Eddi University of Canberra, Canberra, Australia

Reeves Nicolas School of Design, University of Quebec in Montreal, Montreal, Canada

Shukla Niranjan Accenture, Canberra, ACT, Australia

St-Onge David Department of Mechanical Engineering, École de Technologie Supérieure, Montréal, Canada

Stower Rebecca Department of Psychology, Université Vincennes-Paris 8, Saint-Denis, France

Varadharajan Vivek Shankar Department of Computer and Software Engineering, Polytechnique Montréal, Montreal, Canada

Wang Jiefei The School of Engineering and Information Technology, University of New South Wales, Canberra, Australia

Part I
Contextual Design

Chapter 1
Genealogy of Artificial Beings: From Ancient Automata to Modern Robotics

Nicolas Reeves and David St-Onge

Learning objectives

- To understand the mythological origins of contemporary robots and automata
- To be able to connect current trends in robotics to the history of artificial beings
- To understand the role of crafts, arts and creation in the evolution of contemporary robotics.

Introduction

This chapter is an extensive overview of the history of automata and robotics from the Hellenistic period, which saw the birth of science and technology, and during which lived the founders of modern engineering, to today. Contemporary robotics is actually a very young field. It was preceded by a 2000-years period in which highly sophisticated automata were built for very different purposes—to entertain, to impress or to amaze—at different times. You will see that the methods and techniques that were used to build these automata, and that largely contributed to the development of robotics, were at times imported from unexpected fields—astronomy, music, weaving, jewellery; and that the impulse that drove automata makers to build their artificial beings was far from rational, but rather rooted in the age-old mythical desire to simulate, and even to realize, an entity from inert materials.

N. Reeves (✉)
School of Design, University of Quebec in Montreal, Montreal, Canada
e-mail: reeves.nicolas@uqam.ca

D. St-Onge
Department of Mechanical Engineering, ÉTS Montréal, Montreal, Canada
e-mail: david.st-onge@etsmtl.ca

© The Author(s) 2022
D. Herath and D. St-Onge (eds.), *Foundations of Robotics*,
https://doi.org/10.1007/978-981-19-1983-1_1

1.1 What is a Robot?

Whereas most of us would think they know what a robot actually is, a closer look at the concept will show that a precise definition of the term is actually not that easy to frame; and that it broadened again in the last decades to encompass a large variety of devices. From the first appearance of the word in the Czech theatre play R.U.R (Capek, 2004), in which it was referring to human beings artificially created to become perfect and servile workers (Fig. 1.1), it is now used for a range of devices as different as robotic arms in factories, battery-operated toys for kids, androids or biomorphic machines. It even came to describe entities that lie at the boundary of technology and biology and that cannot anymore be described as fully artificial.

This evolution is less paradoxical than it seems. As opposed to a common idea, Capek's robots where not strictly speaking artificial machines: they were created with organic materials synthesized by chemical processes. In the scenario, the core of the project was to build teams of workers that were free from everything that was not essential to the implementation of their tasks—feelings, emotions, sensibility. Their role was that of robots, but they were still living biological organisms, which makes them quite different from the highly sophisticated technological devices that come to mind when thinking of contemporary robots. They were in a sense much more related to automata, a word that Capek has actually used in a previous play, but that was completely replaced by "robots" in this one.

This last point is worth noticing. At the time R.U.R. was written, the word "robot" was a neologism forged from slavish roots referring to work, chore, forced labour. Nothing in its original meaning implied that a robot should be a machine: an automaton created to help human beings in the implementation of some task

Fig. 1.1 Scene from Capek's play Rossum's universal robots, with three robots on the right; 1920

becomes a robot. In that respect, it actually contradicts the original meaning of the word "automaton", which, etymologically, refers to an animated device which acts *by itself*. The word decomposes in the Greek roots *auto*, which precisely means "by itself", whose origin, strangely enough, is unknown; and *matos*, "thinking" or "endowed with will", from the older proto-Indo-European * *mens*, "to think". It therefore designates an animated artificial being that is able to make decisions and to act autonomously, whereas "robot" is frequently associated with a machine that has been designed for the sole purpose of blindly executing sets of instructions crafted by a human being—the opposite of an autonomous entity. What is hardly disputable however is the fact that every contemporary robot finds its place in the age-old genealogy of automata. It might also be interesting to note that the oldest origins of the root "rob", from which "robot" was created, evokes the fact of being orphan, which corresponds surprisingly well to these artificial beings which, as a matter of fact, never had a biological father or a mother.

Since its first occurrence, the meaning and signification of "robot" have extended well beyond this gloomy etymology. A lot of robots are today created for research, experimentation or entertainment, without any practical use; but current roboticists do not yet agree on a single definition. Two elements however strike out as reaching a broad consensus: first, the device must present some form of *intelligence*; second, it must be embodied. As it is well known, "intelligence" is in itself a tricky notion to define. In this context, it does not indeed refer to human intelligence, less again does it correspond to the common perception of artificial intelligence, better represented by the concept of machine learning. Intelligence for a robot is only about taking decision on its own, based on the limited information it has from its context or from its internal states. Here again, etymology comes to help: the word comes from the Latin *inter ligere*—«to link between». The links can be elementary—a bumper sensing a wall makes the robot wheels stop—or more complex—the robot takes a decision by comparing information coming from multiple sources. The concept of embodiment refers to the physicality of the robot, as opposed to software "bots." On its side, «automaton» today refers indifferently to a hardware or a software device.

For the sake of the present chapter, we will tighten the meaning the word "robot" in order to encompass essentially hardware automata fulfilling two criteria: first, they must be dedicated to the autonomous implementation of DDD (Dangerous, Dull or Dirty) tasks, or to facilitate the implementation of such tasks for human beings; second, they must be able to take decisions through some form of interaction with their context. As we will see below, this definition itself has undergone several variations in the last decades, but we will keep it for the moment.

1.2 A Mythical Origin

The genealogy of robots, as well as the history of robotics, are then intimately linked to that of automata. An extensive recapitulation of this history would be far beyond the scope of this chapter, all the more since several books have already been written on it (among many others: Demson & Clason, 2020; Mayor, 2018; Nocks, 2008; Foulkes, 2017 …). However, an efficient way to understand the fundamentals of human motivation and fascination for robots and robotic systems is to recapitulate some of its main chapters, and to locate in time the bifurcations that progressively separated robots from automata: the evolution of historical trends in robotics is of the greatest help to understand why some aspects of robotic research are better known, and better developed, than others.

The first and likely most important point to consider is that the roots of robotics are not anchored in technology or science, but rather in a mythological ground that extends far beyond these fields, and that can be broadly divided in two layers. The first one is concerned by the myth of a being with supernatural power and unpredictable intentions, an image that still hovers over any robot or automaton. The second involves all the attempts that have been made along history to replicate through artificial mechanisms two natural phenomena that escape human understanding, namely life and cosmological events.

These two layers are intricated at many levels, but they differ by their basic intentions. The first one is most likely at the origin of all humanoid or animal-like automata. It led to the pursuit of creating artificial beings whose power surpass those of human beings: autonomous entities that can be made insensitive to pain, fear, boredom, and to any form of emotion, less again empathy. A lot of examples of inventors who try to build such entities can be found in tales, science-fiction stories, movies and video games, covering all the spectrum of intentions towards mankind—from help, assistance and protection to destruction and domination. However, once it is built, because it should possess, as an automaton, a kind of free will, it can become uncontrollable and behave in unpredictable ways, even for its creator. This is illustrated by Capek's play, but also by the wealth of works that has emanated from the Jewish myth of the Golem, first mentioned in the Talmudic literature. Being an artificial creature made of clay, the Golem was an embryonic form of life created for the sole purpose of helping or protecting his creator. It should be noted that historically speaking, Golem is most likely the first entity that corresponds to the above definition of a robot: an artificial entity built specifically for a implementing a practical task or function.

Despite the highly functional and technological nature of most contemporary robotic systems, the evolution of automata and the emergence of robots cannot be fully understood without realizing that most of them originate from the will to simulate life; that automata makers have been developed highly advanced skills, and have been spending tremendous amounts of time and resources, in order to achieve this goal with the highest possible precision; and last but not least, that in every automaton maker rests the secret and hidden dream of seeing one day his own inanimate creatures come to life—a dream to which, in a previous work, we gave the name «Geppetto syndrome» (Reeves, 1992).

From their very beginnings, automata were created to simulate. Their main—and often only—objective was to dissimulate what that they actually were: assemblages of inanimate matter pretending to act by themselves. It is not a coincidence if the first automata appeared at time during which a first, elementary understanding of physics was slowly emerging. Since only a tiny part of the population had access to it, its mastering was often perceived as magical by common people. Even if one of the main objectives of the new-born Greek science was to explain natural events by natural causes, that is, to get rid of supernatural explanation, its power could easily be confused with that of entities found in myths, tales or religious texts. Several works exploited it in order to create devices whose purpose was either to entertain, or to siderate crowds by simulating the intervention of supernatural forces. Automata built for practical purposes were virtually non-existent.

This was not always obvious. At first glance, the perpetual clepsydra built by Ctesibios from Alexandria (Fig. 1.2), of whom we will talk below, could claim to be a primordial robot, since it has the function of giving the time of the day. An ordinary clepsydra cannot be considered as an automaton: it is akin to that of an hourglass that uses water instead of sand, and as such, it does not feature any autonomous component. But Ctesibios' device, built three centuries BC, was coupled with a mechanism that refilled its tanks every day with water coming from a source, and that reconfigured its internal states in order to indicate the time for each day of the year. Being completely autonomous, it qualifies as an automaton. Since it

Fig. 1.2 Ctesibio's clepsydra, circa 250 B.C, as represented in the French translation of Vitruvius' treatise "ten books on architecture" by Claude Perrault (1864)

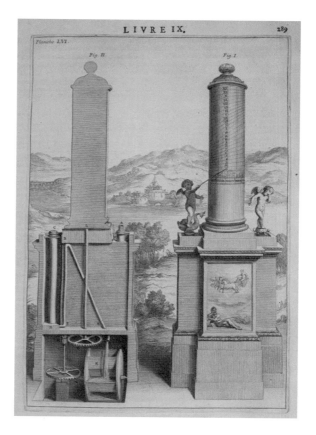

was built for a practical purpose, since it incorporates some kind of *intelligence* by reacting to the amount of water in its tanks, and since it was embodied, it could claim to be a first instantiation of a robot in the modern sense of the term. But this interpretation only holds when considering it with our contemporary eyes. Like most time-measuring devices, Ctesibios' clepsydra was more an astronomical model than a clock: it transposed the movements of the Sun into an autonomous mechanism. Just like humanoid or zoomorphic automata were trying to describe, comprehend and replicate the functions and behaviour of living beings, the first clocks, up to the beginning of the scientific revolution, were mainly planetary or cosmological models built to translate a partial understanding of celestial mechanics.[1] Vitruvius himself, while referring to Ctesibios' clepsydra in the 10th book of its treatise *De Architectura*, does not attribute to it any practical function. The design and building of such instruments usually requested workers and craftsmen that were the best skilled of their generation. The technological challenges implied by such mechanisms triggered the development of fully new technological and theoretical knowledge, and often requested massive amounts of money that could be provided only by the wealthiest members of the society. They became symbols of prestige, and testified for the level of expertise achieved in their country of origin. Even today, building a clock with a very long revolution period is everything but a simple venture. It took more than fifteen years to design and twelve years to build the astronomical clock located at the Copenhagen City Hall, completed in 1955; its slowest gear completes one full revolution in 25,735 years (Mortensen, 1957).[2]

All these examples, as well as many others, show that the impulse for creating automata is not originally driven by practical needs. It comes from the mythical desire to understand some of the deepest mechanisms at the origin of life and cosmological events, a desire that stands at the origin of major developments in mechanical science and in technology, and especially those at the origin of modern robotics. To qualify as an automaton, an artificial being does not need to be useful; it does not even need to move, or to do anything: it just has to be able to provide a convincing enough illusion of life (Reeves & St-Onge, 2016).

---

[1] In the first mechanical clocks, such as the one built by Richard of Wellington around 1330, the great astronomical orloj in Prague, or the very rare heliocentric clock at Olomouc, also in Czech Republic, counting the hours was only one of many different functions: the indication of time becomes almost anecdotical. Many other dials indicated the sidereal time, the signs of the Zodiac, the phases and position of the Moon, the movements of the Sun and of the Planets, the solstices and the equinoxes, the hours of the tides … Some needed several decades to accomplish a single revolution.

[2] Later devices, such as the eighteenth century Peacock clock in the Hermitage museum in Saint Petersburg, intimately associates the simulation of life with the measure of time (Zek et al., 2006). In this incredible piece of mechanics, once a week, a large peacock extends its wings, deploys its tail and moves its head; a rooster sings; an owl turns its head, blinks its eyes and rings a chime. A small dial, almost lost in the rest of the device, gives the time of the day: its presence is inconspicuous. The presence of time and the cycle of the days are mainly evoked by the three animals: the owl is a symbol of night, the rooster a symbol for the day, the peacock a symbol of rebirth.

An Industry Perspective

Charles Deguire, President and CEO

Kinova inc.

I like to think that I was born an entrepreneur. Both my parents were entrepreneurs, as some of our family members, and from the day I had to decide what I was going to become, I knew the path I wanted to follow. But as in every business case, you need THE idea. In my case, I was raised with the idea … When I was younger, I had three uncles living with muscular dystrophy, all power wheelchair users, and very limited upper-body mobility. The challenges they faced never stopped them, they even founded a private company dedicated to the transportation of people with special needs. This concept evolved to become the public-adapted transport system of Montreal.

One of my uncles, Jacques Forest, had only one finger that he could move. He was challenged by the idea to develop an arm that could be controlled by his active finger to allow him to become independent in his functionality and able to grasp and manipulate objects in his surroundings without external assistance. He generated various innovative technical ideas for such devices that were based on his own experience and intuition. The gripping device he succeeded to build was made from a desk lamp frame and ended by a hot dog pincer. The manipulator is built by every member of the family. It was put in motion by bicycle cables attached to windshield wipers motors that were assembled on plywood and located at the back of his wheelchair. Motors were activated through 14 electronic switches that he controlled through his unique moving finger.

While I was studying to become an engineer, I came across all kinds of new technologies that were all extraordinary. But I realized how having an astronaut doing remote manipulation with a space robot arm could be an aberration when people in wheelchairs could not even pour themselves a glass of water alone. As I was already aware of the reality of people living with physical disabilities, I decided I would dedicate my life to solving those problems, starting with a robotic assistive device built from the ground up, specifically for wheelchair users.

We move problems through a funnel. We start very wide, sort of chaotic. We look internally and externally, within our own industry and other industries, and ask, What process can I use to solve this? Once we've selected a few approaches that we believe have potential, we drill down and get really focused on executing each of them.

We robotize tasks. We did that for people using wheelchairs, expanding their reach. In surgery, we expand the capabilities of the surgeon. In hazardous material handling, we robotized the manipulation of toxic or nuclear waste. But it's always the same process, providing better tools to humans.

Creativity is one of Canada's greatest resources. This is what supports the growth of Kinova and which propels our Canadian manufacture to the international scope.

1.3   Early Automata

Ctesibios is considered as the founder of the Greek school of mechanics. After him, four characters stood up in the nascent field of automata around the Eastern part of the Mediterranean Sea: Philon of Byzantium; Vitruvius in Rome; Heron of Alexandria; and later, Ismail Al-Jaziri from Anatolia. By looking at a few examples of their work, we will see that working in the field of illusion and simulation did not prevent them to produce a major corpus of knowledge on the behaviour of real physical systems, to contribute with large instalments in their area and to leave technical writings that became major sources of inspiration for generations of engineers and scientists. The machines and automata they conceived are nothing less than technological wonders of their time.

Philon of Byzantium lived around the third century B.C. He left a number of treatises that give a very precise account of the technological level of his country. He invented an automated waitress that was serving wine and water, and that is generally considered as the first real humanoid robot in history. About three centuries later, mathematician and engineer Heron of Alexandria designed a series of about eighty mechanical devices, one of which being considered as the first steam machine, some others being moved by the sole force of the wind. Since his researches were unknown to Western scholars for more than a millennium, and since most his machines were destroyed during the fire of the Alexandria library, the count of his invention can only be estimated; many may never have been realized. None of them were dedicated to the implementation of practical tasks: he fostered his knowledge of physics and mathematics (mostly geometry) in order to impress or to trigger mythological fascination through mechanisms whose description can be find in his treatises *Automata*

Fig. 1.3 Drawings extracted from Chapuis, 1658 of devices made by Dionysus by Heron of Alexandria, first century A.D (left) and by Ismail Al-Jazari, 1206 (right)

(Murphy, 1995) and *Pneumatica* (Woodcroft, 1851). In what is known as the first example of building automation, the doors of a temple would open after a sacrifice only if the visitors ignited a fire in a receptacle; the fire heated a hidden water reservoir; the accumulated pressure caused a part of the water to be transferred in a second reservoir suspended to a cable and pulleys system attached to the doors; since this reservoir became progressively heavier, it began to go down, which caused the doors to open.[3]

Heron also designed a large animated sculpture of Dionysus (Fig. 1.3, left) in which water flowing from a reservoir to another triggered a sequence of actions: pouring "wine" (red-coloured water) from Dionysus' glass; pouring "milk" (white-coloured water) from his spear; rotating Dionysus central statue; rotating the statue of an angel over that of Dionysus; and finally pouring again wine and milk from opposite outputs. Some versions of the corresponding plans and diagrams include a group of dancers circling around the main statue, as well as a fire that was ignited automatically by a lighting device. Another of his treatise, *Dioptra* (Coulton, 2002), is

---

[3] This mechanism, as well as a number of the automata designed by Heron, have been reconstructed by the Kotsanas Museum of Ancient Greek Technology. They can be seen in function on a video produced by the Museum at http://kotsanas.com/gb/exh.php?exhibit=0301001 (accessed Dec 30 21).

key to modern roboticists, since it describes several instruments with practical aims, such as the measure of distances and angles. It includes the first odometer, a device that worked by counting the rotations of the wheels of a chariot. It was tailored to the Roman mile unit, which was obtained by adding 400 rotations of a 4-feet wheel; a series of gears slowly opened a hatch to release one pebble for each Roman mile. Such a device obviously does not qualify as a robot nor as an automaton; but the very idea of gathering information from the external world through a measuring device is key to modern robotics. It is worth noticing that such a device actually converts information coming from a continuous phenomenon—the rotation of a wheel—into a discrete one—the number of pebbles. Odometry is nowadays often computed from optical encoders fixed to motor wheels, but the measurement concept is similar to what Heron had imagined two thousand years ago.

Another of Heron's achievements is an automated puppet theatre. It represents an impressive example of the level of skills and technological knowledge that was put to use for the implementation of a device meant only for entertainment. It is also the first known historical example of a programmable mechanism: the movements of the puppets were controlled by wires and wheels whose movements followed a pre-recorded sequence. They were actuated by the movement of a weight suspended to a wire, just like for the German cuckoo clocks that appeared two millenia later. Any computer programs that is used today for about every imaginable task is a remote descendant of this machine that was built only to amuse or to surprise people. It is all the more stunning to realize that for centuries, the efforts put to work to achieve such a goal far exceeded those dedicated to the creation of practically useful robots, a situation that lasted up to the middle of the twentieth century; and that this energy has led to intellectual and technological achievements that sometimes did not find any other application for extended periods of time.

About ten centuries later, Ismail Al-Jazari, an engineer and mathematician living in Anatolia, fulfilled numerous contracts for different monarchs; he was hired to invent apparatuses aiming at impressing crowds during public parades (Fig. 1.3, right). By a clever use of hydraulics, levers and weight transfers, he designed several mechanisms whose parts would move autonomously. In his most famous treatise, *The Book of Knowledge of Ingenious Mechanical Devices* (Al-Jazari, 1974), he details systems ranging from a hydraulic alarm clock that generates a smooth flute sound to awake the owner after a timed nap, to a musical instrument based on cams that bumped into levers to trigger percussions. The cams could be modified in order to generate different percussive sequences, which constituted, ten centuries after Heron, another implementation of a programmable automaton.

It is to be noted that other devices, such as the Antikythera Machine, an astronomical calculator dated second century B.C. and whose inventor is unknown, has sometimes been regarded as an automaton; however, according to historians and scholars, it was operated by a crank, and thus does not meet the autonomy criterion. It remains nonetheless related to the first automata, and in particular to the first mechanical clocks that appeared almost fifteen years later, by the fact that it does represent, somewhat like a mechanical clock, a scaled model of a planetary system, executed with stunning precision and skills for the time.

1.4 Anatomical Analogies: Understanding Through Replication

1.4.1 Leonardo Da Vinci

It is impossible to recapitulate the history of automata without referring to Leonardo da Vinci (1452–1519). Some of the works of this visionary artist and inventor are also heavily grounded into the age-old mythological fascination for the simulation of human beings. In order to implement them, he explored extensively the anatomy and kinematics of the human body; but as it is well known, his work spanned about all the existing disciplines of his time. It would be difficult to say which of his endeavours had the greatest impact on modern-day arts and sciences. His inventions and practical treatises on mechanisms triggered and propelled the first industrial revolution that came more than three centuries later. Some of the pieces and assemblages he managed to manufacture thanks to his unique craftsmanship skills, such as gear heads and pulleys, are now mass-produced by complex industrial equipment, but they remain informed by the same design principles.

For roboticists, the inventions that are most related to contemporary projects are his mechanical knight on one side, and his self-propelled cart, also sometimes referred to as *Leonardo's Fiat*, most likely the first autonomous vehicle, on the other. The cart included a differential drive propulsion system with programmable steering for travel. The whole mechanism was originally seen as powered by wound up springs. In 1997, researchers understood that their real use was not to propel the cart, but to regulate its driving mechanism. In 2006, a first working replica, built at scale 1:3, was successfully made in Florence; all previous attempts have failed because of this misunderstanding (Gorvett, 2016).

The mechanical knight on its side is a complex machine (Fig. 1.4). It involves tens of pulleys and gears which allegedly allow him to sit, stand, move its arm and legs; it was however unable to walk. It is not until 2004 that a first prototype was implemented. It confirmed the possibility of all these actions, as well as several others: jaw actuation, neck rotation, visor movement. Way ahead of his time, while still rooted in the ancient mythology of artificial beings, Da Vinci's mechanical knight is connected to the very essence of the automaton. It stands as an ancestor to several recent humanoids, and its role in the original design of the NASA's Robonaut is said to have been influential.

It is not yet possible to account for all of Da Vinci's robotic endeavours, partly because many of them have been lost to history. Additionally, as previously stated, not all of his surviving designs are complete. In some cases, key components regarding machinery or function are missing; in others, as it was the case with his cart, some of his designs are simply to complex, and are not yet fully understood.

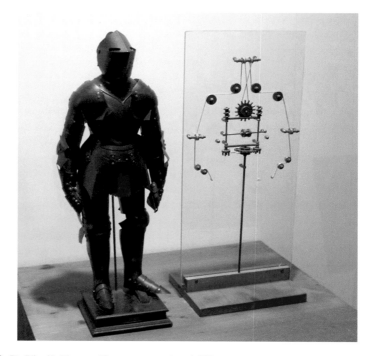

Fig. 1.4 Da Vinci's Humanoid automaton; circa 1495

1.4.2 *The* **Canard Digérateur,** *the Writer, the Musician and the Drawer*

As can be seen from these first examples, the will to simulate living beings is everywhere present in the history of automata and robots. All of these entities try to replicate the main characteristics of life, and to produce, deliberately or not, the illusion, that they managed to extricated themselves from the nothingness of inert matter. The efforts and energy invested to generate this illusion implied technologies that not only systematically accounted for the most advanced of its time, but also widely contributed to the evolution of these technologies. Beyond a simple simulation, the automaton was trying to reach the status of an *explanative device* endowed with descriptive virtues, making it possible to unveil the secrets of life. So it is with Vaucanson's duck, called the digesting duck (*canard digérateur*) by its inventor, built at the end of the seventeenth century (Fig. 1.5). As its inventor says (Vaucanson, 1738):

> This whole machine plays without you touching it when you set it up once. I forgot to tell you that the animal drinks, dabbles in water, croaks like the natural duck. Finally, I tried to make him do all the gestures according to those of the living animal, which I considered with attention.

Fig. 1.5 Vaucanson's "canard digérateur" (digesting duck), 1738. This picture is a fantasy reproduction published by the scientific American magazine (1899). Very few original pictures of Vaucanson's duck have been found

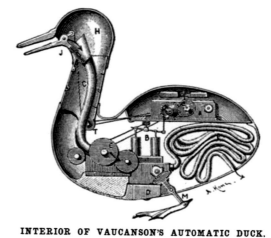

INTERIOR OF VAUCANSON'S AUTOMATIC DUCK.

A, clockwork; *B*, pump; *C*, mill for grinding grain; *F*, intestinal tube; *J*, bill; *H*, head; *M*, feet.

Later in the same text, Vaucanson mentions the most unexpected feature of his automaton, namely the fact that it digests and defecates:

There, in a small space, was built a small chemical laboratory, to break down the main integral parts, and to bring it out at will, by convolutions of pipes, at one end of its body quite opposite.

The simulation of the excretive function is clever: very few people would deliberately implement it for the sake of art or illusion. The very idea seems so unusual that it can only arise, for those who observe it, as a consequence of the will to create an entity that is to the perfect like of a living duck, including all its metabolic processes. One is at times left with the impression that the inventor surrenders to the illusion that the perfect formal simulation of the basic organs of life will fool life itself, so it will appear and animate the entity. The "small chemical laboratory" wants to be the equivalent of a digestive system, by which the food absorbed by the beak would be decomposed into nutritive substances on one side, and on useless substances evacuated through the cloaca on the other.

As can be expected, it was later revealed that Vaucanson's duck was a hoax. Nonetheless, the fact remains that following the Cartesian model, which sees the Universe moved by a great watchmaker, and living beings as nothing more than sophisticated mechanics, such attempts exemplify the tendency to systematically associate living organisms to the most advanced technologies of the time.[4]

---

[4] Interestingly enough, the idea of evoking life through its less prestigious functions finds a contemporary instantiation in his installation series «Cloaca» by Wim Delvoye (Regine, 2008), which reproduces all the phases of human digestion, from chewing to excretion, through successive chambers in which the food is processed by some of the enzymes, bacteria and biochemicals found in the digestive system. The installation must be fed twice a day. By observing the device in operation, it is easy to remain under the impression that the artist, helped by a team of biologists, has perfectly

They also mark the beginning of a slow bifurcation by which the evocation and simulation of life left the domain of formal analogy to join, by a long process, that of information flows and transfers. Here again, this separation was initiated by the model of human beings that prevailed at the end of the seventeenth century, a model that distinguished the body—the material component—from the soul—the driving and decision-making force. Descartes himself considered man as made from these two components. It is generally admitted that his model of the animal-machine (Descartes, 1637) was induced to him when he learned about the existence of a simple automaton, an idea later extended by La Mettrie's concept of man–machine (La Mettrie, 1748): this may be a glimpse on the process by which an object, initially built as a *simple formal simulation* of a given phenomenon, can become a model meant to *describe and explain* that phenomenon.

The concept of man and of animals as sophisticated mechanisms has led to the design of more and more sophisticated automata, with a gradual increase in the complexity of their functions. About a century after Descartes, the automata built by the Jacquet-Droz family initiated the separation between matter and information (Fig. 1.6). Not only were they driven by the equivalent of programs that were advanced versions of those created by Ctesibios and Heron of Alexandria, but the program themselves, recorded in rolls, cams or discs, could be changed, thus modifying the internal states of the automaton: they became independent of its material moving components. Changing the program opened spaces of possibilities that remained limited, but nonetheless real (Carrera et al., 1979). One of the given automata, the Musician Player, could play five pieces of music; the second one, the Drawer, could create four different drawings; the last one, the Writer, was the most complex. It can draw forty different characters; the text to be written is encoded on dented wheels, which makes it a fully programmable automaton. By looking at these delicate and impressive technological pieces, on can only regret the almost complete disappearance of automata arts since the nineteenth century. Fortunately, a few passionate artists still maintain this practice today; some of their most recent works, such as François Junod's *Fée Ondine*, are nothing less than jewellery pieces in movement. And as can be expected, Junod's studio is located close the Swiss town of La Chaux de Fond, the first town ever planned around the activities of the watchmakers.

---

grasped the mechanism of several vital functions; but here again there is an illusion, at least partial. The use of biological substances and living organisms such as bacteria prevents the device to meet the definition of an automaton, since it does not use only inert materials; it thus cannot claim to testify for a full understanding of the phenomena involved—which was not anyway the explicit intention of its author. However, despite all the explanations provided and in spite of the highly technological appearance of the work, the visitor cannot help feeling the presence of a strange animal plunged into the torpor of a heavy digestion, like a beast after a too copious meal; and to ask himself whether or not it presents a risk once awaken: the mere mention of the digestive function is enough for the visitor to readily accept the image of a living being.

Fig. 1.6 Jacquet-Droz's automata: drawer, musician, writer; 1767–1774

1.4.3 Babbage and the Computer-Robot Schism

The bifurcation that made the automaton and its controlling program two distinct entities cannot be located at a single moment in time. As we have seen, it can be traced back to the devices created by early Greek engineers and to Al-Jazari's percussive automaton; but several other steps intervened since; and the trajectory leading to contemporary computer programming has taken an unexpected detour through music and textiles. Seventy-five years after the Jacquet-Droz's automata, Henri Lecoultre created a musical box in which the melodies were recorded on interchangeable rolls. Barrel organs, which first appeared during the sixteenth century, could play melodies that were pre-recorded on rolls, discs, cards or ribbons perforated with holes that determined the melodies to be played—such instruments were actually called *automatophones.*

This principle was almost immediately transposed to create the first Jacquard loom by Basile Bouchon, the son of an organ maker, and by his assistant Jean-Baptiste Falcon (Fig. 1.7); they adapted musical boxes mechanisms from his manufacture to create the card readers that controlled the patterns to be woven (Eymard, 1863). It is worth noticing that the Jacquard loom also used the cylinder developed by Vaucanson, in another illustration that the technologies required to implement machines with practical uses often originated from the artistic realm, where they were developed with completely different motivations.

The perforated card system lasted for more than two centuries. It was extensively used for the programming of the first generations of computers. It played an essential role in the Manhattan project during which the first atomic bomb was created,

Fig. 1.7 A Jacquard loom, 1801

establishing an odd and peculiar connection between the delicacy of the melody played by a musical box, the patterns on a cotton fabric and the thundering apocalypse of a nuclear explosion. It was also by observing the Jacquard loom that Charles Babbage had the idea to design his *Analytical Engine*, today considered as the first full computer in history (Fig. 1.8). This huge machine included all the main elements of a modern computer: an input device that separated data and instructions, thanks to two punch card readers; a mechanical "driver" that prepared and organized the data for processing; a "mill", made of hundreds of gears that performed the operations—the mechanical equivalent of a CPU; a "memory" which stored intermediate and final results; and an output device in the form of a printer.

The Analytical Engine was never completed, due to problems of financing and manufacturing precision. It however remains, along with the Jacquard loom, the first example of a device that fully and completely separates the flow of information from the material processing unit. It is also remarkable for another reason: Ada Lovelace, the daughter of the poet Byron, was fascinated by mathematics. She wrote for the Babbage's machine the first known mathematical algorithm, a sequence of instructions for computing Bernoulli numbers,[5] which makes her the first programmer in history. Her clairvoyance and insights were actually nothing less than visionary. She

---

[5] Bernoulli numbers, named from Swiss mathematician Jakob Bernoulli, were identified in 1713 during the study of sums of powers of integers. If $S_m(n)$ represents the sum of the n first integers individually raised to power m, then the value of this sum is given by:

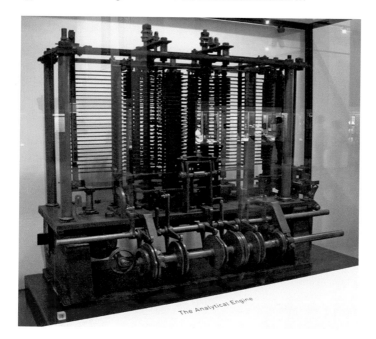

Fig. 1.8 Uncompleted prototype of Babbage's analytical engine, exhibited at the London museum of science; 1871

foresaw the possibility for such devices to perform not only numerical operations, but also symbolic calculations, and to use them to associate letters and signs in order to produce results that had nothing to do with mathematics, such as the composition of musical pieces, in another loop that reconnected the machine with its musical box origins (Lovelace, 1843). This is also probably the first known evocation of a form of artificial creativity, a characteristic which, perhaps more than for many other automata, testifies for the impulse to bring machines closer to human beings: art at that time was seen as the prerogative of the human species, an idea still largely preponderant today. The question of the relations between arts, robots and automata will be discussed more in detail in Sect. 6.2.

$$S_m(n) = \frac{1}{m+1} \sum_{k=0}^{m} \binom{m+1}{k} B_k n^{m-k+1}$$

In which coefficients B^k are Bernoulli number. They can be obtained through the following generating function:

$$\frac{x}{e^x-1} = \sum_{k=0}^{\infty} \frac{B_k x^k}{k!}$$

Ada Lovelace's algorithm was derived from this function.

1.5   Industrial (R)evolutions

Technological progress took a new pace over the course of the last two centuries, as the Western world underwent what we refer to as the "industrial revolutions"; the plural form is used here because at least four revolutions have been identified (Marr, 2016). The first major change intervened as a result of the use of steam and water to generate power. The second corresponded to the emergence of mass production and division of labour, and to the discovery of electricity as a power source. The third took place at the end of the sixties, with automated production and the exponential development of computing and electronics. The fourth can also be called the «digital revolution», and stands as a result of a merging of technologies that broke down the limits between the digital, physical, and biological spheres. The field of application of this last revolution is often referred to as "Industry 4.0."

The first revolution is a direct implementation, at larger scales, of several early contributions that were mentioned above, but it was also grounded on many works by Leonardo Da Vinci. It is only in the third one that the first robotic systems (industrial automata) were widely adopted, though it seems obvious that the second revolution paved the way for it. This third revolution exploited the discoveries and breakthroughs made by several inventors, among which Nikola Tesla is certainly not the least. It is during the third revolution that the lineage of robots branched from the main trunk of the genealogy of automata. For the first time, artificial entities endowed with a degree of autonomy were put to work, becoming nothing less than automated servants or slaves insensitive to fatigue, not vulnerable to health hazards, and hopefully more robust and durable than human workers. The fourth revolution will not be discussed in this book, as history is still being written on the impacts of the changes that it brought, but it will be referred to in the last section of this chapter.

The rise of industrial robots during the twentieth century required several scientific breakthroughs in power (electric, pneumatic, hydraulic), power transport and tele-operation (remote control). Nikola Tesla [1856–1943] was an engineer and inventor who referred to himself as a "discoverer". He solved most of the requirements and constraints needed by the third industrial revolution, and stands out, with about three hundred patents, as of the most proficient inventor of his time. He is widely known for his contributions on electricity transport and alternative current. These works had obviously a major impact on robotics; but we will focus here on his contributions to the use of radio waves.

In November 1898, Tesla demonstrated that a small autonomous boat could be remotely operated, from distances up to several feet (Fig. 1.9). The instructions were sent by coded pulses of electromagnetic waves. On demand of his audience, he instructed the ship to turn left or right, or to stop. This was the first demonstration of a remotely operated vehicle. It was not a robot in the full sense of the term, but it was, according to its inventor, "borrowing the mind" of the human operator so that future, advanced versions could fulfil mission together. A handful of patents, such as the one on advanced "individualized" (protected) multi-band wireless transmission, followed this demonstration; another one concerned the first "AND" circuit, a device that combined two radio frequencies to minimize the risks of interferences.

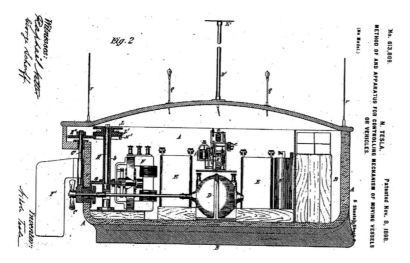

Fig. 1.9 Nikola Tesla radio-operated vessel plan from his US patent 613809Al, 1898

Tesla's boat would be hardly more than a toy today; at the time, he was nothing less than the forerunner of all remotely controlled devices and systems. He tried to write a list all of its potential application. By reading it, one cannot help to find him a little bit optimistic about the consequences of the military ones:

> Vessels or vehicles of any suitable kind may be used, as life, dispatch, or pilot boats or the like, or for carrying letters, packages, provisions, instruments, objects, or materials of any description, for establishing communication with inaccessible regions and exploring the conditions existing in the same, for killing or capturing whales or other animals of the sea, and for many other scientific, engineering, or commercial purposes; but the greatest value of my invention will result from its effect upon warfare and armaments, for by reason of its certain and unlimited destructiveness it will tend to bring about and maintain permanent peace among nations.

As for every new technological breakthrough, the militaries were quick to foresee the uses they could make for this invention. They massively funded the research on related technologies and quickly deployed remotely operated equipment in operation fields—without, as could be expected, helping in any noticeable way the pacification of conflict areas. As he foresaw at the time, the most advanced robotics research ventures and developments are still funded by the military industry, which is still the first to deploy these new technologies. From a mythological and largely poetic origin, robotics became within a few decades a field in which sophisticated war machines were developed.

Still, while most of the works done in this domain are not publicly available, some initiatives do contribute to the general advancement of the field. Nowadays, the United States Defence Advanced Research Projects Agency (DARPA) is hosting several robotics challenges: autonomous vehicle races (2004–2007), humanoid emergency response (2012–2015), heterogenous robotics swarms' tactics (2019–2022)

Fig. 1.10 Unimate robotic
arm deployed at general
electrics facility to handle
pick-and-place of heavy
parts; 1963

and subterranean exploration (2018–2021). Some of the competitors of these challenges are funded millions of dollars by the DARPA to push the boundaries of their research.

If we go back to the industrial realm, mass production in the third revolution has resulted in a lot of repetitive tasks in manufacturing processes. Most of them were perfectly fit for simple robust automation: the sixties welcomed the first industrial robotic arm, the Unimate (Fig. 1.10), designed by Georges Evol. Even if some early version of digital switches (vacuum tubes) and digital encoders were commercially available at the time, none of the off-the-shelf parts would fit his design, so every single component of the first set of Unimates was specifically manufactured for it. It was deployed at General Motors in 1961, and was the object of the first on-site study for market, integration, ease of use and safety of industrial robots.

Several lessons were learned from it; two of them proved essential. The first one is that robot obsolescence is likely to strike well before utter wear-out. It led to the conclusion that the life of an industrial robot depends on its robustness (ability to hold together) as well as on its versatility (ability to evolve and to adapt to new jobs). The second one relates to the fact that the complexity of a robot is so high that it becomes difficult to guarantee its reliability, a criterium that depends on the owner programming skills, on the production system into which the robot is integrated, and on the quality of its maintenance. It is however important to note that, after the Unimate was used for about a decade, several owners agreed that the financial benefits of replacing human workers with it were not significant, but they still wanted to go along with it because it kept their workers away from industrial accidents and health hazards.

The Unimate featured up to six axes, one of them prismatic (translation), and a payload of 225 kg. The first one was sold at a loss, but after six years, the company,

Unimation Inc., started to do profits; it later changed its name for Staubli. Others then joined the market, such as ASEA with its IRB series. The first commercialized IRB, the IRB6, had five axes and a payload of six kilograms. ASEA focused on the ease of integration of its product, whose overall mass was 112 kg, and whose integrated control electronics, including its DC actuators, was fully integrated within the enclosure. It then merged with Brown, Boveri & Cie to become ABB, competing with Staubli to become one of the main robotic arms manufacturers in the world.[6]

1.6 Modern Robotics

During the last decades, while the industry was trying, through several attempts and test sites, to robotize manufacturing processes, tremendous progresses on robotic systems design, kinematics, sensing and control were achieved. The corpus of knowledge on advanced robotic systems resulting from these breakthroughs constitute the fundamentals of modern robotics, a field that explores the possibility to deploy reliable robots in unknown dynamic environments. One of the most important phenomena of this period is certainly the progressive convergence between biological and robotic systems that can be observed since the end the 70's, during which the age-old attempts of simulating life through formal analogies gave place to new experiments that tried to reproduce the dynamic aspects of biological processes.

1.6.1 Coping with the Unknown

Managing complex tasks or missions autonomously in unknown, changing contexts requires a high level of performance in perception, decision-making and agile

---

[6] It may be worth noticing that Staubli and Brown Boveri are both Swiss enterprises; Swiss is widely recognized as the country where watchmaking was born. It is for more than four centuries the country in which the research and development of mechanical clocks of all scales remains the most active in the world. It can legitimately be assumed that the unique expertise thus developed in the field of micro-mechanisms was essential for the development of robotics, leading to the emergence of a cutting-edge robotic industry. What is less known is that this situation originated, rather paradoxically, from religious concerns: when Swiss became a protestant country after the Reform, in the sixteenth century, Calvin banned the wearing of all ornamental objects. Goldsmiths and jewelers had to find another way to use their skills. They applied them to the realization of watches and clocks, which, because they had a practical function that could be used as an alibi, could become miniature artworks and allow people to wear expensive devices that looked like jewels without incurring the wrath of the church. Watchmakers established themselves in several cities, most of them being located in an area called the «Jurassian Arc», not far from France, in the very area where the Jacquet-Droz family built its famous automata. Here again, by a strange detour, expertise coming from an artistic realm—jewelry—becomes the historical origin of one of the most important developments in robotics and in the robotic industry.

motion control, all elements that can be observed in a wide variety of configurations and biological strategies in nature; this is one of the main reasons why living systems quickly became a source of inspiration for roboticists. Among the first fully autonomous robots are a handful of prototypes realized in 1948 by William Grey Walter, a neurophysiologist fascinated by the complexity of emerging behaviours manifested by simple biological systems. He was convinced of the possibility to transpose such strategies in the field of robotics by using elementary devices. In order to prove his hypothesis, he designed a wheeled robot of the steering tricycle type which was able to detect light directly through a frontal photodiode sensor, without any programming (Fig. 1.11, right). It was then instructed by simple electronic logics to actuate the wheels in order to head towards the strongest light source in its environment. This very simple instruction led to an emerging behaviour—a behaviour that was not planned nor programmed, by which it could autonomously avoid obstacles; emerging behaviours represent one of the essential characteristics of living beings. When the battery level was getting low, the robots behaviour switched in order for it to seek the darkest spot around, as if it was trying to burrow in its lair. The protective shell over Walter's robots, as well as their slow velocity, led people to christen them turtles, or tortoises. The latter name was kept by their creator, most likely because, as mentioned in Alice in Wonderlands, tortoise are wise teachers.

Interestingly enough, the relations of roboticist with turtles extended far beyond Walter's prototypes. In the sixties, a new teaching approach, called Logo, was developed. It was based on recent cognition and learning researches and implemented into programming languages. One implementation made Logo history: it consisted in a method to teach the basics of procedural thinking and programming to children. Kids would learn either by instructing a turtle icon to move on the screen of a computer monitor, or a turtle-like robot to move on the floor. Logo remained one of

Fig. 1.11   (left) Stanford Shakey robot, circa 1960; (right) Walter Tortoise (1948–1949)

the only toolsets for the teaching of procedural programming and thinking until the late nineties, in primary schools as well as in high schools.

Walter's tortoises inspired a great deal of other robotic works. Twenty years later, in 2010, two employees from Willow Garage, Tully Foote and Melonee Wise, started working on the newly released Microsoft Kinect camera to integrate it with an iRobot Create platform.[7] The result was an affordable, easy to use robot, perfectly fit for teaching and training, to which they gave the name «TurtleBot». Its popularity is closely intertwined with the one of the Robotic Operating System, or ROS (discussed in Chap. 5). One of the most important conclusions of these experiments is that platforms with heavy limitations on sensing abilities and processing power can develop complex behaviours that mimic those of insects (ants, bees, termites, etc.), birds or fishes; and in particular those of animal societies in which groups of individuals can implement complex tasks that are out or reach of a single element. This paved the way for the field of swarm robotics, discussed in Chap. 11.

Since they were using light as their only source of information, the artificial tortoises became very sensitive to the calibration of their sensor, as well as to their context; they required a very controlled environment to perform adequately. A first step in exploring unknown contexts was accomplished by a Stanford-designed robot named Shakey (1966–1972). Shakey (Fig. 1.11, left) was the first robot able to reason about its own actions: it could make decisions based on the combination of inputs from several sensors in order to fulfil a given task (explore, push an object, go to a location …). The platform itself consisted in a differential drive actuated vehicle equipped with cameras, range finders, encoders and bump detectors. Its "brain" computer was a SDS 940 the size of a room, with which it communicated over a radio link. Shakey vision system was able to detect and track baseboards, which allowed it to navigate in its large playground. Working with Shakey allowed the researchers to produce essential contributions, such as the A* path planning algorithm and the visibility graph, both introduced in Chap. 8, as well as the Hough transform in computer vision.

Right after Shakey, Stanford contribution to modern robotics continued with another autonomous vehicle, called the Stanford Cart (1973–1979). Originally designed to mimic a lunar rover operated from Earth, which implies a 2.6 s delay in the transmissions of instructions, it quickly became obvious that such a setup had only two options to choose between: move really slowly, or make the steering and navigation autonomous. To detect obstacles, the Cart was equipped with the first stereovision system (3D imagery). To plan safely its path, it would take a fifteen minutes break and scan its surrounding after each metre travelled. In 1979, using this strategy, it managed to cross autonomously in five hours a twenty-metre room filled with chairs, without any collision.

---

[7] The iRobot Create comes from the same manufacturer that today sells the Roomba vacuum cleaner robots.

These robots, as well many others that we could have presented here, constitute major milestones in the recent history of technology. As opposed to most of the automata from which they descend, they have the possibility to move by themselves, and to adjust their internal sets and behaviour according to the data coming from their sensors—an elementary form of exteroception. They directly lead to the current state of research and development in self-driving vehicles and drones. Altogether, they pave the way to service robots outside of the industrial realm that are able to cope with challenging dynamic unknown environments.

1.6.2 Robots in Arts and Research–Creation

As anyone may guess, research in robotics is an extremely active field. What is less known is that robotic arts are also very dynamics. As for many technological developments, it didn't take long for artists to take hold of the new knowledge, methods and tools coming from this rapidly expanding field. This should not come as a surprise since, as we have seen, automata of all kinds have always maintained a close relationship with arts. Whereas scientists and engineers were, and still are, concerned on *how* robots should be built, artists, as well as researchers from human sciences, ask the question of *why* they should be developed. Many of their works invite us to evaluate the risks, stakes and potential linked to the emergence of more and more sophisticated machines. As you will see, the border between research and creation in robotics can be very porous, and sometimes completely blurred. Within the recent field of research–creation that lies precisely at the intersection of arts, science and technology, are conceived robotic works that trigger the production of new knowledge and new technological developments in these three domains.

Just like automata arts, robotic arts do not produce robots with practical purposes. They nonetheless managed to trigger a wealth of developments and breakthroughs in mechanics as well as in mechatronics and programming. The impulse that drives them presents no major differences from the one that drove the Jacquet-Droz family to build his Writer or his Musician, or Vaucanson to build his duck, by using some of the most advanced techniques of their time. Furthermore, the often-quoted leitmotiv stating that the first robotic artists were *playing* with their contraptions, instead of *working* to make them useful, should be seen as a positive statement rather than a deprecating one: research in any field is first and foremost a ludic activity, driven by the curiosity and desire for exploration that are inherent to the human nature.

Artists cannot rely, like university researchers, on established research infrastructures; nor do they have access to the same level of human and material resources. But as a counterpart, they have a freedom of research and action that would not be possible in an institutional environment. Not being limited by any calendar constraint, research trend or industrial need, robotic artists are free to explore unexpected research tracks. Not being incited by their peers to work at the edge of technology, they can investigate the potential of low-tech devices with personal sets of motivations, which adds to the specificity and unicity of their work. This has two consequences. First, major results have been obtained by people with limited technological expertise and very limited means and resources, at times verging on *arte povera*, demonstrating, if necessary, that essential breakthroughs can be achieved from elementary devices.[8] Second, the association of artists with university researchers, or with industrial partners, is likely to produce results that could not be possible for artists or researchers alone.

A quick look at artworks from the domain shows that robotic arts are essentially of hybrid nature. From 1920 on, artificial humanoids began to appear in theatre plays and performances. They were most of the time remotely controlled, and thus had no degree of autonomy. It is now commonly accepted that art robots should supprimer be able to interact in some ways with the audience, or with its environment, so that their behaviour can change according to the context in which they are presented.

The very concept of interaction is actually related to a potential dialogue with an artificial being. The occurrence of this dialogue depends on the elements that are used by the robot to communicate, which is why robotic arts have also played an important role in the development of intuitive human–robot interfaces. The first computer-controlled robotic art piece was the Senster (Benthall, 1971). It was equipped with an interface that gave him a pseudo-human behaviour, in the sense that it was attracted towards soft movements and sounds, but repelled by sudden gestures and loud noises. The range and level of technologies that were used to implement it (microphones, Doppler radars, hydraulic rams, plus an 8 K memory P9201 computer from Phillips, whose price at the time exceeded that of a three bedrooms apartment in London) made it impossible to afford by an independent artist; it was actually commissioned by the Philips company.

The Senster, who looked somewhat like a three-legged, four metres giraffe whose movements were derived from that of a lobster's arm, can be seen as pioneering the field of research–creation: its main objective was artistic, but its implementation required a collaboration with experts from several fields and disciplines. Since it was sensitive to the general ambiance of its context, it was able to trigger emotions in the people that interacted with it. It looked like worried when the environment became too agitated or too noisy, which incited people to act so as to make it "feel better" or "more worried". This empathic attitude can be observed in many later works that were designed precisely to trigger it. The Hysterical Machines family by Bill Vorn were octopus-like mechanical robots hanging from the ceiling. When the visitors came too close to them, they become extremely agitated, even showing signs of panics through rapid light effects and frantic movements of their metal tentacles. In front of such reactions, most of the viewers felt sorry for them and were incited to walk back to calm them down (Vorn, 2010). The Aerostabiles project by Reeves and St-Onge consisted in large robotic cubes levitating in wide internal spaces (Reeves & St-Onge, 2016). They could remain still in the air thanks to sensors, actuators and ducted fans (Fig. 1.12). A micro-computer permanently readjusted their position, producing slow oscillations. Despite their high-tech appearance, far remote from

---

[8] This is obviously not limited to artists, as shown by Walter's tortoises, which are among the simplest robots that can be imagined; or by a software automaton like the Life Game by John Conway, a quasi-elementary system that triggered the birth and evolution nothing less than artificial

Fig. 1.12 Three Aerostabiles, flying cubic automata by Reeves and Saint-Onge (Moscow, 2010)

that of any living being, they managed to trigger intense emotions, since their very soft movements were interpreted as a form of hesitation, or breathing; they were seen by some visitors as large, floating animals that were prisoners in some way of their technological envelope.

This connexion between the artificial movements of a robot and the emotions felt by the visitor is of outmost importance on three points. First, it demonstrates again, if needed, that the essence and potential of any automaton lie in its simulation abilities. Second, it shows that, even for living beings, powerful impressions and emotions can be communicated even while considering only the formal components of movements, displacements and gestures. Third, as a consequence of the second point, it opens the possibility to develop strictly formal or mechanical vocabularies for triggering and controlling human impressions and emotions, with all the risks and potentialities that such a project implies.

Several other aspects of early automata can be observed in robotic artworks. The puppet theatre built by Heron of Alexandria finds contemporary counterparts in Szajner's "The Owl and the Robot" or "Petit Nicolas", two interactive, theatrical computer-controlled automata scenes; in Vorn and Demers' "No Man's Land", which involved more than fifty robots of nine different *species* detecting the presence of viewers and reacting to it (Demers & Vorn, 1995); or in Rogers' «Ballet Robotique», a movie showing large industrial robots choreographed so as to evoke animals or

---

life, a new science that has since the 60's produced a wealth of theoretical and technological results in several disciplines.

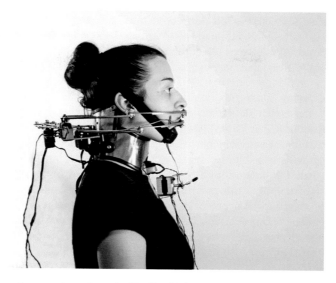

Fig. 1.13 Human speaker experiment by artist Nataliya Petkova, 2017

plants. "The Robotic Church" by Chico McMurphy involves forty different robots that play their individual sound-producing sequence (McMurphie, 1989).[9]

The level of interaction in these pieces is rather elementary, but they still demonstrate the importance for robotic artists of attempting a dialogue between the robots and the viewers; or at least, to trigger a reaction or an emotion from the latter. The next step consisted in conceiving works in which human and robots would act together in installation or performance scenes, trying to maximize the integration and the collaboration between human and robotic performers. Among the pioneers of such projects, Stelarc stands out as the first artist to have experimented robots as advanced prosthesis of his own body. In his seminal piece "The Third Hand", he tried to control a robotic arm affixed to his right forearm through his own muscular impulses, in order to make it write the same thing as his right hand (Stelarc, 1981). He also designed pieces in which he reversed the mutual roles of the human and of the robots: in his "Ping Body" piece, distant viewers located in three different cities could trigger his body movements through a muscular stimulation device (Stelarc, 1995). A less known but maybe more radical piece, "The Human Speaker Experiment" (Fig. 1.13), presents a performer whose tongue, throat, cheeks and lips are actuated by mechanic and electric devices, so as to allow a computer to make her pronounce different words and sounds (St-Onge et al., 2017). Such installations convert the body into passive objects whose only role is to follow the instructions sent by the computer, like human

---

[9] McMurphie's installation strangely evokes a famous low-tech automated piece from the outsider arts category, "Le manège de Petit Pierre", a life-size mechanical fair created and built by Pierre Avezard, a handicapped farm boy, and which differs from more sophisticated automata theatres only by the precarity of its materials (Piquemal & Merlin, 2005).

interfaces. Just like the self-destructive multi-machine performances in the 80's by the Survival Research Laboratories (Pauline, Heckert and Werner), they convey strikingly powerful messages about the risks linked to the expansion of robotic devices in our daily lives, and the possibility for them to escape all human control (Ballet, 2019).

One cannot evoke robotic arts without mentioning another category of pieces, namely those that deliberately try to give inanimate objects the appearance of life. "Robotic Chair" by Max Dean is an ordinary looking chair that disassemble and reassembles autonomously (Gérin, 2008); Boursier-Mougenot's Grand Pianos slowly move in an exhibition space, sometimes bumping into each other (Bianchini & Quinz, 2016); Mike Phillips Sloth Bot is a white abstract prism, several metres high, which imperceptibly moves in the atrium of a public building, getting closer and closer from groups of people who end up noticing his ominous presence and quickly getting out of its way (Phillips, 2007). Paul Granjon's sexed robots live in an enclosure called the "Robotarium", in which their only concern and objective is to mate with each other. They are also inspired by Walter tortoises in several ways; for instance, when low in battery, they seek the darkest spot as their nest (Pitrus, 2013). Such works are often infused with a dose of humour, which does not prevent them to carry strong statements about the potential futures of robotics, and the necessity for us to carefully evaluate the risks involved in some specific development axis.

Other artists propose works that directly address these notions, by entering active discussions and controversies surrounding the research and development of killer robots. The ethical problems raised by such machines are nothing less than overwhelming. In a 2021 piece called Spot's Rampage, Brooklyn collective MSCHF has purchased one of the famous yellow dog-shaped robots from Boston Dynamics, which used to be displayed playing and jumping on videos that became viral. They equipped it with a paint gun and offered to anyone the possibility to pilot it online, so as to make more concrete the possibility of armed police robots wandering in the street of large cities (MSCHF, 2019).

One common point of the works mentioned in the present section is that they can hardly be relegated to a single domain: all of them are nurtured by data and information coming from the three fields of art, science and technology. They are inherently trans-disciplinary—some authors even qualify them as post-disciplinary, since a robotic artist can navigate between theatre, performance, music, video, installation, sculpture, bio-arts, visual arts, and many others, producing equally valuable works in each of these fields. They can be characterized by the fact that they constantly cross boundaries between all fields and domains, and by the way they manage to thrive on these boundaries. Just like former automata makers, researchers-creators in robotic arts are dedicated to the creation of artworks which constitute their final objective. Just like them, through the process of conceiving and implementing them, they develop advanced new skills, expertise and knowledge that can be then transferred to several other fields; and just like them, the mechanisms they imagine can be seen as models for some hidden or ill-explained aspects of reality, and help understand these aspects. And last but not least, the interactive nature of most robotic artworks directly connects with the age-old impulse to create works that simulate features of living beings.

1.7 Social Robotics

While several challenges still need to be addressed for robots to be able to robustly navigate any cluttered terrain, vacuum robots and robotic pets are getting common in household. Robots in our daily routine can have a significant impact on our lives and no enough study were conducted yet on this topic. However, opening the door to psychology, education and sociology over the past decades of research in robotics also contributed to promote robots as potential good artificial companions (Fig. 1.14). A handful of companies hit that market with innovative products, but very few succeeded, in a surprising contradiction to the success of AI start-ups. The often-quoted refrain in the industry is that "robotics is hard."

If you think engaging Alexa or Siri in a natural conversation is difficult, just try building a robotic humanoid that can function in any capacity similar to a human. Simply put, initiatives in social robots such as Rethink Robotics, Jibo, Nao and Mayfield Robotics helped to grow and spawn an industry only to find that more nimble competitors, in the shape of robotic assistants with no mobile components,

Fig. 1.14 A small pack of Nao's humanoid robots from Jaume I University. Nao is one of the most popular robotic platform for human–robot interaction in psychology and it has made its place in the child education market

outcompeted them. For whatever reasons, the venture investors determined that these market forces were more important than any longer-term vision that the robotics company had and decided not to continue funding it. Anki CEO and co-founder Boris Sofman gives a clue of the reasons behind that state of affair:

> You cannot sell a robot for $800 or $1000 that has capabilities of less than an Alexa.

Roboticist Guy Hoffman adds:

> When designers will start their own social robotics companies and hire engineers, rather than the other way around, we will finally discover what the hidden need for home robots was in the first place.[10]

This does not mean that social robots have no role to play whatsoever. Many things that are not directly connected with robotics as such can be learned from each of these experiments. Jibo, for instance, is a major case study for the first large-scale human grief and mourning for robotic systems, with hundreds of owners sharing their distress and psychological state after its end of life was announced. There is obviously a major field of research, centred around the emotions that can be triggered by an artificial being, to be investigated here. As we have seen, the field of robotic arts has been considering and exploring these phenomena for several decades now. It is not unreasonable to suppose that joint research–creation ventures involving human scientists, psychologists, artists and engineers will be ideally equipped, theoretically and technologically, to address these questions.

1.8   Robotic Futures and Transrobotics

Throughout this chapter, we encountered several examples of a sequence in which an entity that evokes more or less precisely the shape of living beings induces the creation of more sophisticated devices intended to bring this evocation to the level of a similitude, then to an assimilation, then to a model: the representation becomes the paradigm. The same situation reoccurred at the procedural level with computers and computer science, where it stands at the origin of a new model of human beings in which the antique separation between body and soul is transposed, through an immediate formal analogy, into a separation between matter and information.

Before exploring the consequences of this model, it should be noticed that the mechanistic paradigm of the human body readily led to the resurgence of another primordial myth, through the hope that immortality was at reach. The idea of the body as a machine implicitly supposes the independence of its various components and the possibility of remedying the failure of an organ by the transplant of an identical one, or by the implantation of a prosthesis. From there was born the vision of human beings gradually transformed into robots through the progressive replacement

---

[10] Stalker and stalked: What Killed Off Jibo, Kuri and Kozmo? in Asian Robotics Review 273, https://asianroboticsreview.com/home273-html (accessed January 30, 2022).

of their biological, ephemeral components by artificial ones; and whose longevity becomes considerable thanks to the use of unalterable materials, such as titanium, gold, palladium. Moreover, such hybrid beings would progressively become able to wander in extremely hostile environments, such as deep space or ocean abysses, and even of surviving intergalactic journeys, indefinitely pushing the borders of territories colonizable by mankind.

Today we know that no material is eternal. A stable element such as gold that can remain unchanged for billions of years, but this remains very far from eternity; no robot can last forever. Information however has no prescribed age limit. Since it can be transposed from one material entity to another, it can theoretically last as long as the Universe itself, which is as close to eternity as it is possible to be. Analogies with certain properties of living matter readily come to mind: if we look today at the fossil of a fern in a museum, we know immediately what we are looking at, because we have seen living ferns quite often in our lives. But this fossil is 300 million years old: the fact that we are able to identify it means that the information that controlled its morphology has remain unchanged since it was living—it lasted longer than the highest mountain ranges of the late Paleozoic. We are thus led to the conclusion that life is the optimal process that Nature has found to preserve information, and to allow it to travel towards the future: being immaterial, it can jump from one individual to its offspring when the materiality of the parent degrades. This life-inspired strategy led to the emergence of a particular class of automata, on which will now focus our attention, and which tries to embed three characteristics of living beings: self-building and healing; replication; evolution.

Automata with such abilities are still in their infancy: they are mainly found in university or industrial labs. However, the development of miniaturized mechatronic components and of new materials, as well as the availability of cheap and powerful microcontrollers, allow to foresee their use for practical applications in a not-so-far future.

Several examples of self-building or self-reconfigurable robotic structures have been proposed in the two last decades. One of the first examples consists in basic cubic "bricks" equipped with an arm on each of their faces. Sets of such cubes can built cubic lattices with various topologies: the cubes can carry one another from one node of the lattice to the next (Yoshida et al., 2003). These modular robots were rather heavy and cumbersome, but they prepare the grounds from miniaturized systems in which such "bricks" could become the basic cells of robots with advanced functions; moreover, a robot built this way could theoretically self-disassemble and reconfigure in a completely different one in order to perform different tasks. Such devices may seem very upstream of potential applications; but their potential is so promising that they are the object of intensive researches in several labs. Many designs have been tried, such as the two-hemispheres ATRON robot by Modular (Jorgensen et al., 2004), chain structured systems such as the CEBOT (Fukuda & Kawauchi, 1990) made of three different cells (wheel mobile, rotation joint, bending joint), Yim's Polypod, made of segments and joints (Yim et al., 1995), truss structured systems such as Hamlin's Tetrobots (Hamlin & Sanderson 2012) or Ramchurn's Ortho-Bot (2006), which remains at the state of a concept.

Self-healing robots are also the object of a lot of attention from researchers. They can be broadly divided into two categories. The first one consists in mechanical robots that are able to repair their own components by using tools that are integrated in their structure, such as the PR2 robot configured at Tokyo University (Murooka et al., 2019). Such robots would theoretically be able to fix themselves after a failure, like a surgeon that performs surgery on himself. The second one includes robots that are made of soft materials («softbots») that self-reconstruct after having accidentally been damaged or ripped by a collision, like a biological skin (Guo et al., 2020).

Replication and evolution on their side are not independent processes. In both cases, the robot must carry the information that represents itself, in order to transmit it to a device that could built an identical copy of itself. This device could be a separated piece of equipment; but in order to stay closer to the analogy with living processes, which can be deemed optimal since they have been elaborating and fine-tuning through billions of years of evolution, the robot itself should be able to produce its own replicas. Directly evolving a physical robot is out of reach of current technologies; it should however be mentioned that the first evolution of digital organisms has been observed in 1984 on a cellular automaton (Langton, 2000; Salzberg & Sayama, 2004), where it appeared, surprisingly enough, as an emergent feature of the system. A cellular automaton is not a robot; but the fact that an evolution process can take place in the memories of a computer means that generations of physical robots, progressively more adapted to a given task, can be successively built along its course.

Lipson and Pollack's Golem project[11] was specifically aimed to create robots with specific performance specifications, without any human input at the level of design (Lipson & Pollack 2000). Their morphology resulted from a digital evolution process whose results were evaluated and selected through computer analysis, simulation and optimization, before reaching a final shape. The only human intervention consisted in affixing the actuators on the various components. The final product was an articulated worm equipped with a triangular arm; it was able to crawl on different surfaces. Such a result may seem disappointing as compared to the sophistication of the process; but this opinion can be relativized when considering that it took hundreds of millions of years to biological life to reach the same result on Earth. Moreover, a close look at the evolution diagrams reveals striking analogies with biological evolution: both underwent stable phases, where they seemed not to be able to produce new proposals or *species*, followed by phases where the number of such proposals literally exploded. Knowing that a computer can evolve robots much faster, and maybe more efficiently, than biological evolution, reminds us that the field of robotics faces us with unlimited possibilities, but also with risks that must be carefully considered for each of its new development axis and trends.

We will end up this chapter with a small tale that will briefly take us back to the first age of automata. Thousand years ago, a craftsman created a human-like automaton for the Chinese emperor. It was so realistic, and behaving so humanly, that it almost became a star in the emperor's court. Everyone wanted to be seen with him. He behaved very elegantly, and with exquisite politeness towards everyone,

---

[11] Note the mythological reference!

especially young women, with whom it even happened to engage in some form of flirt. Unfortunately for him, he made the error of flirting with the emperor's favourite spouse. The wrath of the emperor was terrible; he feared that the automaton and his wife could become lovers; he ordered the automaton to be executed, which was done immediately. The automaton has made the error of entering a territory which was exclusively reserved to the emperor, namely that of its succession, threatening the perpetuation of his life and heritage.

Today, the situation is completely reversed. Despite all the mythical worries associated with such as project, building an automaton or a robot that could reproduce itself and evolve by following lifelike processes is an objective that is looked for rather than feared; the first team to accomplish such a feat would be immediately acclaimed at the international level. This is illustrated by a very recent project by Kriegman and Bongard (Kriegman et al., 2021), in which small entities made of skin cells of frogs are dubbed «biological robots» in the media, a name that looks like a contradiction in itself, but that translates the perplexity of contemporary commentators in front of such researches.[12] These microscopic entities are able to replicate themselves, not by regular cellular division (mitosis), but by assembling other cells freely floating in their environment—the new ones are biological constructions, rather than offspring of biological «parents».

Most of the robots we know today are dedicated to practical tasks. In that respect, one can wonder to which extensive research about bio-inspired robots, lifelike robots or biorobots so remote from our daily concerns can be relevant. The answer lies in two points. The first one is the observation of the optimal efficiency of living processes for about all imaginable tasks, and the hope that this efficiency can be one day transposed in artificial entities. The second one is linked to the fact that after thousands of years of evolution, the most advanced researches on automata and robots remain deeply connected with the myths and fears that led to the creation of the very first ones, thousands of years ago. As shown by Kriegman and Bongard's experiments, the convergence with living beings, once seen as an illusory attempt, is now stronger than ever; and the meaning of the term «robot», as well of that of the suffix «bot», has expanded far beyond its original significance. New knowledge about biological and genetic processes led to the emergence of life-inspired automata and robots, which in turn ended up bringing new knowledge and models for some of these processes.

It is still too early to know which of these attempts will become successful, and which ones will remain as milestones in the ongoing genealogy of automata; but we can legitimately suppose that the future of robotics lies in a more and more pronounced convergence between artificial and biological entities at all scales, from a whole organism to cells and molecules; and that we will soon see hybrid robots involving more and more biological or life-inspired components going out of the lab to enter industrial and domestic environments.

---

[12] The authors gave the name «Xenobots» to their creatures.

Chapter Summary

After examining the difficulties linked to the precise definition of the word "robot", the mythical origin of all robots and automata was exposed; it was regularly reminded in the following sections. Early automata built by the Greek founders of mechanics, namely Ctesibios, Philon of Tarentum, Heron of Alexandria, were described. They were followed by the presentation of works from the Renaissance to the Classical Age, in which automata that tried to simulate life and life processes by replicating as precisely as possible the form and/or anatomy of living beings. From there, the genealogy of robotics bifurcated. A new branch appeared, in which the machine and the information controlling it became fully separated. It can be seen as the origin of modern computers and robots. Some early automata from Antiquity could be programmed to modify their behaviour, through different mechanisms; but surprisingly enough, the ancestors of modern programming are to be found in musical boxes and in the Jacquard loom. They also led to Babbage's Analytical Engine (1843), the first device that featured all the components of a modern computer, for which was written the very first algorithm. The industrial era saw an almost complete loss of interest for automata. The expressed needs of large-scale manufacturing paved the way to the implementation of the first industrial robots. It was simultaneously realized that mobile robots, able to cope with unknown, changing environments, could find a wealth of potential applications in several fields. Robots became more and more autonomous; their sensors became more and more efficient; computer and mechatronics equipment became smaller and smaller. Robots could begin to take decisions on their own by comparing information from different sources and by using processes inspired from biological organisms. From there, a marked convergence was established, and is still going on, between artificial and natural beings. Some of the latest robots developed in research labs use materials and strategies coming both from biology and technology. Their potential, as well as the interest they raise, allows to see them as harbingers of the next phases of robotics. The possible applications of such machines are impressive, and we can legitimately be fascinated by such technological achievements. But we must also consider the risks raised by the introduction, in our daily life as in industry, of autonomous artificial entities increasingly close to living beings, and whose abilities and power expand almost exponentially with time.

Revisions Questions

1. *How do the arts contribute to the development of robotics?*
   There may be one or more correct answers, please choose them all:

   A. *By allowing studies to be carried out free from the laboratory environment and the constraints of research*
   B. *By prioritizing aesthetic considerations*
   C. *By making researchers popular*
   D. *By questioning the present and future emotional implications of technologies*

2. Which of the following historical figures is recognized for having produced the first programmable automaton?
3. Who is known to have written the first machine algorithm?
4. Can you identify the main impulse(s) that drove the first automata makers to build their works

   There may be one or more correct answers, please choose them all:

   A. *To demonstrate their knowledge and skills*
   B. *To simulate and/or replicate living organisms*
   C. *To help understand phenomena such as life or celestial mechanics through explicative models*
   D. *To impress, entertain or amaze crowds*

Further Reading

Demson M, Clason C R (2020)

Romantic Automata: Exhibitions, Figures, Organisms. Bucknell University Press

A brilliant collection of essays, most of them based on the eighteenth century literature about robots and automata, which describe the contradictory feelings that emerged in the Romantic times, when it was realized that the construction of life-like artificial entities, once seen as a technological achievement, could actually lead to the emergence of mechanical beings deprived of the qualities that are inherent to humanity such as empathy and compassion. A source of fascination and entertainment for centuries, automata began, in a short period of time, to trigger less positive emotions such as dread and fear, in a pivotal moment that is cleverly apprehended by the authors.

Herath D, Kroos C, Stelarc (2016)

Robots and Arts: Exploring an Unlikely Symbiosis. Springer, Berlin

A pioneer book about robotic arts of all kinds, Robots and Arts presents, through some of the most emblematic projects of the field, a thorough and in-depth reflexion about the role, status and future of robots in a world where these artificial beings are progressively becoming daily companions and partners. It constitutes an eloquent demonstration of the essential contribution of artists to the general discourse on the evolution of technologies. The argument is elaborated through a trans-disciplinary compendium of texts by artists, scientists and engineers. Though this is not the main objective of the book, the different contributions also make the case for the importance of research–creation, by showing the wide number of disciplines, expertises and skills that are required to produce even simple robotic art pieces, and the necessity to promote such fruitful collaborations in university labs as well as in artists' studios and technological research centres.

Foulkes N (2017)

Automata. Xavier Barral, Paris

The epic history of automata, from the oldest to the most recent, in a book profusely illustrated with documents from all periods. The intimate links of automata with clocks watchmaking, their parallel evolution with that of technologies, their links with magic and myths, are clearly exposed, as well as the different roles they have occupied throughout history, in several regions of the world. A well-argumented book that can be used as an introduction as well as a reference.

Mayor A (2018)

Gods and Robots: Myths, Machines and Ancient Dreams of Technology. Princeton University Press

A historical account of the links between the fantastic characters of the earliest myths in history, recorded in Crete, the Roman Empire, Greece, India and China, and the first instantiations of these artificial beings and mechanisms that are the ancestors of all robots and automata that have been designed since. Perhaps one of the clearest evocations of the origin of automata, all born from this obsession to breathe life into inanimate beings, and an irrefutable demonstration of the fundamental role of art, imagination and legends for the greatest scientific and technical developments.

Nocks L (2008)

The Robot: The Life Story of a Technology. John Hopkins, Baltimore

A history of robots mainly centred on the technological aspects of the field. The argument remains generally more factual than for the previous ones and gives less prominence to the non-technological roots of automata; but it takes on its full importance in the light of several elements which will serve as a useful reference: a glossary, a timeline, an abundant bibliography, as well as information on the state of research and development of contemporary robotics through statistics on currently operating laboratories, firms and companies.

Wilson S (2003)

Information Arts : Intersections of Art, Science, and Technology. MIT Press, Cambridge

A reference book in the field, Information Arts presents itself as the first international survey of these artists who prefigured the development of research–creation by exploiting data and concepts from all scientific fields, as well as the results of a large number of technological advances. Soundly argumented from a theoretical point of view, this essential work, based among other things on the visual and bibliographical analysis of major artistic approaches, shows here again that the artist does not limit himself to staging these concepts and their developments: through the positions it takes in front of their social and cultural consequences, it participates in the discourse on their evolution and becomes a full player in the determination of future research programs.

References

Al-Jazari, A. (1974). *The book of knowledge of ingenious mechanical devices* (D. R. Hill Trans.). D. Reidel.

Ballet, N. (2019). Survival research laboratories: A dystopian industrial performance art. *Arts, 8*(1), 17. https://doi.org/10.3390/arts8010017

Benthall, J. (1971, November). Edward Inhatowicz's senster. In *Studio International* (p. 174).

Bianchini, S., & Quinz, E. (2016). *Behavioral objects | A case study: Céleste Boursier-Mougenot.* MIT Press.

Capek, K. (2004). R.U.R., Penguin Classics (tr. C. Novack-Jones).

Carrera, L., Loiseau, D., & Roux, O. (1979). *Les automates des Jacquet-Droz.* Sciptar—F.M. Ricci

Coulton, J. J. (2002). The dioptra of Heron of Alexandria. In L. Wolpert, J. Tuplin, & T. E. Rihl (eds.), *Science and mathematics in ancient Greek culture,* Oxford University Press (pp. 150–164)

Demers, L. P., & Vorn, B. (1995). Real artificial life as an immersive media. In *5th Biennial Symposium on Arts and Technology* (pp. 190–203).

Demson, M., & Clason, C. R. (2020). *Romantic automata: Exhibitions, figures.* Bucknell University Press.

Descartes, R. (1637). *Discourse on the method of rightly conducting the reason, and seeking truth in the sciences,* part V. Project Gutenberg, https://gutenberg.org/files/59/59-h/59-h.htm#part5. Accessed 31 Dec 2021.

Eymard, P. (1863). *Historique du métier Jacquard.* Imprimerie de Barret.

Foulkes, N. (2017). *Automata.* Xavier Barral.

Fukuda, T., & Kawauchi, Y. (1990). Cellular robotic system (CEBOT) as one of the realization of self-organizing intelligent universal manipulator. *Proceedings of the IEEE International Conference on Robotics and Automation, 1,* 662–667. https://doi.org/10.1109/ROBOT.1990.126059

Gérin, A. (2008). The robotic chair: Entropy and sustainability. *Espace Sculpture, 83,* 40–40.

Gorvett, Z. (2016). *Leonardo da Vinci's lessons in design genius,* BBC Future, https://www.bbc.com/future/article/20160727-leon

Guo, H., Tan, Y. J., & Chen, G. et al. (2020). Artificially innervated self-healing foams as synthetic piezo-impedance sensor skins. *Nature Communication, 11,* 5747. https://doi.org/10.1038/s41467-020-19531-0. Accessed 30 Dec 2021.

Hamlin, G. J., & Sandersen, A. C. (2012). *Tetrobot: A modular approach to reconfigurable parallel robotics.* Springer Verlag.

Herath, D., Kroos, C., & Stelarc. (2016). Robots and arts: Exploring an unlikely symbiosis. Springer.

Jorgensen, M. W., Ostergaard, E. H., & Lund, H. H. (2004). Modular ATRON: Modules for a self-reconfigurable robot. In *2004 IEEE/RSJ International Conference on Intelligent Robots and Systems (IROS)* (Vol. 2, pp. 2068–2073). https://doi.org/10.1109/IROS.2004.1389702

Kriegman, S., Blackiston, D., Levin, M., & Bongard, J. (2021). Kinematic self-replication in reconfigurable organisms. *PNAS, 118*(49), e211267211. https://doi.org/10.1073/pnas.2112672118. Accessed 30 Dec 2021.

La Mettrie, J. O. (1748). *L'homme machine.* Elie Luzac Fils.

Langton, C. G. (2000). Evolving physical creatures. In M.A. Bedeau, J.S. McCaskill, N.H. Packard, & S. Rasmussen (eds.), *Artificial Life VII: Proceedings of the Seventh Artificial Life Conference* (pp. 282–287). MIT Press.

Lovelace, A. (1843). Notes on Luigi Menabrea's paper, autograph letter to Charles Babbage. Add MS 37192 folios 362v–363, British Library.

Marr, B. (2016). Why everyone must get ready for the 4th industrial revolution. https://www.forbes.com/sites/bernardmarr/2016/04/05/why-everyone-must-get-ready-for-4th-industrial-revolution/?sh=6849f19d3f90

Mayor, A. (2018). *Gods and robots: Myths.* Princeton University Press.

McMurphy, C. (1989). *The robotic church.* In web site Amorphic Robot Works. http://amorphicrobotworks.org/the-robotic-church. Accessed 30 Dec 2021

MSCHF. (2019). Spot's Rampage. https://spotsrampage.com. Accessed 30 Dec 2021

Mortensen, O. (1957). *Jens Olsen's clock: A technical description*. Technological Institute.

Murphy, S. (1995). Heron of Alexandria's "on automaton-making." *History of Technology, 17*, 1–44.

Murooka, T., Okada, K., & Inaba, M. (2019). Self-repair and self-extension by tightening screws based on precise calculation of screw pose of self-body with CAD data and graph search with regrasping a driver. In *2019 IEEE-RAS 19th International Conference on Humanoid Robots (Humanoids)* (pp. 79–84). https://doi.org/10.1109/Humanoids43949.2019.9035045

Nocks, L. (2008) *The robot. The life story of a technology*. John Hopkins.

Phillips, M. (2007). Sloth-Bot. https://arch-os.com/projects/slothbot/. Accessed 30 Dec 2021

Piquemal, M., & Merlin, C. (2005). *Le manège de Petit Pierre*. Albin Michel Jeunesse.

Pitrus, A. (2013). No longer Transhuman: Handmade machines by Paul Granjon. *International Journal of Cultural Research, 3*(12), 129–133.

Pollack, J. B., & Lipson, H. (2000). The GOLEM project: Evolving hardware bodies and brains. In *Proceedings. The Second NASA/DoD Workshop on Evolvable Hardware* (pp. 37–42). https://doi.org/10.1109/EH.2000.869340

Ramchurn, V., Richardson, R. C., & Nutter, P. (2006). ORTHO-BOT: A modular reconfigurable space robot concept. In M.O. Tokhi, G.S. Virk, & M.A. Hossain (eds.), *Climbing and walking robots* (pp. 659–666). Springer. https://doi.org/10.1007/3-540-26415-9_79

Regine. (2008). Cloaca 2000–2007, We Make Money Not Art, 19/01/2008. https://we-make-money-not-art.com/wim_delvoye_cloaca_20002007/. Accessed 30 Jan 2022.

Reeves, N. (1992). Syndrome de Geppetto et machine de Türing. *Agone, 8–9*, 139–156.

Reeves, N., & St-Onge, D. (2016). Still and useless: The ultimate automaton. In D. Herath, C. Kroos, & Stelarc (eds.), *Robots and art: Exploring an unlikely symbiosis*. Springer.

Salzberg, S., Sayama, H. (2004). Complex genetic evolution of artificial self-replicators in cellular automata. *Complexity, 10*(2), 33–39

St-Onge, D., Reeves, N., & Petkova, N. (2017). Robot-Human interaction: A human speaker experiment. In *HRI '17: Proceedings of the Companion of the 2017 ACM/IEEE International Conference on Human-Robot Interaction* (pp. 30–38). https://doi.org/10.1145/3029798.3034785

Stelarc. (1995). Ping Body. http://www.medienkunstnetz.de/works/ping-body/. Accessed 30 Dec 2021.

Stelarc. (1981). Third Hand. http://stelarc.org/?catID=20265. Accessed 30 Dec 2021.

Vaucanson, J. (1738). *Le mécanisme du flûteur automate*. Jacques Guérin.

Vorn, B. (2010). Mega hysterical machine. Google Arts & Culture. https://artsandculture.google.com/asset/mega-hysterical-machine-bill-vorn/twEoqSJUmM0i7A. Accessed 30 Dec 2021.

Wilson, S. (2003). *Information arts, intersections of art, science, and technology*. MIT Press.

Woodcroft, B. (1851). *The pneumatics of Heron of Alexandria from the original greek*. Taylor Walton and Maberly.

Yim, M., Lacombe, J. C., Cutkosky, M., & Kathib, O. (1995). Locomotion with a unit-modular reconfigurable robot. Dissertation, Stanford University.

Yoshida, E., Murata, S., Kamimura, A., Tomita, K., Kurokawa, H., & Kokaji, S. (2003). Research on self-reconfigurable modular robot system. *JSME International Journal, 4*(46), 1490–1496.

Zek, Y., Balina, A., Guryev, M., & Semionov, Y. (2006). The Peacock clock. https://web.archive.org/web/20080202131950/http://www.hermitagemuseum.org/html_En/12/2006/hm12_1_22.html. Accessed 12 Dec 20210.

Nicolas Reeves is Full Professor at the School of Design at University of Quebec in Montreal. A graduate of U. Montreal, U. Plymouth and MIT, trained in architecture and physics, he has been developing for thirty years a research and an art practice in the field of science-art/technological arts. His work is characterized by a highly poetic use of science and technology. Founding member, then scientific director of the Hexagram Institute (2001–2012), vice-president of the Montreal Society for Technological Arts for ten years, he heads the NXI Gestatio Design Lab which explores the impact of digital technologies in all fields related to creation. Several of his works have had a major media and public impact: Cloud Harp, Aérostabiles (flying cubic automata capable of developing autonomous behaviors), Point d.Origine (real-time musical transposition of remarkable architectures) … Winner of several awards and grants, he presented his work and gave lectures on four continents.

David St-Onge (Ph.D., Mech. Eng.) is an Associate Professor in the Mechanical Engineering Department at the École de technologie supérieure and director of the INIT Robots Lab (initrobots.ca). David's research focuses on human-swarm collaboration more specifically with respect to operators' cognitive load and motion-based interactions. He has over 10 years' experience in the field of interactive media (structure, automatization and sensing) as workshop production director and as R&D engineer. He is an active member of national clusters centered on human-robot interaction (REPARTI) and art-science collaborations (Hexagram). He participates in national training programs for highly qualified personnel for drone services (UTILI), as well as for the deployment of industrial cobots (CoRoM). He led the team effort to present the first large-scale symbiotic integration of robotic art at the IEEE International Conference on Robotics and Automation (ICRA 2019).

Chapter 2
Teaching and Learning Robotics: A Pedagogical Perspective

Eleni Petraki and Damith Herath

2.1 Learning Objectives

By the end of this chapter, you will be able to:

- Understand the current challenges in robotics course design in higher education
- Analyse current teaching practices and innovations in robotics teaching
- Reflect on the link between learning theories and pedagogies for designing robotics education
- Select and assemble suitable pedagogies and techniques for self-directed learning and development in the field of robotics.

2.2 Introduction

The previous chapter outlined technological developments and growth in the robotics field. The advancements and proliferation of robotics applications have had an enormous impact on our daily lives and have changed the skills and competencies of the emerging workforce (Ahmed & La, 2019). Ahmed and La (2019) argue for robotics integration into all levels of education to prepare the future workforce for a technologically advanced society. Considering the growth of robotics applications and the increase in robotics courses in academia, it is vital that the curricula of higher education be carefully designed to address graduate workplace demands. In that domain, there is an absence of systematic discussion and examination of robotics education,

E. Petraki (✉)
Faculty of Education, University of Canberra, Canberra, Australia
e-mail: eleni.petraki@canberra.edu.au

D. Herath
Collaborative Robotics Lab, University of Canberra, Canberra, Australia
e-mail: Damith.Herath@Canberra.edu.au

© The Author(s) 2022
D. Herath and D. St-Onge (eds.), *Foundations of Robotics*,
https://doi.org/10.1007/978-981-19-1983-1_2

both of the syllabus and the pedagogies for addressing graduate student needs at tertiary level. This systematic discussion of teaching and learning practices is an imperative dictated not only from an education renewal perspective but also from the design and product development perspective in the newly developed industries that will have lasting and far-reaching societal implications.

This chapter aims to review current evidence-based research studies on robotics in higher education. Due to the expansion of robotics application in numerous fields, such as mechanical engineering, mechatronics, information technology, artificial intelligence to name a few, we reviewed research investigating teaching and assessment practices in robotics courses primarily in the last 10 years. This time frame will capture the current developments and innovations in the field and will provide a comprehensive understanding of effective teaching practices. These teaching practices will then be explained in the context of well-established educational theories and philosophies in adult learning with the goal of assisting teachers and academics in the design and selection of pedagogies and learning principles to suit robotics education.

In writing this chapter, we have two primary audiences in mind. First, we hope this discussion is applicable to teachers, academics and course designers of higher education robotics courses as it will introduce a bank of resources which they can use to design effective, pedagogically appropriate and industry-relevant curricula. Guided by learner-centred educational philosophies, and with an understanding of the link between educational theories and practices, it will contribute to a principled approach to the design, reflection and improvement in current educational practices, pedagogies and assessment in robotics education.

Second, the pedagogical discussion will be immensely valuable to students who are enrolled in robotics courses or who might want to advance their knowledge and skills in the field. It will provide them with a comprehensive understanding of the theories and pedagogies underpinning course design and a clear insight into interdisciplinary nature of the field. Knowledge and awareness of effective practices will empower and propel students to pursue their own learning and endow them with an array of strategies to learn autonomously and enhance their self-directed learning. Constructivist, constructionist and connectivist education theories (Bower, 2017) discussed in Sect. 6 in more detail, regard teachers as facilitators and guides of student learning and learning is seen as a continuous co-construction between learners and teachers. We hope that this chapter will provide them with an incentive and inspiration to continue their engagement in robotics, develop lifelong learning skills and exploit opportunities outside the university walls.

2.3   Defining the Body of Knowledge of the Robotics Field

An important starting point for designing an appropriate and relevant curriculum for any course is clearly delineation of articulating the body (mass) of knowledge, along with the skills and learning outcomes of any course. This process is guided by curriculum design principles, which view curriculum design as dynamic, comprising a series of interconnected stages: theoretical and epistemological beliefs about the nature of learning, needs analysis, definition of aims and learning outcomes, syllabus design and assessment, methodologies and pedagogies for implementation and the evaluation plan (Richards, 2017). This process suggests that each of these stages is not acting independently, but is mutually dependent on one another. In order to address the research gap in the educational robotics literature and guide the development of robotics courses in higher education, this chapter will survey the literature to identify the body of knowledge expected of graduates of robotics and review the current pedagogies and practices in the robotics field, with a view to suggesting a more holistic approach to robotics education that transcends the traditional boundaries and domains.

Despite the wealth of research in the robotics field, there have been few attempts at describing the body of knowledge expected for those working in the field. To date, we trace the most recent discussion of the body of knowledge and skills for robotics to two reviews in 2007 and 2009 which we summarise here in an effort to describe the state of the art in the field and further illustrate the challenges facing academics today (Gennert & Tryggvason, 2009; McKee, 2007).

While robotics is a field that is taught in various courses and disciplines such as engineering, computer science, information technology, it is common knowledge among researchers that the field is highly diverse and draws on a variety of disciplines (Berry, 2017; McKee, 2007; Wang et al., 2020). According to McKee (2007), this knowledge goes beyond traditional fields of study such as mathematical modelling and machine learning but includes key theoretical and practical dimensions that reflect the diversity in the field: it can cover areas such as mathematics, computing, control engineering, electronic systems, computing systems, programming and algorithms, robotics systems and practice, artificial and computational intelligence, human–computer interaction, artificial intelligence, algorithmic and mathematical modelling, machine learning (McKee, 2007). The multidisciplinary nature of robotics poses several challenges for curriculum developers in the field and calls for a systematic and theory-driven approach to the design of tertiary curricula. In the second study, Gennert and Tryggvason (2009) highlight the importance of defining the body of knowledge necessary for robotics education and preparing ardent prospective robotics engineers to handle the complex nature of robotics applications. They argue that robotics education must not simply attempt to transfer knowledge but attempt to "educate innovators who will have the imagination to shape our world" (p. 20). Discussing their difficulties in their own course design, they identify certain gaps in robotics education:

- Robotics engineering does not seem to have a firm intellectual basis, which is necessary for defining the knowledge and skills required for undergraduate courses in robotics.
- Robotics engineering is not an accredited programme of study and the authors recommend that researchers identify the body of knowledge expected.

- Robotics engineering should bridge the gaps between the scientific, theoretical knowledge and hands-on industrial knowledge.
- There is insufficient research on appropriate curricula and syllabi for robotics engineering education.

Besides the interdisciplinary nature and skills needed in the design of robotics courses, other compounding factors include the role of robotics courses in different disciplines, schools and faculties, and the selection of content to meet the level of prerequisite knowledge expected of students when enrolling in a robotics course (Berry, 2017; McKee, 2007). These concerns are further compounded by the challenges of balancing theory and practice (Jung, 2013), the appropriateness of selection of teaching methods in robotics courses and the design of assessment that evaluates students' achievement of skills in practical and theoretical understanding (Jung, 2013).

A comprehensive inspection of the educational literature on robotics reveals that the current teaching of robotics has not changed dramatically, since the studies in 2007 and 2009, despite the wide applications and developments in the research space (Berry, 2017; Jung, 2013). This is the point of departure for the present chapter which will review a series of studies that pioneer innovative pedagogies and assessment in robotics and which will guide our subsequent theoretical discussion and recommendations for pedagogical approaches in the robotics field.

2.4 Review of Research on Pedagogies and Practices in Robotics Education

Due to the STEM integration in school years, robotics engineering has widespread appeal among university students (Berengual et al., 2016; Gennert & Tryggvason, 2009; Hamann et al., 2018; McKee, 2007; Wang et al., 2020) and this appeal has captured the attention of educators. Educational practitioners and researchers in the field highlight the need to shift away from traditional modes of delivering robotics education (McKee, 2007) to encapsulate the diverse applications of automata, integrate interdisciplinary research and resolve some of the aforementioned tensions. Given the technological advancements, innovations have been introduced in the delivery of courses which include virtual learning environments, virtual robotic laboratories and mobile robotics education to support distance and online courses in robotics (Gabriele et al., 2012; Khamis et al., 2006).

This section reviews current research on educational robotics and reports on innovative pedagogies and content selection employed in the design and teaching of robotics courses, especially in the last 10 years. The research studies originate in courses which received favourable student evaluations and led to improved learning outcomes (Gabriele et al., 2012; Jung, 2013; Wang et al., 2020). The presentation will pave the way for revolutionising higher education robotics courses and assist students and teachers in identifying pedagogical tools for autonomous learning development and teacher curriculum development.

2.4.1 Adaptation of Content from Different Disciplines

One of the key challenges is the selection of suitable content for robotics courses that target the needs and knowledge of different disciplines and subfields. For instance, Gennert and Tryggvason (2009) discuss the design of their robotics undergraduate course in a Polytechnic university aiming to teach the basic fundamentals to students in mechanical engineering, computer science and electrical engineering. In addressing the different student background knowledge, the syllabus integrated a unique range of modules on areas such as power, sensing, manipulation, and navigation, adjusting and incorporating content from each of the students' disciplines. In another study discussing the review of a robotics course in the faculty of mechatronics at a Korean university, Jung (2013) raised the need to combine theory and practice by integrating knowledge in Manipulator robots with hands-on experiences in laboratory practicals. The course incorporated interdisciplinary theoretical content covering robot kinematics, dynamics, path planning and control, while the laboratory practical experience made use of a range of robot applications, experimental kits, Lego robots and humanoid robots to develop student skills in motor control. Wang et al. (2020) and Hamann et al. (2018) share these views and stress that, because of the popularity of robotics as a discipline and its cross-disciplinary nature, new methodologies and content need to be developed to allow students to combine hardware and software implementation and to prepare future engineers to handle unfamiliar and complex problems. This complies with current educational curriculum principles, which recommend a thorough analysis of the context and student needs in the courses to design relevant and student-centred courses.

The development and redesign of new robotics courses and the increasing diversity of contexts of robotics have led to the emergence and necessity of new pedagogies to engage students in the field and to design appropriate content effectively (Martínez-Tenor et al., 2019). Similarly, Wang et al. (2020) argue that new methodologies need to be developed to allow students to combine hardware and software implementations.

2.4.2 Constructivist Approaches to Learning

An important consideration emerging in this research is the importance of educational theory in underpinning curriculum design and assessment. Few studies identified the role of combining instructivist or didactic and constructivist paradigms in course design (Johnson, 2009; Martínez-Tenor et al., 2019). Instructivist pedagogies are associated with traditional forms of learning such as lectures, videos and examinations where learners aim to gain knowledge. Constructivist modes of instruction focus on student engagement in active participation and problem solving, where teachers are facilitators and enablers of student learning. The constructivist paradigm is typically associated with activities and pedagogies such as task-based learning, collaborative activities, group tasks in which students engage in problem solving and learning through collaboration and exchange. A revision of a recent master's course (Martínez-Tenor et al., 2019) on cognitive robotics led to the integration of two approaches using Lego Mindstorm. Students were first exposed to instructional videos on machine learning and reinforcement learning as a preparation for their engagement in interactive sessions using reinforcement learning working on two decision-making problems. Students' evaluation of the teaching methods in the course showed that students appreciated and benefitted from autonomous learning and collaborative learning activities and found the possibility of programming a robot intensely motivating. They also offered suggestions for improvement, which could be considered in future courses. These comprise time allocation for analysis and reflection on the experiments, addition of problem-solving activities, increasing opportunities for collaboration, reflection and retention by students. Martinez-Tenor et al. (2019) echo Johnson's suggestion (2009) for a carefully designed programme that combines instructivist and constructivist approaches to teaching to address diversity in learning styles.

2.4.3 Situated Learning Methodology

Wang et al. (2020) discuss the implementation of an innovative pedagogy, which they name situated learning methodology combined with the development of a hands-on, project-oriented robotics curriculum in an undergraduate and postgraduate unit for computing students. To address the challenge of combining theory and practice, the course employed a situated learning-based robotics education pedagogy, guided by four central principles: content, context, community and participation (Stein, 1998). The situated learning methodology assumes that learning is a process of participation and practice for solving real-life authentic problems (Lave & Wagner, 1991). Based on the belief that knowledge and skills are developed effectively in the context of real life, situated learning allowed students to work on a real-life application: interacting with a multimodal collaborative robot who is employed as the students' classmate. A classroom-based learning community is established with groups working on solutions to different hands-on tasks. The situated learning approach could be regarded as a technique belonging to the constructivist education paradigm that promotes collaboration and co-construction of learning in authentic real life environments (Selby et al., 2021).

2.4.4 Flipped Classroom

Another novel method introduced in a mobile robotics course in a US university was the flipped classroom (Berry, 2017). This method was adopted to address time limitations in explaining the theoretical components of robotics and encourage more student participation (Berry, 2017). The flipped classroom is a new pedagogical method which distinctively combines instructivist and constructivist approaches to learning. The term "flipped classroom", often referred to as "reversed instruction", incorporates a switch between in-class and out-of-class time, thus fostering more interaction between teachers and students during class time. Students spend most of the time engaged in experiential activities, problem solving and diversified platforms (Nouri, 2016). A meta-analysis of flipped classroom research has demonstrated the effectiveness of this model over traditional learning on student achievement and learning motivation (Bergmann & Sams, 2012). The flipped approach was utilised in the course to allow students to focus on their development of technical skills in controlling robots, designing and experimentation with the real mobile robots for laboratory experiments. This model has enormous potential for addressing the challenges of balancing theory and practice in a university course and allowing adequate time for problem solving, self-paced learning activities and student negotiation.

2.4.5 Gamification

Another area of increasing interest is the role of gamification in robotics education, which refers to the addition of play-based elements such as games as a method of instruction to increase student engagement. Hamann et al. (2018) discuss the gamification in teaching swarm robotics to first-year undergraduate students in computer science, with a focus on teaching/learning theory and practice. Videogames allowed student immersion in a simulated environment and inspired student creativity. Students were presented with several robot manipulator challenges, engaged in designing fully working prototype robots and models from the start with a gradual increase in their functionality and complexity. The curriculum integrated robot-based videogames and student competitions, thus building students' teamwork skills and triggering their imagination and engagement. Simultaneously, these learner-centred methods offer students flexibility in learning and enhance their autonomy in problem solving and engineering.

2.4.6 Online Interactive Tools

The advances in educational technologies have impacted education worldwide by creating a variety of online tools and technological affordances. The educational domain experienced a boom in online learning and hybrid learning modes which led to the creation of several online and virtual tools. To facilitate online delivery of robotics courses, virtual laboratories were used engaging students in building and guiding robots remotely with a range of tools. For instance, Berengual et al. (2016) employed an array of interactive tools which they defined as "a set of graphics windows whose components are active, dynamic and clickable ones" in order to practice the theoretical aspects of the course. The "Mobile Robot Interactive Tool" (MRIT) aimed at teaching students about robot navigation, allowing students to explore a variety of parameters, such as robot kinematics, path planning algorithm, the shape of the obstacles. It assisted students in understanding the basis of mobile robot navigation and allowed them to modify different characteristics, such as robot kinematics, path planning algorithm and the shape of the obstacles. The second interactive software tool, the slip interactive tool (slip-IT) was used to teach the concept of slip in off-road mobile robots and last for the teaching of robotics manipulation MATLAB/SIMULINK and robotics toolbox for conducting robot simulations. The courses integrated two robots, some of which could be controlled remotely or offline through Internet connection to the labs allowing students to work remotely. In addition to the simulation activities, the adoption of a real robot for demonstration and implementation was a fundamental aspect of the course. Another interactive tool, called ROBOT DRAW, was discussed by Robinette and Manseur (2001), which has been widely used in robotics education. The tool was designed to enable students to easily visualise robots in various configurations and evaluate the effect of a parameter variation on the robot. Among others, a popular online platform (https://www.theconstructsim.com/) provides a range of online robot manipulation tools and can be used by both students and teachers for autonomous practical learning. It consists of virtual laboratories allowing students to experiment with manipulating, building real and virtual robots online using a range of tools. Exposure and interaction with a range of tools build students' technological competencies and problem-solving skills.

2.5   Assessment Practices

Changes in pedagogies and methods in teaching are closely intertwined with transformative assessment practices that match the learning–teaching philosophies of these methods. Traditional methods of assessment have been embedded in many higher education courses and comprised examination-based assessment or/and experimental work. A few attempts have been made to modify assessment practices to reflect changes in pedagogical approaches in robotics.

2.5.1 Collaborative and Individual Project-Based Assessment

The majority of new assessment tasks integrated into some courses comprise project-based assessment and competition reward systems. Group and individual projects provide opportunities for authentic and collaborative learning experiences and enhance student motivation and problem solving. In the design of courses reviewed by Hamann et al. (2018) and Jung (2013), student assessment consisted of a group project using competition-based learning, in which students had to engage and collaborate through a series of tasks in a boxing match, using humanoid robots. Students found the competition-based assessment a valuable and motivating experience in applying many theoretical robotics skills although they acknowledged the challenges of the time requirement of the competitions (Jung, 2013). Similarly, Wang et al. (2020) employed project-based assessment allowing students to create a complete robot control architecture in software and hardware during laboratory sessions. This form of assessment enabled a classroom-based learning community with groups working on solutions to different hands-on tasks. Consistent with the situated teaching methodology, project-based learning was adopted: each student was equipped with a robotics development kit containing ultrasonic sensors, an Arduino board and other robotics electronic accessories. The practical hands-on application, combined with the step-by-step progression part of the syllabus and the teaching methodology, led to student satisfaction and the effectiveness of this approach in the development of students' learning outcomes. Berengual et al. (2016) equally employed a project-based group assessment expecting students to build, programme and navigate a robot, and a series of online reflections on theory and laboratory participation in a range of tasks that assisted with the group project. Students identified the project task as one of the most vital educational experiences that developed their technical and engineering skills. Last, using a simple to complex curriculum design model, Hamann et al. (2018) report on the use of group project allowing students to progress the robot applications through a series of phases from simulation to real robots leading to a battle royale game. The adoption of games and competitions both as sources of learning and assessment offer students opportunities for collaboration, development of student autonomy in problem solving and engineering and allow students to see and test the effects of their programming and engineering.

2.5.2 Competition-Based Assessment

As mentioned previously, competition-based assessment can be a powerful tool in engaging students in collaborative assessment. It was integrated into Martínez-Tenor et al. (2019) and Jung (2013) course design studies and contributed to rich learning and increase in student engagement and motivation. Some courses used project-based learning to generate conference presentations which offered multiple opportunities for student academic development, rich learning and networking with industry.

2.5.3 Reflective Learning

To foster deep processing of learning, reflective writing in the form of continuous assessment such as reflective posts was also introduced in some robotics courses. The use of reflective activities is often combined with other forms of assessment such as group projects which integrate experimental work with reflective writing where students explain and focus on consolidation of theoretical knowledge. Wang et al. (2020) designed project-based assessment expecting students to work towards creating a complete robot control architecture in software and hardware during laboratory sessions. Assessment was redesigned to include weekly literature reflections, online quizzes on the theory and staged group project assessment conducted in laboratories consisting of three graded components: a demonstration, a technical memo and a code submission. Martínez-Tenor et al. (2019) also incorporated reflections as part of the group/project assessment focusing on robot manipulation, which resulted in a valuable learning experience for students. Individual reflections also allow for flexibility and self-paced learning and when shared publicly in online learning platforms offer rich learning opportunities for all students in the course.

The aforementioned discussion identified some attempts at transforming teaching/learning practices and assessment in robotics higher education courses based on a review of educational research in the last decade. To truly transform education practices and to identify effective teaching pedagogies in robotics education and beyond, it is vital for teachers and students to develop an advanced awareness of the relationship between education theories, curriculum design principles and methods of learning and teaching. Equipped with these skills, academics, teachers and students can make systematic and theory-driven selections to revise, adapt and refine robotics education.

2.6 Paving the Way for Innovative Pedagogies and Assessment in Robotics Education

To address the call for more diverse and current educational practices, to tackle the current diverse applications of robotics and the growth of the industry, it is important that robotics education prepares future engineers adequately to cope with arising challenge in the field (Wang et al., 2020). This section will provide a guide to novel pedagogical practices and assessment in teaching robotics, relying on research in educational literature and the challenges facing robotics education at the academic level. Important caveats for applying these suggestions will be discussed at the end of this section.

First, we will begin with a discussion of educational theories/epistemologies that drive pedagogical practices, as this is an integral aspect of any teaching and curriculum design process (Richards, 2017). Research on adult learning and education theory is well-established, highly researched and has undergone many transformations. Educational theories and ideologies are defined as a set of epistemological beliefs concerning the nature and value of learning, teaching and the role of education and serve as a justification for particular approaches, pedagogies and methods to teaching (Richards, 2017).

Historically, one of the first theories which influenced educational processes was behaviourism which viewed learning as habitual behaviour, that is, observable, conditioned upon a stimulus-reward action and reinforced through habitual learning (Skinner, 1974). Influenced by a series of experiments on dogs, Skinner (1974) concluded that learning is observable through actions and is shaped by the environment and instructional design. He continued to suggest that learning can be achieved through a series of teacher questions and student responses, where positive and negative feedbacks determined the learning process. The behaviourist learning theory influenced educational design, by emphasising that teaching is an objective body of knowledge that is to be delivered and measured though performance measures and outcomes (Bower, 2017; Howell, 2012). The behaviourist approach is associated with the transmission-based model of teaching placing teachers as the authority of knowledge, organisers and planners of learning and learners as passive recipients of this knowledge. This is evident in traditional and authoritative models of teaching and classical forms of assessment such as examinations, quizzes, not acknowledging the role of the learners in the process or other environmental or psychological factors (Bower, 2017). Despite the early successes of the behaviourist paradigms, one of its drawbacks was the lack of consideration of the complexity of human cognition and the individual learner processes.

In addressing the limitations with the behaviourist theory, another group of researchers examined the role of mental and information processing in the learning process, which LD to the development of cognitivism. Within the theory of cognitivism, learning is an internal mental process of storing, receiving, consolidating and reorganising information and knowledge structures or schemata (Bower, 2017). Cognitivism could be seen as an extension of behaviourism, with attention to the workings of the brain. Proponents and researchers in the field focused on aspects of selection, organisation and retrieval of information and used some of this research to design a curriculum with learner conditions in mind. These included aspects of knowledge sequencing, information load, staged instruction to improve learning comprehension and consolidation. However, within cognitivism the transmission model of education and the focus on demonstration of learning outcomes prevailed.

This gave way to the theory of constructivism, one of the most influential paradigms that focused on learning as a process rather than learning as a product. Constructivist paradigms have dominated modern educational practices at all education levels (Jones & Brader-Araje, 2002). The paradigm is based on the idea that learning is not static but dynamic and is a process of reflection, negotiation and individual or collaborative discussion through interaction with other learners, interaction with social and cultural influences. Individual constructivism was pioneered by Piaget (1970), who considered learning as a result of processes of assimilation and accommodation of new knowledge to existing knowledge, while social constructivism, introduced by Vygotsky (1978) focused on sociocultural influences on learners and their learning. Within Vygotsky's social constructivism (1978), group activities and collaborative learning are preconditions and must precede any individual learning. Learning is regarded as a continuous interplay between others and the self through internal assimilation and extension/addition of new knowledge. Intrinsic to the social constructivist model, which has had tremendous impact on learning, is the idea of scaffolding, which is defined as additional assistance and support which can gradually be removed after the learner has gained independence. Based on the constructivist perspective, the teachers are considered guides and facilitators and providers of the conditions, tools and prompts enabling students to discover principles and engage in knowledge construction by themselves (Bruner, 1990). The constructivist paradigm gave birth to several teaching methodologies that promote co-construction, negotiation of learning and self-discovery, comprising students' engagement in self-directed learning but also and most importantly collaborative learning, project-based learning and competitions-games and tournament tasks (Jones & Brader-Araje, 2002).

Constructionism is regarded as an extension of constructivism which considered the impact of technologies and artefacts on the learning process. The origins of this theory can be traced to Papert (1980) who observed that learners create their own reflections through experimentation with tangible objects, which were initially referred to Lego, Logo and Mindstorms. It was suggested that learning takes place when people are active during their creation of tangible objects in the real world. It further assumes that learning is reinforced through engagement in authentic tasks, creation of tangible objects, collaborative learning or other design activities in the real world such as authentic and situated learning experiences (Howell, 2012; Papert & Harel, 1991).

With similar roots to constructionism and inspired by the digital networking, researchers introduced connectivism as the new epistemology based on the dominance of digital learning. Connectivism subscribes to the views that learning takes place in an organic fashion and is a result of building connections and skills in connecting the digital world, technologies and platforms with social networks, knowledge and information (Siemens, 2005). It centres on the metaphor of networks with nodes and connections as the basis for learning. Influenced by constructivist principles, connectivism is a novel approach, adopted in technology-enhanced learning and online learning, and aims to develop students' skills in critical thinking, connecting and collaborating through interactions with technologies and connectivist learning environments (Bower, 2017; Howell, 2012; Siemens, 2005).

It is evident in the above review that there has been exponential growth in educational theory, which in turn generated new methods and pedagogies that could be integrated into robotics education. Some of these new methods employed in the course design literature identified in Sect. 4 were influenced by constructivist, constructionist and connectivist ideologies and were considered effective. Given the role of robotics education in preparing the undergraduate students in handling complex real-life problems, curriculum design in the field could benefit from integrating such novel methodologies.

While traditional didactic learning is an integral aspect of acquiring key knowledge, admittedly, to align with current research developments in learning theories and to address today's global challenges and to develop competitive and multi-skilled graduates, it is vital that robotics education be enriched to bring about more educational benefits. Instructivist, behaviourist and cognitivist methods have dominated the delivery and implementation of higher education courses but they are limited and inadequate in improving learning outcomes. This section will highlight novel and evidence-based pedagogies that could improve robotics course design and facilitate graduates' self-directed learning.

Some of the most effective pedagogies that are consistent with constructivist and constructionism theories are collaborative learning, project-based learning and competition-framed tasks. These methods should play a significant role in the delivery of robotics education in academic as well as other educational levels. There is abundant research to suggest that social engagement and collaboration with peers have positive impact on individual development, problem solving as well as social collaboration skills, skills and attributes expected of university graduates (Zheng et al., 2020). Collaborative learning can be enhanced through discussion forums, web-conferencing systems, virtual worlds, project-based learning during experimental work. Collaborative learning allows students to treat their collaborators as resources and guides for their own growth and development. It also provides opportunities for scaffolding by allowing for information exchange and learning from one another and teamwork skills on problem-solving activities. It needs to be mentioned that project-based learning comprising group collaboration comes with several challenges. These challenges can be frustrating for students, but with sufficient guidance, they can empower students, help them develop student independence, creativity and equip them with innovative problem-solving skills.

Project-based learning can sometimes take the form of problem-based learning and design-based learning, which all align with constructivist and constructionist principles. Design-based learning is a novel learning approach encouraging students to work collaboratively on authentic real-life design tasks with the aim of advancing their design skills, problem-solving abilities, reasoning and critical thinking skills and develop attitudes to continuously tackle emerging challenges (Howell, 2012; Kim et al., 2015). Problem-based learning is a pedagogical technique that provides students with an authentic problem, with the aim of advancing student engagement and motivation and supporting student-centeredness, self-regulation, development of cognitive and metacognitive strategies, autonomy and student independence (Stefanou et al., 2013). It has also been suggested that project-based learning is easily combined with other methods such as flipped classroom models, inquiry-based learning, collaborative learning, and the combination of such methods maximises the effectiveness on student learning (Zheng et al., 2020).

Last but not least, competitions, games, tournaments combined with or incorporated in collaborative projects enhance students' motivation and interest to learn and encourage independence and further learning. Games are built on constructivist principles and promote cognitive and social interaction, and build risk-taking, strategic negotiation, problem solving, collaboration, reflection and lateral thinking (Gee, 2005). They can increase student engagement, motivation and promote a high sense of achievement and competition (Stefanou et al., 2013). Gamification principles could be used as learning approaches or as assessment tools and have the potential to increase students' continuous engagement and excitement in the course and the range of activities (Hwang & Chang, 2016).

Changes in learning methods and pedagogies implicate changes in assessment practices. An effective curriculum expects consistency between the syllabus, pedagogies and assessment practices, a notion known as "constructive alignment" (Biggs, 2014, p. 5). The aforementioned literature has paved the way for integrating a wide range of assessment items that align with constructivist and project-based approaches to learning.

Educational research points to the significance of project-based assessment, as it offers authentic learning experiences for students, builds their collaborative skills and develops their problem-solving skills. It is consistent with the new pedagogies promoted in the previous review and would also endow students with skills for the real world where teams work together to build, design and manipulate robots.

Due to the multidisciplinary aspects of robotics and its contribution to a range of fields, robotics courses could benefit from online reflections on the literature and theory. This was assumed and encouraged in the early work by Papert (1980) who suggested that knowledge is created through reflection and engagement with people and artefacts. These online reflections could be used as formative assessments to engage students' reflective, critical learning skills and problem solving abilities (Merlo-Espino et al., 2018). Reflective activities and discussions can also be integrated into project-related work to assist students in resolving these challenges and offer a mechanism of getting support from lecturers (Serrano et al., 2018).

Admittedly, authentic assessment should be an indispensable component of robotics assessment in higher education. Authentic learning is a suitable pedagogy that operates within the theory of constructionism, hypothesising that learning takes place during students' interaction with practical tasks and robots. Authentic assessment, therefore, refers to assessment requiring students to build/design/create artefacts or robotics applications and provides them with opportunities to develop real-world skills. Gulikers et al. (2004) highlight a number of aspects of authenticity in assessment: the task, the physical, virtual and social context, the artefact produced (or behaviour assessed) or/and, the criteria and expected standard. Authentic assessment assists the students with developing competencies appropriate for the workforce and is often requirements for meeting professional accreditation standards. Project-based assessment that enables students to design a robot-based application is paramount to developing students' real-life skills and foster effective human–robot interaction (Gurung et al., 2021). They further enhance situated learning/learning by doing (Wang et al., 2020) as they provide the environment for students to learn from one another and develop collaborative skills.

An important caveat needs to be mentioned here. The choice of assessment tasks, formative, summative, group and/or individual need to be closely linked with the pedagogy and epistemology of the course, syllabus and the teaching, something known as epistemological alignment to improve the course success. There must be an effective triadic relationship between epistemology (the nature of learning), pedagogy an assessment for the course to be successful and meet its objectives (Knight et al., 2014).

It is important to highlight that these suggestions are pertinent to students who are studying in robotics and robotics adjacent fields. Students interested in advancing their knowledge and skills can seek opportunities, extra-curricular and industry opportunities to be involved in authentic projects, collaborative activities and pursue conference or industry presentations. Reflective learning activities and participation in discussions can create valuable learning opportunities for students to advance their skills and be competitive in the field (Fig. 2.1).

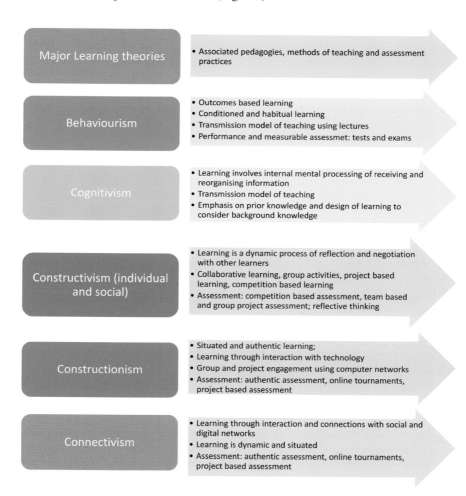

Fig. 2.1 Learning theories, principles and pedagogies

2.7 Chapter Summary

In addressing the absence of systematic reviews of research and recommendations in teaching robotics, this chapter offered an overview of the current challenges in teaching and learning robotics and reviewed pedagogical trends in robotics education at higher education institutions. The need for a systematic presentation of current educational practices is further enhanced when considering that the purpose of the book is to introduce the theory, design and applications of robotics for students and academics, and to advance students' skills to handle complex problems. This chapter first highlighted several challenges facing designers of robotics courses which include lack of systematic research in robotics education and the complex network of disciplines which need to be synthesised to design robotics courses. Next, it reviewed current innovations in higher education course design and pedagogy, specifically focusing on the last ten years, which were found to lead to improved learning outcomes. This aimed to raise students' awareness of the history and theoretical principles underlying the teaching of robotics at the academic level. To address the challenges and complexities in designing appropriate syllabus and instruction, and the need to shift away from traditional forms of learning, the last section offered a comprehensive understanding of learning theories and relevant pedagogies that have the potential to improve educational practices and lead to learning benefits if used appropriately in robotics education.

To shape the future of robotics education, it is imperative that academics, teachers and industry practitioners work collaboratively and be involved in negotiating and co-designing the syllabus and assessment of academic robotics courses. In addressing the chasm in the knowledge, we hope this chapter developed their in-depth awareness of the theoretical basis of teaching pedagogies and advances in learning theory which should guide course design, syllabus and assessment. Learner-centred, constructivist and connectivist learning theories should be the basis for selecting suitable methods which address the challenges embedded in the multidisciplinary nature of robotics, and the diverse skills engineers need in today's technologically advanced society. These pedagogies comprise project-based learning, problem-based and collaborative learning, reflective writing and authentic assessment, to name a few.

Revolutionising robotics education and building work-ready graduates are not simple tasks. Recognising the complexity of the robotics field and the diversity in educational processes is a starting point which can assist in our definition of roles, responsibilities and identities as learners and teachers. It requires changes in beliefs and practices that both students and teachers implement and manage effectively. Zhou et al. (2020) argue that students' dissatisfaction in academic courses is often ascribed to their lack of understanding of their role in the learning process and, of the epistemological beliefs underpinning learning and assessment (Zhou et al., 2020). Teachers should be willing to adopt such roles as guides, facilitators, moderators of learning and enablers of change, and invite students in negotiations and co-constructions of their learning experiences. Armed with tools and strategies to improve their learning, students should be co-creators and active participants of classroom realities (Harmer, 2015). Students need to engage in sociocultural and professional practices in robotics, shaping and negotiating their identities and social relations in this academic community of practice (Saltmarsh & Saltmarsh, 2008). It is hoped that with the discussion in this chapter, students are empowered and inspired in taking charge of their own learning and armed with a multitude of tools to continue their professional development and lifelong learning.

2.8 Quiz

According to this chapter,

- What are some key challenges facing robotics education course design?
- What were some of the pedagogical innovations discussed and reviewed in the robotics literature in this chapter?
- Name some interactive tools which have been incorporated in teaching robotics in higher education.
- What is the learning theory which espoused the idea that knowledge is built when we interact, experiment and reflect on our experience by building and creating artefacts?
- What are some methods that you can employ to advance your skills in robotics?

Acknowledgement The contribution of the first author is funded by the Australian Research Council Discovery Grant DP200101211.

References

Ahmed, H., & La, H. M. (2019). Education-robotics symbiosis: An evaluation of challenges and proposed recommendations. In *IEEE Integrated STEM Education Conference (ISEC)* (pp. 222–229). https://doi.org/10.1109/ISECon.2019.8881995

Berenguel, M., Rodríguez, F., Moreno, J. C., Guzmán, J. L., & González, R. (2016). Tools and methodologies for teaching robotics in computer science and engineering studies. *Computer Applications in Engineering Education,24*(2), 202–214. https://doi.org/10.1002/cae.21698

Bergmann, J., & Sams, A. (2012). *Flip your classroom: Reach every student in every class every day.* Internal Society for Technology in Education.

Berry, C. A. (2017). Robotics education online flipping a traditional mobile robotics classroom. *IEEE Frontiers in Education Conference (FIE),2017*, 1–6. https://doi.org/10.1109/FIE.2017.819 0719

Biggs, J. (2014). Constructive alignment in university teaching, *HERDSA Review of Higher Education, 1,* 5–22 .

Bower, M. (2017). *Design of technology-enhanced learning: Integrating research and practice.* Emerald Publishing Limited.

Bruner, J. (1990). *Acts of meaning.* Cambridge, MA: Harvard University Press.

Gabriele, L., Tavernise, A., & Bertacchini, F. (2012). Active learning in a robotics laboratory with university students. In C. Wankel & P. Blessinger (Eds.), *Increasing student engagement and retention using immersive interfaces: Virtual worlds, gaming, and simulation, Cutting-edge technologies in higher education* (Vol. 6 Part C, pp. 315–339). Emerald Group Publishing Limited, Bingley. https://doi.org/10.1108/S2044-9968(2012)000006C014

Gee, J. P. (2005). Good video games and good learning. Paper presented at the Phi Kappa Phi Forum.

Gennert, M. A., & Tryggvason, G. (2009). Robotics engineering: A discipline whose time has come [education]. *IEEE Robotics & Automation Magazine, 16*(2), 18–20. https://doi.org/10.1109/MRA. 2009.932611

Gulikers, J. T. M., Bastiaens, T. J., & Kirschner, P. A. (2004). A five-dimensional framework for authentic assessment. *Educational Technology Research and Development, 52*(3), 67–86.

Gurung, N., Herath, D., & Grant, J. (2021, March 8–11). Feeling safe: A study on trust with an interactive robotic art installation. *HRI '21 Companion.* Boulder, CO, USA.

Hamann, H., Pinciroli, C., & Mammen, S. V. (2018). A gamification concept for teaching swarm robotics. In *12th European Workshop on Microelectronics Education (EWME)* (pp. 83–88). https://doi.org/10.1109/EWME.2018.8629397

Harmer, J. (2015). *The practice of English language teaching* (5th ed.). Longman.

Howell, J. (2012). *Teaching with ICT: Digital pedagogies for collaboration and creativity.* Oxford University Press.

Hwang, G.-J., & Chang, S.-C. (2016). Effects of a peer competition-based mobile learning approach on students' affective domain exhibition in social studies courses. *British Journal of Educational Technology, 47*(6), 1217–1231.

Johnson, G. M. (2009). Instructionism and constructivism: Reconciling two very good ideas. *International Journal of Special Education, 24*(3), 90–98.

Jones, M. G., & Brader-Araje, L. (2002). The impact of constructivism on education: Language, discourse, and meaning. *American Communication Journal, 5*(3).

Jung, S. (2013). Experiences in developing an experimental robotics course program for undergraduate education. *IEEE Transactions on Education, 56*(1), 129–136. https://doi.org/10.1109/TE.2012.2213601

Khamis, A., Rodriguez, F., Barber, R and Salichs, M. (2006). An approach for building innovative educational environments for mobile robotics. *Special Issue on Robotics Education, International Journal of Engineering Education, 22*(4), 732–742.

Kim, P., Suh, S., & Song, S. (2015). Development of a design-based learning curriculum through design-based research for a technology enabled science classroom. *Educational Technology Research Development, 63*(4), 575–602.

Knight, S. B., Shum, S., & Littleton, K. (2014). Epistemology, assessment, pedagogy: where learning meets analytics in the middle space. *Journal of Learning Analytics, 1*(2), 23–47.

Lave, J., & Wenger, E. (1991). *Situated learning: Legitimate peripheral participation.* Cambridge University Press.

Martínez-Tenor, A., Cruz-Martín, A., & Fernández-Madrigal, H-A. (2019). Teaching machine learning in robotics interactively: The case of reinforcement learning with Lego® Mindstorms. *Interactive Learning Environments, 27*(3), 293–306. https://doi.org/10.1080/10494820.2018.152 5411.

McKee, G. T. (2007). The robotics body of knowledge [Education]. *IEEE Robotics & Automation Magazine, 14*(1), 18–19. https://doi.org/10.1109/MRA.2007.339621

Merlo-Espino, R. D., Villareal-Rodgríguez, M., Morita-Aleander, A., Rodríguez-Reséndiz, J., Pérez-Soto, G. I., & Camarillo-Gómez, K. A. (2018). Educational robotics and its impact in the development of critical thinking in higher education students. In *2018 XX Congreso Mexicano de Robótica (COMRob)* (pp. 1–4). https://doi.org/10.1109/COMROB.2018.8689122

Nouri, J. (2016). The flipped classroom: For active, effective and increased learning—especially for low achievers. *International Journal of Educational Technology in Higher Education,13*, 33. https://doi.org/10.1186/s41239-016-0032-z

Papert, S. (1980). *Mindstorms: Children, computers and powerful ideas.* Basic Books Publishers.

Papert, S., & Harel, I. (1991). Situating constructionism. *Constructionism, 36*, 1–11.

Piaget, J. (1970). *The science of education and the psychology of the child.* Grossman.

Richards, J. (2017). *Curriculum development in language teaching.* CUP.

Robinette, M. F., & Manseur, R. (2001). Robot-draw, an Internet-based visualization tool for robotics education. *IEEE Transactions on Education,44*(1), 29–34. https://doi.org/10.1109/13.912707

Saltmarsh, D., & Saltmarsh, S. (2008). Has anyone read the reading? Using assessment to promote academic literacies and learning cultures. *Teaching in Higher Education,13*(6), 621–632.

Selby, N. S., Ng, J., Stump, G. S., Westerman, G., Traweek, C., & Harry Asada, H. (2021). TeachBot: Towards teaching robotics fundamentals for human-robot collaboration at work. *Heliyon, 7*(7). https://doi.org/10.1016/j.heliyon.2021.e07583

Siemens, G. (2005). Connectivism: A learning theory for the digital age. *International Journal of Instructional Technology and Distance Learning, 2*(1), 3–10.

Skinner, B. F. (1974). *About behaviourism.* Penguin.

Stefanou, C., Stolk, J.D., Prince, M., Chen, J.C., & Lord, S.M. (2013). Self-regulation and autonomy in problem- and project-based learning environments. *Active Learning in Higher Education, 14*(2), 109–122. https://doi.org/10.1177/1469787413481132

Stein, D. (1998). *Situated learning in adult education.* ERIC Clearinghouse on Adult, Career, and Vocational Education.

Vygotsky, L. S. (1978). Tool and symbol in child development. In M. Cole, V. John-Steiner, S. Scribner, & E. Souberman (Eds.), *Mind in society: The development of higher psychological processes.* Harvard University Press.

Wang, W., Coutras, C., & Zhu, M. (2020). Situated learning-based robotics education. In *2020 IEEE Frontiers in Education Conference (FIE)* (pp. 1–3). https://doi.org/10.1109/FIE44824.2020.9274168

Zheng, L., Bhagat, K. K., Zhen, Y., & Zhang, X. (2020). The effectiveness of the flipped classroom on students' learning achievement and learning motivation: A meta-analysis. *Educational Technology & Society,23*(1), 1–15.

Zhou, J., Zhao, K., & Dawson, P. (2020). How first-year students perceive and experience assessment of academic literacies. *Assessment & Evaluation in Higher Education,45*(2), 266–278. https://doi.org/10.1080/02602938.2019.1637513

Eleni Petraki is an Associate Professor at the University of Canberra. She is an applied linguist with close to three decades of experience in language teaching, discourse analysis and intercultural communication. Her experience in teaching English has been accumulated in different countries including Vietnam, Greece, UK, USA and Australia. In addition to her research in these fields, she has evolved a research program on artificial intelligence, where she is applying educational curriculum theories and pedagogies to new fields including machine education.

Damith Herath is an Associate Professor in Robotics and Art at the University of Canberra. He is a multi-award winning entrepreneur and a roboticist with extensive experience leading multidisciplinary research teams on complex robotic integration, industrial and research projects for over two decades. He founded Australia's first collaborative robotics start-up in 2011 and was named one of the most innovative young tech companies in Australia in 2014. Teams he led in 2015 and 2016 consecutively became finalists and, in 2016, a top-ten category winner in the coveted Amazon Robotics Challenge—an industry-focused competition among the robotics research elite. In addition, he has chaired several international workshops on Robots and Art and is the lead editor of the book "Robots and Art: Exploring an Unlikely Symbiosis"—the first significant work to feature leading roboticists and artists together in the field of robotic art.

Chapter 3
Design Thinking: From Empathy to Evaluation

Fanke Peng

3.1 Learning Objectives

This chapter introduces methods and approaches for design thinking as the main drivers in developing the ability to identify critical problems in a given situation. This problem identification represents the opportunities for design intervention and creative solutions to a range of possible scenarios and practical applications. The chapter also develops the students' understanding of design as an iterative process involving empathy, ideation and prototypes to test and evaluate concepts and solutions to a wide variety of identified problems.

By the end of this chapter, you will be able to:

- Discover the history of the "designerly way of thinking" as the origin of design thinking
- Understand what design thinking is and why it is so important
- Reflect on a human-centred design (HCD) process through empathy, collaboration and creative thinking
- Select and assemble suitable design thinking models and tools for self-directed learning and problem-based learning.

3.2 Introduction

The need for design thinking in robotics is becoming the catalyst for digital transformation (Automeme, n.d.). Design thinking applies from the origin of a robotic system for industry through interactive robotic art and ongoing research. It helps

F. Peng (✉)
UniSA Creative, University of South Australia, Canberra, Australia
e-mail: Fanke.Peng@unisa.edu.au

© The Author(s) 2022
D. Herath and D. St-Onge (eds.), *Foundations of Robotics*,
https://doi.org/10.1007/978-981-19-1983-1_3

designers and non-designers empathise, learn, develop and deliver creative possibilities. To understand the importance of design thinking in robotics, we need first to understand what design thinking is and why it is so important?

3.2.1 What Is Design Thinking

Design thinking was introduced in the 1960s to the "design science decade" (Cross, 2001, 62). The theories evolved from the understanding that wicked problems are at the centre of design thinking. Buchanan's (1992) article about "wicked problems" in design has become a foundational reference for the discourse about design thinking and the whole design area. When designers engage in design processes, Buchanan (1992) stated that they face wicked and indeterminate problems. The designer is not merely discovering, uncovering and explaining the phenomenon in question (which is undeterminate) but is also suggesting other possibilities and creating and transforming the matter. Dewey (1938) defined the process of inquiry as a transformation process beginning from an indeterminate problem. Inquiry is a process that begins with doubt and ends with knowledge and a set of beliefs so concrete that they can be acted upon, either overtly or in one's imagination (Dewey, 1938). To engage in this process, one must ask questions and seek answers to eliminate the initial doubt.

'These complex and multidimensional problems require a collaborative methodology that involves gaining a deep understanding of humans' (Dam & Siang, 2020, par 7). Nonetheless, the main strength of this design process is that it can introduce novel approaches that the key stakeholders directly inform.

3.2.2 Design Thinking Models (Double Diamond Model, IDEO Design Thinking and d.school Methods)

The design thinking as a process model has an established ground for both divergent and convergent thinking. Various design thinking models divide the design process into different stages (see Table 3.1). According to Kueh and Thom's review, there are 15 design thinking models. For example, according to the Double Diamond design framework developed by the British Design Council, there are four steps in the creative process—Discover, Define, Develop and Deliver (Design Council, n.d.). Like this, the Hasso Plattner Institute of Design at Stanford d.school encourages empathising, defining, ideating, prototyping and testing in a completed design process. Ambrose and Harris (2009) divided the design process into seven stages: Define, Research, Ideate, Prototype, Select, Implement and Learn. IDEO Education (2012), a leader in design thinking techniques, breaks the design process into five steps: Discovery, Interpretation, Ideation, Experimentation and Evolution. Brown (2009) opined that design thinking covers three stages: inspiration-identifying

Table 3.1 Comparison of design thinking models (Kueh & Thom, 2018)

Model	Steps in the process						
Human Centred Design Toolkit (IDEO, n.d.)	Hear	Create	Deliver				
Acumen HCD Workshop (Acumen Fund, n.d.)	Discover	Ideate	Prototype				
Design thinking - Business Innovation (Vianna, Vianna, Adler, Lucena, & Russo, 2012)	Immersion	In-depth Immersion	Analysis and synthesis	Ideation	Prototyping		
Design thinking (Cross, 2011)	Quantify problem	Generate concepts	Refine concepts	Select a concept	Design	Present	
Design thinking for Educators (IDEO, 2012)	Discover	Interpretation	Ideation	Experimentations	Evolution		
Basics Design 08 Design Thinking (Ambrose, 2010)	Define	Research	Ideate	Prototype	Select	Implement	Learn
Double Diamond (Design Council, 2015)	Discover	Define	Develop	Deliver			
IDEO (Myerson, 2001)	Observations	Brainstorming	Rapid Prototyping	Refining	Implementation		
Leading Public Sector Innovation (Bason, 2010)	Knowing	Analysing	Synthesising	Creating			
Service Design (Stickdorn & Schneider, 2011)	Exploration	Creation	Reflection	Implementation			
Collective Action toolkit (Frog, 2013)	Seek	Imagine	Make	Plan	Build		
Bootleg Bootcamp (dschool, n.d.)	Empathise	Define	Ideate	Prototype	Test		
dSchool (dSchool, 2009)	Understand	Observe	Point of View	Ideate	Prototype	Test	
Designing for growth (Liedtka & Ogilvie, 2011)	What is?	What if?	What wows?	What works?			
Business Model Generation (Osterwalder, Pigneur, & Clark, 2010)	Mobilise	Understand	Design	Implement	Manage		

☐ Context Framing Phase ☐ Ideation Phase ◼ Prototyping Phase ◼ Implementation Phase ■ Reframing Phase

a problem/an opportunity; ideation-conceive general concepts and solutions; and implementing, producing and launching the final solutions (products or services). Kueh and Thom (2018) reviewed the design processes that are most commonly used and summarised that there are five main phases: 1. Context or problem framing phase; 2. Ideation generation phase; 3. Prototyping phase; 4. Implementation phase; 5. Reframing phase.

It is of value to point out that none of the design thinking models represents a linear process. "Cyclical icons" (as seen in Fig. 3.1) are always added to design thinking models, meaning that you could shift back and forth between these states, generating the new, analysing it, shifting and often, starting the whole process again. Our mode of thinking shifts among design stages and mental states: divergent and convergent thinking, and analysis and synthesis (Brown, 2008, 2009). No matter which model is adopted for the design practice, each step in the design process leads to a creative solution that addresses a known or otherwise unknown problem. For this chapter, we use the Double Diamond model (Fig. 3.1) as an example to demonstrate the process, from information extraction to decision-making.

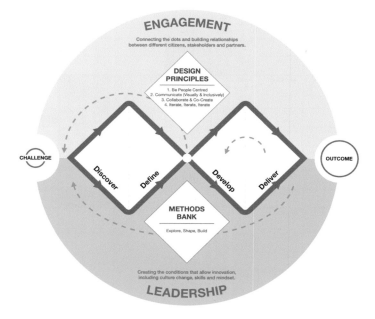

Fig. 3.1 Double Diamond model (Design Council, 2019)

3.2.3 Design 1.0–4.0 and Its Alignment with Robotics

A design approach and mindset to learning encourage understanding the complexity of a given situation. According to Jones (2013, 23–28) and Jones and VanPatter (2009), there are four levels of the design approach that are aligned with the levels of complexity in problems:

- **Design 1.0 Traditional "form-giving" Design**: This design approach focuses on creating design solutions in the form of websites, logos and posters. This deals mainly with a discrete problem that can be solved with an obvious solution. It aligns with embodied design in robotics and robotic product design.
- **Design 2.0 Service and Product Design**: This design approach seeks to explore complicated problems associated with human experiences through products and services. Designers often seek collaboration with stakeholders to explore possibilities in innovating experiences. Design 2.0 also aligns with embodied design in robotics and robotic product design.
- **Design 3.0 Organisational Transformation Design**: Commonly engaged in complex organisational challenges, designers engage in activities such as co-design of change processes for organisations and business systems. Challenges that are facing designers here are bounded by systems and strategies. Co-creation is the focus to achieve change-making processes in organisations.

- **Design 4.0 Social Transformation Design**: This design approach focuses on ill-defined wicked problems and can be challenging to solve. Design activities include iteration of prototyping interventions, observing their impact on the community and reframing the design problem. Projects in this phase involve social and systemic challenges that are difficult to define. Design 3.0 and 4.0 seem to align with the broader question of robots transforming human lives outside of industrial environments, such as caregiving robots and hospital robots—these social robots might displace human workers. This helps to understand automation in its broader context—the impact of automation and loss of work, ethics in design and broad acceptance.

Design approaches and mindsets that focus on the levels of complexity allow people to cultivate the attitude of questioning challenging situations and experimentation with opportunities. This attitude is different from the "problem-solving" mindset that was appropriate in producing products. According to Medley and Kueh (2015), the "problem-solving" approach focuses on the simple and discrete problem that sees designers being detached from stakeholder's needs, while the "experimental approach" allows designers to emphasise on empathic and reflective exploration that would contribute to more complex problems in design levels 3.0 and 4.0. Therefore, an experimental design paradigm is an approach that encourages students to understand complexity in a holistic manner. An experimental design mindset encourages students to see outcomes as interventions applied in a more extensive system.

An Industry Perspective

Alexandre Picard
Mechanical Designer, Senior

Kinova Inc.

I have a technical degree in composite material transformation and a mechanical engineering bachelor's degree. I got into the robotic industry by total coincidence. I spent the first years of my career as a product designer for a design firm playing with anything ranging from airplane components to household products. Eventually moved on to designing patient simulators (aka manikins) for the healthcare industry. About three years later, and with a baby on the way, I got sick of spending three hours a day stuck in traffic so I decided it was time for something new. I started looking for an opportunity that checked all the boxes in terms of my professional interests without the transportation hassle. I was lucky enough to stumble upon a small robotic company's job post, hi-tech designs, dynamic team, free coffee and robots! Why not? So yeah, I got the job and I've been there ever since … In short, I stumbled upon robotics because of a baby and traffic jams.

I think the most challenging portion of designing robots, and probably any product, is the constant "compromise negotiation" that is taking place between all the parties involved. It always starts with the idea of a product that can do anything at a budget price and, for fiscal reasons, that said product has to be completed and sold within a fixed timeframe. In a list of wishes and requirements, often the most rigid ones are linked to money and/or time. When designing you just have to deal with it and find ways of meeting the needs in a satisfactory manner without all the sparkles and refinements you initially had in mind. In my career, I think the most obvious example is when we designed a robot that needed to be dirt cheap compared to the competition but still at a professional quality grade. Of course, the initial drafts and requirements did not give a good perspective of achievability but, the "compromise negotiation" eventually led to what I believe was the first professional robot with a structure entirely made of plastic even with one-piece articulated fingers!

From what I see, with the design and prototyping tools expanding it will get much easier to iterate through ideas and concepts, especially for parts requiring complex or expensive production processes. It is already possible to test plastic components out of 3D printers prior to investing in tooling, and in some cases, it has become more cost-effective if the part remained printed. Also, in recent years, we have been using metal laser sintering (metal 3D printing) to produce entire robots out of aluminium to use as fully functional prototypes. I imagine that as these technologies continue to evolve and the materials offering expands, we will eventually be able to print robots using robots.

3.3   Design Thinking Process: Discover, Define, Develop and Deliver

Numerous design methods could be adopted and applied to the design thinking process to support this iterative process. This section will unfold the concept and definition of each design stage. Among the different design thinking models, we choose the Double Diamond model as a framework to demonstrate the critical concept and methods of design thinking. We will also introduce practical design methods for each stage in the design thinking process. You should know what these models and stages are, why they are helpful, and how to implement these methods at each stage.

3.3.1 What Is the Discover Mode, Why Empathise and How

According to the Double Diamond model, the discover mode is the first step in the design thinking process. The first step helps designers and non-designers understand and empathise, rather than simply assume, what the problem is (Design Council, n.d.). Empathy is the foundation of the discover stage and the core for a human-centred design (HCD) process. HCD is a systematic approach to problem-solving that focuses on empathy and encourages its practitioners to explore and understand the key stakeholders' emotions, needs and desires for which they are developing their solutions (Matheson et al., 2015). In order to empathise, you can observe, engage and immerse (d.school, n.d.).

- Observe: Observe your users and understand their behaviour in the context of their daily lives.
- Engage: Interact with your users through scheduled and short "intercept" encounters, such as interviews, focus groups and co-design workshops.
- Immerse: Put yourself into the shoes of your users and gain an "immersive" experience of what your users experience.

In order to design for the users, human-centred designers need to build empathy for who they are and what is important to them. The design tools help remove bias from the design process and help the team build a shared understanding of the users.

HCD denotes that the professionals involved consider the users' needs when designing a product. HCD is a form of innovation occasioned by developing a knowledge of people and then creating a product specifically for them, with the designer driving the process involved (Desmet & Pohlmeyer, 2013). In addition, HCD has much evidence in providing a solid approach to robotics.

Good HCD is generated from deep insights into human behaviour and a solid understanding of the users' beliefs and values. However, learning to recognise those insights, beliefs and values isn't easy. This is partly due to our minds automatically filtering out much information in ways we aren't even aware of (d.school, n.d.). To achieve this "enlightenment", you need to learn to put yourself into the users' shoes and see things "with a fresh set of eyes". Design tools for empathy, along with a human-centred mindset, could help you to tackle the problems with those fresh eyes (d.school).

Through discovering and empathise, you could engage others to

- uncover needs that people have which they may or may not be aware of
- guide innovative efforts
- identify the right users to design for

- discover the emotions that guide behaviours.

As you learn more and more about our users and their needs, ideas or possible solutions would then spring to mind. You document these ideas to make the process more tangible and generate conversation with users and stakeholders about solutions (DHW Lab, 2017).

3.3.1.1 Design Tools and Methods for Discover Mode: To Translate Ideas into Action

As identified in the framework of "Design tools and methods in the design thinking process" (Table 3.2), there are many design tools to guide innovative mind at the discover stage, including Empathy Mapping, Personas, Cultural Probes, Feedback Stations and Photo Boards. Due to the length of this chapter, we selected two essential design tools and methods for this section, they are 1. Visualising empathy and 2. Persona.

Visualising empathy

Brown (2009) and Vianna et al. (2012) identified a key element of design as having empathy and understanding for those affected by the problem. To tackle complex challenges, designers must identify, understand, reflect upon, challenge and possibly change their frame of reference, and habits of thinking. There are various empathy mapping canvases you can use, such as d.school's four-quadrant layout "Say, Do, Think and Feel" (d.school, n.d.) and Grey's "empathy mapping template" (Gray, 2017) (Table 3.3).

A simple "traditional" empathy map has a four-quadrant layout (Say, Do, Feel and Think). Table 3.1 gives a detailed explanation of the four traits. It's also an analysis

Table 3.2 Design tools and methods in the design thinking process (Double Diamond model)

Discover	Define	Develop	Deliver
Project brief	How might we?	Tomorrow's narratives	Decision matrix
Empathy mapping	Theming and coding	Science fiction prototypes	Low volume production
Personas	Design principles	Low-fi prototypes	Feedback station
Visual probes	Journey mapping	Hi-fi prototypes	Beta testing
Cultural probes	User goals	Role-play	Quantitative evaluation
Feedback stations	Rose, bud, thorn	CAD models	Full-scale testing
Photo boards	Comparing notes	Review survey	Role-play

Table 3.3 A traditional empathy mapping tool (adapted from d.school, n.d.)

SAY	DO
What are some quotes and defining words your user said?	What actions and behaviours did you notice?
FEEL	THINK
What might your user be thinking? What does this tell you about his or her beliefs?	What emotions might your subject be feeling?

tool to review your primary data from your user workshop, interview and fieldwork (Fig. 3.2).

Personas: composite character profile

The information you collected through the empathy mapping will help to create personas. What are personas? Personas are reference models, representing a subgroup of users. Technically, they can be called behavioural archetypes when they focus on capturing the different behaviours (e.g. "the conscious chooser") without expressing a defined personality or socio-demographics. The more the archetypes assume a realistic feeling (e.g. name, age, household composition, etc.), the more they become real personas, fully expressing the needs, desires, habits and cultural backgrounds of specific groups of users. Creating personas help designers to get inspired by their specific life and challenges (sdt, 2021) (Fig. 3.3).

EMPATHY MAP

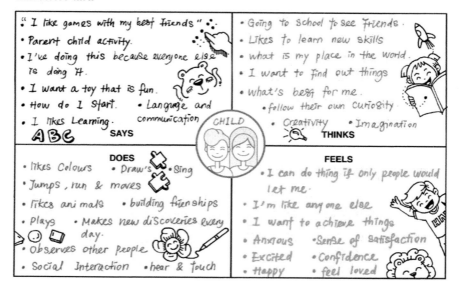

Fig. 3.2 Empathy map example (Master of Design Strategies student's coursework by Boon Khun Ooi)

Exploration of Service Design Solutions

Persona Profile 1

MAX

Max was born blind due to genetic mutations. He is an only child. His parents want a braille learning toy, for parent-child activity at home. At the same time they can improve Max's spelling and communication skills.

SOCIAL
· Socially active
· Loves family
· Loves new discoveries

GOALS
· Building friendships
· Be independent
· Having fun with friends and parents
· Wants a fun learning toy

PERSONALITY & FEELINGS
· Likes animals
· Loves stories
· Loves toys and drawing
· Likes to learn new skills

FRUSTRATIONS
· Want to achieve things
· Struggles with socialising with new people
· Very dependent on his parents
· Sensory sensitive
· Not enough braille toys in the market
· Complicated and expensive learning toys

CLINICAL
· Born blind due to genetic mutations
· Regular practitioner visit

Age : 6 years old
Gender: Male
Living Condition:
Lives with parents
School: Kindergarten
Hobbies:
Loves animals and Learning

Persona Profile 2

EMELIA

Emelia was around two years when she was diagnosed as having Retinopathy of Prematurity. She lives in a regional city with both parents and two older siblings. She interacts well with her siblings but is uncomfortable in social situations outside the family unit.

SOCIAL
· Loves family
· Likes colouring outlines
· Loves new discoveries

GOALS
· Having fun with friends and parents
· Want a fun learning toy
· Want to achieve things
· Opportunity to develop learning abilities

PERSONALITY & FEELINGS
· Likes bright colours
· Likes to dance around
· Loves toys and drawing
· Likes social interaction

FRUSTRATIONS
· Struggles with socialising with new people
· Very dependent on her parents
· Avoidant personality disorder
· Can see only light and dark
· Not confident with technology
· Complicated and expensive learning toys

CLINICAL
· Regular practitioner visit

Age : 5 years old
Gender: Female
Living Condition:
Lives with parents
School: Kindergarten
Hobbies:
Loves School and Learning

Fig. 3.3 Personas examples (Master of Design Strategies student's coursework by Boon Khun Ooi)

Quiz: key questions to ask for reflective designers at this stage

- What problem are you solving? What solutions already exist?
- What are your assumptions about the problem?
- Whom are you designing for? What types of users are involved?
- What are the constraints of the project?
- Who are the stakeholders could be involved?
- What are the needs, pain points and desires of different users?
- How might this idea solve problems or pain points for different users?

3.3.2   What Is the Define Mode, Why Ideate and How

Data collected through research and investigation during the discover phase helps us build a clearer picture of the problem. The design team group, theme and distil qualitative and quantitative findings into insights that will guide the development of design solutions.

The define mode is "convergent thinking" rather than "divergent thinking". Two goals of the define mode are 1. To develop a deep understanding of your users and the design space and 2. Based on those deep insights into human behaviour and a solid understanding of their beliefs and values, to develop an actionable problem statement. The problem statement focuses on targeted users, insights and needs uncovered during the discover mode.

At this mode, you understand the "why" is the key to addressing the "wicked problems" and provide the insights that be leveraged in design concepts to create a "how" towards a successful solution.

3.3.2.1   Design Tools and Methods for Define: To Translate Ideas into Action

Possible design tools at this stage include: Design Principles, User Journey Mapping, Theming and Coding; How Might We? Card Sorting; Hypothesis Generation.

Design principles

Design principles are fundamental laws, guidelines and strategies to solve a design challenge independent of a specific solution (d.school, n.d.). You can articulate these principles, translating your findings into design directives, such as needs and insights. These principles represent the accumulated wisdom and knowledge in design and related disciplines, including behavioural science, sociology, physics, occupational therapy and ergonomics. Many well-established design principles are critical to defining your problem-based learning. From simple to complicated, Common Principles of Design & Global Health (Design for Health, n.d.) are principles where the Bill & Melinda Gates foundation attempts to build a shared understanding, language and a shared sense of purpose between designers and global health practitioners.

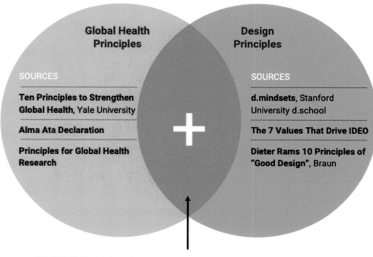

COMMON PRINCIPLES OF DESIGN & GLOBAL HEALTH

Fig. 3.4 Common principles of design & global health (Design for Health, n.d.)

This set of simple statements, some more aspirational than others, demonstrates the alignment and commitment by designers to longstanding global health principles and values. This resource outlines a code of practice for design in global health (Fig. 3.4).

User journey mapping

The journey map is a synthetic representation that describes step-by-step how a user interacts with a service. The process is mapped from the user perspective, describing what happens at each stage of the interaction, what touchpoints are involved, what obstacles and barriers they may encounter. The journey map is often integrated with additional layers representing the level of positive/negative emotions experienced throughout the interaction (sdt, 2021) (Fig. 3.5).

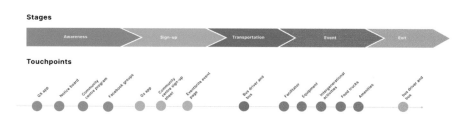

Fig. 3.5 A touchpoint diagram is a graphical representation of how the user interacts with the service (Master of Design Strategies student's coursework by Jordan Mckibbin)

Recap: key questions to ask for reflective practitioners at this stage

- What are the common needs or pain points for users?
- Where in the journey are they experienced or desired?
- How did users or stakeholders respond to ideas presented?
- Who might benefit most from the ideas presented?

3.3.3 What Is the Develop Mode, Why Ideate and Prototype and How

Once you've defined your insights and identified areas to improve the user experience, you begin developing design concepts explored during discover mode or generate further ideas in response to our insights. There are two key concepts in the develop mode: 1. Ideate and 2. Prototype.

Ideation is a mode of divergent thinking rather than convergent thinking. You ideate to generate radical design ideas, concepts and alternatives. The goal of ideation is to explore both a large number of ideas and a diversity among those ideas (d.school).

To further develop the diverse and large quantity of ideas during ideation, prototypes are built to test with users from this vast depository of ideas. Prototypes are "any representation of a design idea, regardless of the medium" (Houde & Hill, 1997, 369). Prototyping is a process of "building, visualising and translating a rough concept into collectively understandable, defined and defendable ideas" (Kocsis, 2020, 61).

> Prototypes traverse from low-fidelity representations in the initial stages (discover and define) of designing to high-fidelity realisations when design outcomes near finalisation (develop and deliver) and can include haptic, oral, digital, spatial, virtual, visual, graphical and also modes beyond a purely technical functional scope through embodied representations of communication such as art, dance and performance. (Kocsis, 2020, 61)

Prototyping facilitates an iterative, interactive communication process. A prototype tests if parts work together for the intended design. This allows further exploration of risks, opportunities and refining of the iterative prototype into the next phase (deliver). "Practices oscillate between creation and feedback: creative hypotheses lead to prototypes, leading to open questions, leading to observations of failures, leading to new ideas and so on" (Dow et al., 2009, 26).

3.3.3.1 Design Tools and Methods for Develop Mode: To Translate Ideas into Action

There are various prototyping tools for this stage, including the low-fi prototype, high-fi prototype, desktop walkthrough, role-play, science fiction prototype and 3D printed prototype. (Chapter 2.7 in the Embodied Design section will discuss 3D Printed Prototypes and CAD in more detail.)

Role-playing

Role-play is a representation tool often used during co-design sessions; it explains a service or product idea by acting out an exemplificatory scenario. Role-playing could be applied at different stages of the design thinking process, not limited to develop mode. Role-playing is a popular technique for building empathy in the discover mode and demonstrating the user experience in the develop mode. It typically requires defining some roles or personas (e.g. Max and Emelia in Fig. 3.3, the service provider, etc.) and preparing rough prototypes (e.g. paper prototypes) or other materials that can facilitate the performance. While a team is acting out their story with given scenarios, the rest of the participants learn about the idea, understand the high-level sequence of actions required, and gain an immersive experience of the actual user experience (sdt, 2021 and Stickdorn & Schneider 2011) (Fig. 3.6).

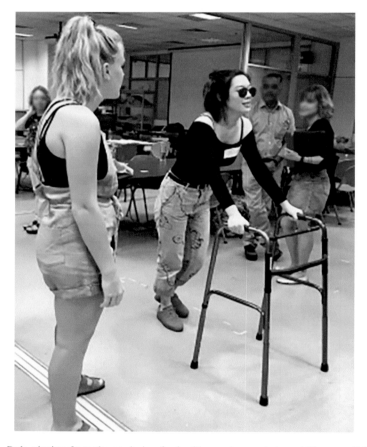

Fig. 3.6 Role-playing from the co-design for healthy ageing workshop at Nanyang Polytechnic 2019

Recap: key questions to ask for reflective practitioners at this stage

- How do users respond or interact with solutions?
- What do users find easy or difficult about our solutions?
- What can we do to improve the prototype?

3.3.4 What Is the Deliver Mode, Why and How

The final stage is delivering the design solutions. Following design development/prototyping, concept testing and review sessions, potential solutions are narrowed down based on assessment criteria. "The process of designing, building and testing continues to go through iterations until you achieve the final solution" (Automeme, n.d.). The process of prototype testing and looping in feedback also provides continuity to create a seamless way forward in the HCD. The final solution (e.g. robot) delivered should be created to empathise with the customer requirements and concerns. The validation and evaluation process is crucial so organisations spend a good chunk of time testing the prototype against business objectives and metrics. Upon completion of detailed design and production, the realised solution will be physically installed or digitally implemented into the business environment, depending on the type of project.

3.3.4.1 Design Tools and Methods for Deliver Mode: To Translate Ideas into Action

Possible design tools and methods at this step: decision matrix, full-scale testing, system map and feedback stations.

Decision matrix

A decision matrix is an analysis tool to compare and evaluate to select the best option between different options. Through the develop mode, you developed several design prototypes and there are several factors you need to consider. Decision matrix can help you to make your final decision. Between more than one option in order to make your final decision.

There are various formats and styles that you can adopt. Using the sample decision matrix as an example, you can list each of the criteria/metrics you evaluate against in the left column of the table. You then place the options available to you across the top row of your table. For the scoring system, you can choose different systems. Table 3.4 chooses the scale of 1–5, with 5 being a good score and 1 being a very poor score. In the bottom row, you can sum all the scores for each option for your decision-making.

Recap: key questions to ask for reflective practitioners at this stage

- What will it cost to manufacture a high-fidelity prototype?

Table 3.4 Simple decision matrix

Criteria	Options		
	Option 1	Option 2	Option 3
Criteria 1	x	x	x
Criteria 2	x	x	x
Criteria 3	x	x	x
Criteria 4	x	x	x
Criteria 5	x	x	x
Total	x	x	x

x: choose the scale of 1–5, with 5 being a good score and 1 being a very poor score

- What additional capability might you need to deliver the design?
- What existing channels can you leverage to implement our solution?
- What is change management required to implement our solution?
- What criteria are you evaluating against?
- What is the best way to measure the success of this solution?

3.4 Conclusion

This chapter provides valuable and practical guidance on design thinking models and tools for people interested in applying design thinking in their projects. Design thinking is an iterative process, which encourages people to empathise, collaborate and prototype. Doing so helps to generate user-centred design to tackle wicked problems in our society.

This chapter covered the history of the "designerly way of thinking" to introduce the origin of design thinking. The development of Design 1.0–4.0, in comparison to the field of robotic, helped provide a context for the past, present and future.

The design thinking process was then deconstructed into different stages to provide a practical toolkit for people from non-design backgrounds to adopt. Many existing design methods can be used for different stages in the design thinking process. Some of them would be applied from the start to the end, such as service blueprint and prototyping. Due to the length of the chapter, we could not include all the existing design methods. However, the key design methods included in this chapter provided a solid ground for the entry level of design thinking. Design thinking in robotics allows practitioners and researchers to seek opportunities through which they can discover, define, develop and deliver value to their stakeholders and additionally, get them engaged, and create ripples of change.

3.5   Quiz

- What is the difference between divergent and convergent thinking?
- What are some key stages in the design thinking process?
- Name some design tools incorporated in achieving iterative processes in design thinking.
- What design methods can you adopt to advance your empath in the discover stage?
- What methods can you employ to test your concepts in the second diamond stages?

References

Ambrose, G., & Harris, P. (2009). *Basic design: Design thinking*. Fairchild Books AVA.

Automeme. (n.d.). Why is design thinking important in robotics automation? Retrieved November 9, 2021, from https://autome.me/why-is-design-thinking-important-in-robotics-automation/#:~: text=The%20impending%20need%20for%20Design,learn%20and%20develop%20amiable% 20personalities

Brown, T. (2008). Design thinking. *Harvard Business Review, 86*(6), 84–92.

Brown, T. (2009). *Change by design*. Harper Collins.

Buchanan, R. (1992). Wicked problems in design thinking. *Design Issues, 8*(2), 5–21.

Cross, N. (2001). Designerly ways of knowing: Design discipline versus design science. *Design Issues, 17*(3), 49–55.

Dam, R., & Siang, T. (2020). What is design thinking and why is it so popular? Retrieved June 9, 2020, from https://www.interaction-design.org/literature/article/what-is-design-thinking-and-why-is-it-so-popular

Design Council. (n.d.). *What is the framework for innovation?* Design Council's evolved Double Diamond (online). Retrieved November 9, 2021, from https://www.designcouncil.org.uk/news-opinion/what-framework-innovation-design-councils-evolved-double-diamond

Design Council. (2019). Double Diamond model. Retrieved May 9, 2022, from https://www.des igncouncil.org.uk/our-work/news-opinion/double-diamond-15-years/

Design for Health. (n.d.). *Common principles of design & global health*. Bill & Melinda Gates foundation.

Desmet, P. M. A., & Pohlmeyer, A. E. (2013). Positive design: An introduction to design for subjective well-being. *International Journal of Design, 7*(3), 5–19.

Dewey, J. (1938). *Logic: The theory of inquiry*. Holt, Rinehart and Winston.

DHW Lab. (2017). *How we design: Better healthcare experiences at Auckland City Hospital*. Design for Health & Wellbeing Lab.

Dow, S. P., Heddleston, K., & Klemmer, R. S. (2009). The efficacy of prototyping under time constraints. In *Proceedings of the Seventh ACM Conference on Creativity and Cognition*.

d.school. (n.d.). *Bootcamp bootleg*. Institute of Design at Stanford.

Frog. (2013). Frog collective action toolkit. Retrieved June 20, 2016, from http://www.frogdesign. com/work/frog-collective-action-toolkit.html

Gray, D. (2017). Empathy map (online). Retrieved November 9, 2021, from Xplane.com

Houde, S., & Hill, C. (1997). What do prototypes prototype? In *Handbook of human-computer interaction* (pp. 367–381). North-Holland.

IDEO Education. (2012). *Design thinking for educators*. IDEO.

Johnson, B. D. (2011). Science fiction prototyping: Designing the future with science fiction. *Synthesis Lectures on Computer Science, 3*(1), 1–190.

Jones, P. H. (2013). *Design for care: Innovating healthcare experience*. Rosenfeld.

Jones, P. H., & VanPatter, G. K. (2009). Design 1.0, 2.0, 3.0, 4.0: The rise of visual sensemaking. NextDesign Leadership Institute.

Kocsis, A. (2020). Prototyping: The journey and the ripple effect of knowledgeability. *Fusion Journal* (18).

Kueh, C., & Thom, R. (2018). Visualising empathy: A framework to teach user-based innovation in design. In S. Griffith, K. Carruthers, & M. Bliemel (Eds.), *Visual tools for developing student capacity for cross-disciplinary collaboration, innovation and entrepreneurship.* Common Ground Publishing.

Liedtka, J., & Ogilvie, T. (2011). *Designing for growth: A design thinking tool kit for managers.* Columbia Business School Pub., Columbia University Press.

Matheson, G. O., Pacione, C., Shultz, R. K., & Klügl, M. (2015). Leveraging human-centred design in chronic disease prevention. *American Journal of Preventive Medicine,48*(4), 472–479. https://doi.org/10.1016/j.amepre.2014.10.014

Medley, S., & Kueh, C. (2015). *Beyond problem solving: A framework to teach design as an experiment in the university environment.* Paper presented at the Ministry of Design: From Cottage Industry to State Enterprise, St Augustine.

Myerson, J. (2001). *Ideo: Masters of innovation.* Laurence King.

Osterwalder, A., Pigneur, Y., & Clark, T. (2010). *Business model generation.* Wiley.

sdt. (2021). Journey map: Describe how the user interact with the service, throughout its touchpoints. Retrieved November 15, 2021, from https://servicedesigntools.org/tools/journey-map

Stickdorn, M., & Schneider, J. (2011). *This is service design thinking.* Wiley.

Vianna, M., Vianna, Y., Adler, I., Lucena, B., & Russo, B. (2012). *Design thinking business innovation.* MJV Press.

Fanke Peng is an Associate Professor and Enterprise Fellow at the University of South Australia. She is an award-winning educator, designer and researcher in design-led innovation, design for health, digital fashion and cross-cultural design. She has been heavily involved in extensive research projects in the UK and Australia, including Australian Council for the Arts project: Home Economix, ACT Government Seniors Grants Programme project: ACT Intergenerational Pen Pal Service, Economic and Social Research Council (ESRC) project: E-Size, Technology Strategy Board (TSB) project: Monetising Fashion Metadata and Fashioning Metadata Production Tools, Engineering and Physical Science Research Council (EPSRC) project: Body Shape Recognition for Online Fashion, and an Arts & Humanities Research Council (AHRC) project: Past Present and Future Craft Practice.

Fanke is passionate about design-led innovation and design for health and wellbeing. To share this passion with students and inspire them. She has developed new courses and units, established innovative work-integrated learning opportunities and international faculty-led programmes (FLP). In 2020, she was awarded Senior Fellow of the Higher Education Academy (SFHEA).

Chapter 4
Software Building Blocks: From Python to Version Control

Damith Herath, Adam Haskard, and Niranjan Shukla

4.1 Learning Objectives

Software is an essential part of robotics. In this chapter, we will be looking at some of the key concepts in programming and several tools we use in robotics. At the end of the chapter, you will be able to:

- Develop a familiarity with common programming languages used in robotics
- Learn about the fundamental programming constructs and apply them using the Python programming language
- Understand the importance of version control and how to use basic commands in Git
- Select appropriate tools and techniques needed to develop and deploy code efficiently

4.2 Introduction

Whether working with an industrial-grade robot or building your hobby robot, it is difficult to avoid coding. Coding or programming is how you instruct a robot to perform a task. In robotics, you will encounter many different programming languages, including programming languages such as C++, Python, and scientific

D. Herath (✉)
Collaborative Robotics Lab, University of Canberra, Canberra, ACT, Australia
e-mail: Damith.Herath@Canberra.edu.au

A. Haskard
Bluerydge, Canberra, ACT, Australia
e-mail: Adam.Haskard@blurydge.com

N. Shukla
Accenture, Canberra, ACT, Australia

© The Author(s) 2022
D. Herath and D. St-Onge (eds.), *Foundations of Robotics*,
https://doi.org/10.1007/978-981-19-1983-1_4

languages like MATLAB®. While many of the examples in this book will utilise Python, there will be instances where we will use code examples in C/C++ or MATLAB®. While we do not assume any prior programming knowledge, previous coding experience will undoubtedly help you advance quicker.

The following section will briefly outline some of the essential programming constructs. By any means, this is neither exhaustive nor comprehensive. It is simply to introduce you to some fundamental programming concepts that will be useful to get started if you do not already have any programming experience. We will begin with a few essential programming tools such as flowcharts and pseudocode and then expand into fundamental building blocks in programming. If you already have some experience in programming, you may skip this section.

In the subsequent sections, we will discuss two important software tools that would be extremely useful in programming robots, version control and containerisation. While these are all great starting points, there is no better way to build your confidence and skills than to practice and dive into coding. So, we will introduce many case studies and provide code snippets throughout the book for you to follow and try and a comprehensive set of projects at the end of the book. Once you have some confidence, you must explore new problems to code to develop your skills.

4.2.1 Thinking About Coding

As you may have already noticed, we use programming and coding interchangeably, and they both mean instructing your robot to do something logically. Before you start programming, it is essential to understand the problem you are going to address and develop an action plan for how to construct the code. Flowcharts and pseudocode are two useful tools that will help you with this planning phase. Once you have the programme's general outline, you will need to select the appropriate programming language for the task. For tasks where execution speed is important or low-level hardware is involved, this is usually a language like C or C++. However, when the intention is rapid prototyping, a language like Python comes in handy. Robotics researchers also tend to use languages like MATLAB® that are oriented towards mathematical programming. MATLAB® is a proprietary language developed by MathWorks[1] and provides a set of toolboxes with commonly used algorithms, data visualisation tools, allowing for testing complex algorithms with minimal coding. In addition to such code-based languages, several visual programming languages such as Max/MSP/Jitter, Simulink, LabVIEW, LEGO NXT-G are regularly used by roboticists, artists and enthusiasts for programming robots and robotic systems. Whatever language you use, the basic programming constructs are the same.

Irrespective of the programming language used, it is common to think of a programme as a set of inputs to be processed to deliver the desired output (Fig. 4.1).

---

[1] https://www.mathworks.com/.

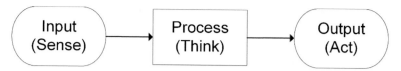

Fig. 4.1 A simple program flows from input to output after processing in the middle

In robotics, a similar framework is used called the sense–think–act loop, which we will explore further in Chap. 7.

4.2.1.1   Flowcharts

Flowcharts are a great way to think about and visualise the flow of your program and the logic. They are geometric shapes connected by arrows (see Figs. 1 and 2. The geometric shapes represent various activities that can be performed, and the arrows indicate the order of operation (*flowline*). Generally, the flowcharts flow from top to bottom and left to right. Flowcharts are a handy tool to have when first starting in programming. They give you a visual representation of the programme without needing to worry about the language-specific syntax. However, they are cumbersome to use in large programmes.

In the following sections, we will explore the meaning of these symbols further.

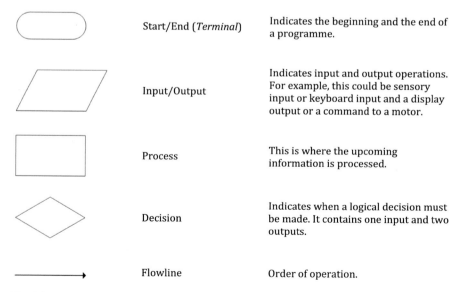

Fig. 4.2 Common flowchart elements

Fig. 4.3 A simple
pseudocode example with a
repetitive read, process,
output loop

Program input_process_output
repeat
read input data
process input data
output the processed data
until user exit

4.2.1.2   Pseudocode

Pseudocode is another tool that you can use to plan your code. You could think of them as simply replacing the geometric shapes discussed in the previous section in flowcharts with instruction based on simple English language statements. As the name suggests, pseudocode is programming code without aligning with a specific programming language. Therefore, pseudocode is a great way to write your programming steps in a code-like manner without referring to any particular language. For example, the input, process, output idea could be presented in simple pseudocode form, as shown in Fig. 4.3. In this example, we have extended the previous program by encompassing the read, process, output block within a repetitive loop structure, discussed later in the chapter. In this variation of the program, the input, process, output sequence repeats continually until the user exits the program. The equivalent flowchart is shown in Fig. 4.4.

4.3   Python and Basics of Programming

First released in the 1990s, Python[2] is a high-level programming language. Python is an interpreted language meaning it is processed while being executed compared to a compiled language which needs to be processed before it is executed. Python has become a popular language for programming robots. This may be due to its easily readable language, the visually uncluttered and dynamically typed nature, and the availability of many ready-to-use libraries that provide common functionalities such as mathematical functions. Python is useful when you want to rapidly prototype as it requires minimal lines of code to realise complex tasks. It also alleviates another major headache for beginner programmers by being a garbage collecting language. Garbage collection is the automatic process by which memory is managed and used by the program.

Python uses indentation (whitespace or a tab inserted at the beginning of a line of code) to identify blocks of code. Unlike languages like C/C++ and Java that uses curly brackets { } to delimit code blocks, it is vital to maintain proper indentation

---

[2] https://www.python.org/.

Fig. 4.4 Flowchart diagram
of a simple read, process,
output loop

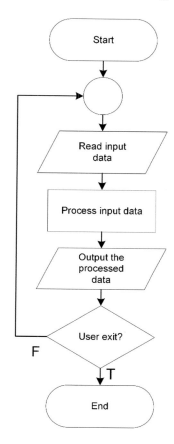

in Python for your code to work correctly. This requirement also improves code readability and aesthetics.

Let us now explore some of the common programming constructs with the help of Python as the example language.

4.3.1 Variables, Strings and Assignment Statements

Python is a dynamically typed language, which means that the variables are not statically typed (e.g. string, float, integer). Therefore, developers do not need to declare variables before using them or declare their type. In Python, all variables are an object.

A typical component of many other programming languages is that variables are declared from the outset with a specific data type, and any value assigned to it during its lifetime must always have that type. One of the accessibility components of Python is that its variables are not subject to this restriction. In Python, a variable may be assigned a value of one type and later reassigned a new value of a different type. Every value in Python has a datatype. Other data types in Python include Numbers, Strings, Dictionary and many more. Variables are quickly declared by any name or even alphabets like a, ab, abc, so on and so forth.

Strings are a useful and widely used data type in Python. We create them by enclosing characters in quotes. Python treats single quotes and double quotes the same. Creating strings is as simple as assigning a value to a variable. For example,

```
var1 = 'Hello World!'
var2 = "Banana Robot"
```

We see two variables notated by the 'var1' and 'var2' labels in the example above. A simple way is to think of a variable as a name attached to a particular object. To create a variable, you just assign it a value and then start using it. The assignment is achieved with a single equal sign (=).

4.3.2 Relational and Logical Operators

To manage the flow of any program and in every programming language, including Python, conditions are required. Relational and logical operators define those conditions.

As an example, and for context, when you are asked if 3 is greater than 2, the response is yes. In programming, the same logic applies.

When the compiler is provided with some condition based on an *expression*, it computes the expression and executes the condition based on the output of the expression. In the case of relational and logical expressions, the answer will always be either *True* or *False.*

Operators are conventional symbols that bring *operands* together to form an expression. Thus, operators and operands are the deciding factors of the output.

Relational operators are used to define the relationship between two operands. Examples are less than, greater than or equal to operators. Python understands these types of operators and accordingly returns the output, which can be either `True` or `False`.

```
1 < 10
True
```

1 is Less Than 10, so the Output Returned is True.

A simple list of the most common operators:

1. **Less than** → used with <
2. **Greater than** → used with >
3. **Equal to** → used with = =
4. **Not equal to** → used with ! =

5. Less than or equal to → used with <=
6. Greater than or equal to → used with >=

Logical operators are used in expressions where the operands are either True or False. The operands in a logical expression can be expressions that return True or False upon evaluation.

There are three basic types of logical operators:

1. **AND**: For AND operation, the result is True if and only if both operands are True. The keyword used for this operator is and.
2. **OR**: For OR operation, the result is True if either of the operands is True. The keyword used for this operator is or.
3. **NOT**: The result is True if the operand is False. The keyword used for this operator is not.

4.3.3 Decision Structures

Decision structures allow a program to evaluate a variable and respond in a scripted manner. At its core, the decision-making process is a response to conditions occurring during the execution of the program, with consequential actions taken according to the conditions. Basic decision structures evaluate a series of expressions that produce TRUE or FALSE as the output. The Python programming language provides you with the following types of decision-making sequences.

1. **if** statements: An if statement consists of a Boolean expression followed by one or more statements.
2. **if...else** statements: An if statement can be followed by an optional else statement, which executes when the Boolean expression is FALSE.
3. **nested if** statements: You can use one if or else if statement inside another if or else if statement(s).

Below is an example of a one-line if clause,

```
# this is a comment (beginning with the # symbol).
# Comments are important documentation element in programming
var = 1300#a variable assignment
if (var == 1300): print "Value of expression is 1300" #decision
structure in a single line
print "Bye!"#display the word Bye!
```

When the above code runs, the following is the output,

```
Value of expression is 1300
Bye!
```

In general, statements are executed sequentially. The first statement in a function is executed first, followed by the second, and so on. It is good to think of code as just a set of instructions, not too different from a favourite cooking recipe. There may be a situation when you need to execute a block of code several times. A loop statement allows us to execute a statement or group of statements multiple times.

4.3.4 Loops

There are typically three ways for executing loops in Python. They all provide similar functionality; however, they differ in their syntax and condition checking time.

1. **While loop**: Repeats a statement or group of statements while a given condition is TRUE. It tests the condition before executing the loop body.
2. **For loop**: Executes a sequence of statements multiple times and abbreviates the code that manages the loop variable.
3. **Nested loops**: You can use one or more loops inside any another while, for or do..while loop.

```
# while loop.
count = 0.
while (count < 3):
    count = count + 1#note the indentation to indicate this section
of the code is inside the loop.
    print("Hello Robot")
```

When the code above is run, we would expect to see the following output.

```
Hello Robot
Hello Robot
Hello Robot
```

4.3.5 Functions

A function is a block of code designed to be reusable which is used to perform a single action. Functions give developers modularity for the application and a high degree of reusable code blocks. A well-built function library lowers development time significantly. For example, Python provides functions like print(), but users can develop their own functions. These functions are called user-defined functions.
e.g.

```
def robot_function():
print("Robot function executed")
# You can then call this function in a different part of your program;
robot_function()
```

When you execute the code, the following will be displayed.

```
Robot function executed
```

You can pass external information to the function as arguments. Arguments are listed inside the parentheses that come after the function name.

e.g.

```
def robot_function(robot_name):
print("Robot function executed for robot named " + robot_name)
```

We have modified the previous function to include an argument called robot_name. When we call the new function, we can now include the name of the robot as an argument:

```
robot_function('R2-D2')
which will result in the following output.
Robot function executed for robot named R2-D2
```

4.3.6 Callback Function

A *callback* function is a special function that can be passed as an argument to another function. The latter function is designed to call the former callback function in its definition. However, the callback function is executed only when it is required. You will find many uses for such functions in robotics. Particularly, when using ROS, you will see the use of callback functions to read and write various information to and from robotic hardware which may happen asynchronously. A simple example illustrates the main elements of a callback function implementation.

```
def callbackFunction(robot_status):
print("Robot's current status is " + robot_status)
def displayRobotStatus(robot_name, callback):
#     This function takes robot_name and a callback function as
arguments
#   The    code to read the robot status (stored in the variable
robot_status) goes here
# the read status is then passed to the callback function
callback(robot_status)
```

You can now call the displayRobotStatus function in your main program.

```
if __name__ == '__main__':
displayRobotStatus ("R2-D2", callbackFunc)
```

4.4 Object-Oriented Programming

Object-oriented programming (OOP) is a programming paradigm based on the concept of 'objects', which may contain data in the form of fields, often known as attributes, and code in the form of procedures, often known as methods. Here is a simple way to think about this idea;

1. A person is an object which has certain properties such as height, gender and age.
2. The person object also has specific methods such as move, talk and run.

Object—The base unit of object-oriented programming that combines data and function as a unit.

Class—Defining a class is defining a blueprint for an object. Describes what the class name means, what an object of the class will consist of and what operations can be performed on such an object. A class sets the blank canvas parameters for an object.

OOP has four basic concepts,

1. **Abstraction**—It provides only essential information and hides their background details. For example, when ordering pizza from an application, the back-end processes for this transaction are not visible to the user.
2. **Encapsulation**—Encapsulation is the process of binding variables and functions into a single unit. It is also a way of restricting access to certain properties or components. The best example for encapsulation is the generation of a new class.
3. **Inheritance**—Creating a new class from an existing class is called inheritance. Using inheritance, we can create a child class from a parent class such that it inherits the properties and methods of the parent class and can have its own additional properties and methods. For example, if we have a class robot with properties like model and type, we can create two classes such as Mobile_robot and Drone_robot from those two properties, and additional properties specific to them such that Mobile_robot has a number of wheels while a Drone_robot has a number of rotors. This also applies to methods.
4. **Polymorphism**—The definition of polymorphism means to have many forms. Polymorphism occurs when there is a hierarchy of classes, and they are related by inheritance.

4.5 Error Handling

A Python program terminates as soon as it encounters an error. In Python, an error can be a syntax (typo) error or an exception. Syntax errors occur when the python parser detects an incorrect statement. Observe the following example:

```
>>> print( 0 / 0 ))
1 ^
SyntaxError: invalid syntax
```

The arrow character points to where the parser has run into a **syntax error**. In this example, there was one bracket too many. When it is removed, the code will run without any error:

```
>>> print( 0 / 0)
Traceback (most recent call last):
File "<stdin>", line 1, in <module>
ZeroDivisionError: integer division or modulo by zero
```

This time, Python has 'thrown' an **exception error**. This type of error occurs whenever correct Python code results in an error. The last line of the message indicated what type of exception error was thrown. In this instance, it was a ZeroDivisionError. Python has built-in exceptions. Additionally, the possibility exists to create user-defined exceptions.

4.6 Secure Coding

Writing secure code is essential for protecting data and maintaining the correct behaviour of the software. Writing secure code is a relatively new discipline, as typically developers have been commissioned to write functions and outputs, not necessarily in a secure manner. However, given the prevalence of exploits, it is important developers build in sound security practices from the outset.

Python development security practices to consider:

1. **Use an up-to-date version of Python:** Out of date versions have since been rectified with vulnerability updates. Not incorporating the updates into the python environment ensures vulnerabilities are available to exploit.
2. **Build the codebase in a sandbox environment:** Using a sandbox environment prevents malicious Python dependencies pushed into production. If malicious packages are present in Python environments, using a virtual environment will prevent having the same packages in the production codebase as it is isolated.
3. **Import packages correctly:** When working with external or internal Python modules, ensure they are imported using the right paths. There are two types of import paths in Python, and they are absolute and relative. Furthermore, there are two types of relative imports, implicit and explicit. Implicit imports do not specify the resource path relative to the current module, while Explicit imports specify the exact path of the module you want to import. Implicit import has been disapproved and removed from Python 3 onwards because if the module specified is found in the system path, it will be imported, and that could be very dangerous, as it is possible for a malicious module with an identical name to be in an open-source library and find its way to the system path. If the malicious module is found before the real module, it will be imported and used to exploit applications in their dependency tree. Ensure either absolute import or explicit relative imports as it guarantees the authentic and intended module.
4. **Use Python HTTP requests carefully:** When you send HTTP requests, it is always advisable to do it carefully by knowing how the library you are using handles security to prevent security issues. When you use a common HTTP request library like Requests, you should not specify the versions down in

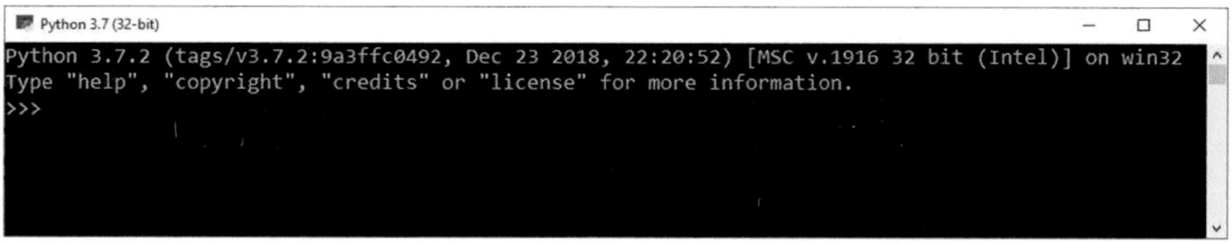

Fig. 4.5 Python command line

your requirements.txt because in time that will install outdated versions of the module. To prevent this, ensure you use the most up-to-date version of the library and confirm if the library is handling the SSL verification of the source.
5. Identify exploited and malicious packages.

Packages save you time as you don't need to build artefacts from scratch each time. Packages can be easily installed through the *Pip* package installer. Python Packages are published to PyPI[3] in most cases, which essentially is code repository for Python Packages which is not subject to security review or check. This means that PyPI can easily publish malicious code.

Verify each Python package you are importing to prevent having exploited packages in your code. Additionally, use security tools in your environment to scan your Python dependencies to screen out exploited packages.

4.7 Case Study—Writing Your First Program in Python

To start experimenting with Python, you can install the current version of the Python program from the Python website.[4] Follow the instruction on this website to download the recommended current version of your operating system. Once installed, you can call the **Python (command line)** shell for an interactive programming environment (see Fig. 4.5).

In any programming language, the Hello World program is a shared bond between all coders. You can go ahead and make your own 'hello world' program. Look at the classic example below. Note that the # symbol is a comment line, which means Python does not read this as code to execute. Instead, it is intended for human audiences, so coders can easily see what each line of code is supposed to do. Commenting well and regularly is key to good collaboration and development hygiene.

```
# This program prints Hello, world!
print('Hello, world!')
```

Output.

```
Hello, world!
```

---

[3] https://pypi.org/.

[4] https://www.python.org/downloads/.

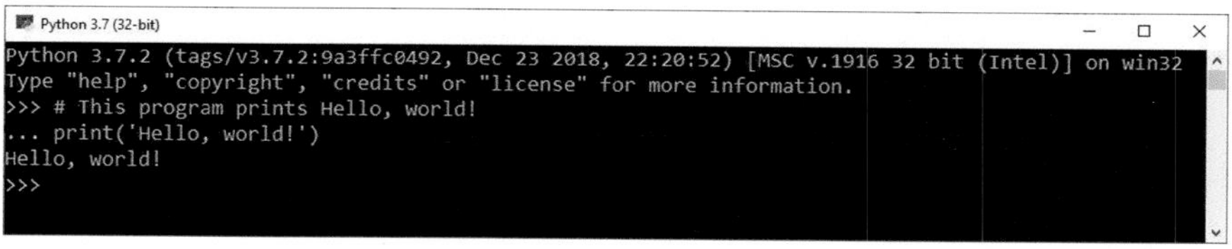

Fig. 4.6 Hello, World program interactively executed in a Python command line window

4.7.1 A Note on Migrating from MATLAB® to Python

As you dwell into robotics programming and writing algorithms, you will notice that many examples are written in MATLAB®, particularly in academia due to previously mentioned reasons. However, there are compelling reasons to use Python instead of a proprietary language like MATLAB. One of the main reasons is the cost of acquiring MATLAB and related toolboxes. Python allows you to easily distribute your code without worrying about your end-users needing to purchase MATLAB® licences to run your code. In addition, Python being a general-purpose programming language offers you a better development environment for projects targeting a wide use and deployment audience.

If you are thinking of migrating any code from MATLAB® to Python, the good news is that the two languages are 'very similar'. This allows for relatively easy transitioning from MATLAB to Python. One of the key reasons for MATLAB's popularity has been its wide array of well-crafted toolboxes by experts in the field. For example, there are several popular toolboxes related to robotics including the *Robotics Toolbox* developed by Peter Corke.[5] These toolboxes provide specific mathematical functions reducing the time it takes to develop new code when building or testing new ideas for your robot. Python also offers a similar mechanism to expand its capabilities through Python packages. For example, one of the powerful elements of MATLAB is its native ability to work with matrices and arrays (side note: matrices and arrays will play a major role in robotics programming!). Python, being a general-purpose language does not have this capability built-in. But a package available in Python called NumPy[6] provides a way to address this through multidimensional arrays allowing you to write fast, efficient, and concise matrix operators comparable to MATLAB. As your knowledge in robotics and programming matures, it would be a worthwhile investment to spend some time to explore the similarities and differences between the two languages and to understand when to utilise one or the other. Figure 4.7 shows our humble Hello world program being executed in a MATLAB® command line window. Can you spot the differences between the syntaxes from our Python example in Fig. 4.6?.

---

[5] https://petercorke.com/toolboxes/robotics-toolbox/.

[6] https://numpy.org/.

```
Command Window                                                            ⊙
New to MATLAB? See resources for Getting Started.                         ×
  >> %Hello, world! example written in MATLAB(r)
  >> disp('Hello, world!');
  Hello, world!
fx >> |
```

Fig. 4.7 Hello, World program interactively executed in a MATLAB command line window

4.8 Version Control Basics

Version control is the practice of managing changes to the codebase over time and potentially between multiple developers working on the same project. It is alternatively called source control. Version control provides a snapshot of development and includes tracking of code *commits*. It also provides features to merge the code contributions arising from multiple sources, including managing merge conflicts.

A version control system (or source control management system) allows the developer to provide a suite of features to track code changes and switch to previous versions of the codebase. Further, it provides a collaborative platform for teamwork while enabling you to work independently until you are ready to commit your work. A version control system aims to help you streamline your work while providing a centralised home for your code. Version control is critical to ensure that the tested and approved code packages are deployed to the production environment.

4.8.1 Git

Git is a powerful open-source distributed version control system.[7] Unlike other version control systems, which think of version control as a list of file-based changes, Git thinks of its data more like a series of snapshots of a miniature filesystem. A snapshot is a representation of what all the files look like at a given moment. Git stores reference to snapshots as part of its version management.

Teams of developers use Git in varying forms because of Git's distributed and accessible model. There is no policy on how a team uses Git. However, projects will generally develop their own processes and policies. The only imperative is that the team understands and commits to the workflow process that maximises their ability to commit code frequently and minimise merge conflicts.

A Git versioned project consists of three areas: the *working tree*, the *staging area* and the *Git directory*.

---

[7] https://git-scm.com/.

As you progress with your work, you typically stage your commits to the staging area, followed by committing them to the Git directory (or repository). At any time, you may *checkout* your changes from the Git directory.

4.8.1.1   Install Git

To check if Git has already been bundled with your OS, run the following command (at the command prompt):

```
git --version
```

To install Git, head over to the download site[8] and select the appropriate version for your operating system and follow the instructions.

4.8.1.2   Setting up a Git Repository

To initialise a Git repository in a project folder on the file system, execute the following command from the root directory of your folder:

```
git init
```

Alternatively, to clone a remote Git repository into your file system, execute the following command:

```
git clone <remote_repository_url>
```

Git repositories provide SSH URLs of the format git@host:user_name/repository_name. git.

Git provides several commands for this syncing with a remote repository:

Git remote: This command enables you to manage connections with a remote repository, i.e. create, view, update, delete connections to remote repositories. Further, it provides you with an alias to reference these connections instead of using their entire URL.

The below command would list the connections to all remote repositories with their URL.

```
git remote -v
```

The below command creates a new connection to a remote repository.

```
git remote add <repo_name> <repo_url>
```

The below command removes a connection to a remote repository.

```
git remote rm <repo_name>
```

---

[8] https://git-scm.com/download/.

The below command renames a remote connection from repo_name_1 to repo_name_2

```
git remote rename <repo_name_1> <repo_name_2>
```

Upon cloning a remote repository, the connection to the remote repository is called origin.

To pull changes from a remote repository, use either *Git fetch* or *git pull.*

To fetch a specific branch from the remote repository, execute the below command:

```
git fetch <repo_url> <branch_name>
```

where repo_url is the name of the remote repository, and branch_name is the name of the branch.

Alternatively, to fetch all branches, use the below command:

```
git fetch —all
```

To pull the changes from the remote repository, execute the following command:

```
git pull <repo_url>
```

The above command will fetch the remote repository's copy of your current branch and will merge the changes into your current branch.

If you would like to view this process in detail, use the verbose flag, as shown below

```
git pull —verbose
```

As git pull uses merge as a default strategy, if you would like to use rebase instead, execute the below command:

```
git pull —rebase <repo_url>
```

To push changes to a remote repository, use git push, as described below:

```
git push <repo_name> <branch_name>
```

Where repo_name is the name of the remote repository, and branch_name is the name of the local branch.

4.8.1.3 Git SSH

An SSH key is an access credential for the secure shell network protocol. SSH uses a pair of keys to initiate a secure handshake between remote parties—a public key and a private key.

SSH keys are generated using a public key cryptography algorithm.

1. To generate an SSH key on Mac, execute the following command:

```
ssh-keygen -t rsa -b 4096 -C "your_email@domain"
```

2.   Upon being promoted to enter the file path, enter a file path to which you would like the key to be stored.
3.   Enter a secure passphrase.
4.   Add the generated SSH key to the ssh-agent

```
ssh-add -K <file_path_from_step_2>
```

4.8.1.4   Git Archive

To export a Git project to an archive, execute the following command:

```
git archive --output=<output_archive_name> --format=tar HEAD
```

The above command generates an archive from the current HEAD of the repository. The HEAD refers to the current commit.

4.8.1.5   Saving Changes

As you make changes to your local codebase, for instance, feature development or bug fixes, you will want to stage them. To do so, please execute the following command for each file you would like to add to the staging area:

```
git add <file_name>
git commit -m <commit_message>"
```

The first command puts your changes to the staging area while the second command creates a snapshot of these changes, which can then be pushed to the remote repository.

If you would like to add all files in one go, consider using the variation of Git add with the—all option.

Once you add the file(s) to the staging area, they are tracked.

4.8.1.6   Syncing

Upon committing changes to the local repository, it is time to update the Git remote repository with the commits from the local repository. Please refer to the syncing commands listed at the start of this section.

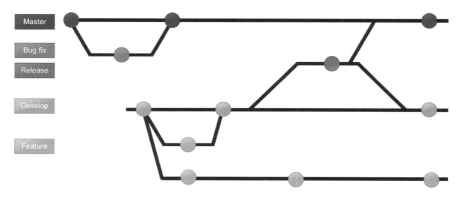

Fig. 4.8 Examples of pull requests

4.8.1.7 Making a Pull Request

A pull request is used to notify the development of changes, such as a new feature or a bug fix so that the development team (or assigned reviewers) can review the code changes (or commits) and either approve/decline them entirely or ask for further changes.

As part of this process:

1. A team member creates a new local branch (or creates their local branch from an existing remote branch) and commits their changes in this branch.
2. Upon finalising the changes, the team member pushes these changes to their own remote branch in the remote repository.
3. The team member creates a pull request via the version control system. As part of this process, they select the source and destination branches and assign some reviewers.
4. The assigned reviewer(s) discuss the code changes in a team, using the collaboration platform that is integrated into the version control system, and ultimately either accepts or declines the changes in full or part.
5. The above step #4 may go through more cycles or reviews.
6. Upon completing the review process, when all changes have been accepted (or approved), the team member merges the remote branch into the code repository, closing the pull request (Fig. 4.8).

4.8.1.8 Common Git Commands

The table lists some commonly used Git commands that are useful to remember. Figure 4.9 depicts the relative execution direction of some of these commands.

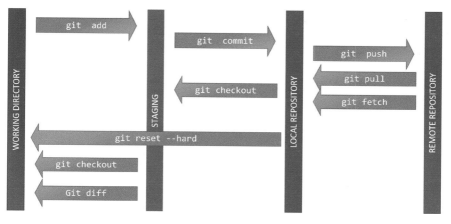

Fig. 4.9 Common git commands and relative execution directions

Configure your username and email address with Git	git config –global user.name "<user_name>"
Initialise a Git repository	git init
Clone a Git repository	git clone <repo_url>
Connect to a remote Git repository	git remote add origin <remote_server>
Add files to a Git repository	git add <file_name>
Check the status of the files	git status
Commit changes to the local repository	git commit –m "<message>"
Push changes to the remote repository	git push origin master
Switch across branches	git checkout –b <branch_name> git checkout<branch_name> git branch git branch –d <branch_name>
Update from the remote repository	git pull git merge <branch_name> git diff
Overwrite local changes	git checkout -- <file_name> git reset --hard origin/master

4.9   Containerising Applications

A minor difference in the version of a library can alter the functionality of your application, resulting in an unintended outcome. Fortunately, containerising an application allows it to execute in the same way regardless of the workspace or computer that it is deployed on. You can think of containerisation as an efficient alternative to virtual machines.

Docker[9] is a great tool to consider for containerisation. A key reason why the development community has adopted Docker is that if you containerise your application and transfer the image to a teammate's environment, the application will have the same performance on both devices. This is because the container includes all the dependencies needed by the application.

4.10   Chapter Summary

The chapter began with an introduction to common constructs found in programming and discussed using Python as an example language. The intention has been to provide a starting point for readers who are not familiar with the basics of programming or as a quick refresher for those picking up coding after some lapse in practice. We also discussed a few useful tools in aiding computational thinking, such as flowcharts and pseudocode. We then covered several important concepts, including OOP, error handling, secure coding and version control. Any robotics programmer worth their salt must be well versed in these aspects. Again, we have aimed to provide you with pointers to essential concepts to explore further and build on. Finally, we discussed containerisation as an efficient way to deploy your code on multiple platforms and operating systems. The projects section of the book will provide further opportunities to practice and explore these ideas further.

4.11   Revision Questions

1.   What are some of the common programming languages used in robotics?
2.   You are required to display the following pattern on a screen. Write the pseudocode of a suitable algorithm for this task.

```
*
*  *
*  *  *
*  *  *  *
*  *  *  *  *
```

3.   Convert the pseudocode developed in 2. above to a Python implementation.
4.   What are the four basic concepts of OOP?
5.   What is Git and why is it important?

---

[9] https://www.docker.com/.

4.12   Further Reading

It is far too numerous to suggest a set of suitable reading for this chapter as there are many online resources as well as excellent books were written on each of the topics covered in this chapter. You may head over to the book's website for a list of up-to-date resources. The following have served as useful online resources in writing this chapter:

- Python programming (Phython.org, 2019; Python Application, 2021; Python Exceptions, 2021; Python Security, 2020; Python Tutorial, 2021)
- Git (Atlassian, 2021)
- Containerisation (Docker, 2013)

References

Atlassian. *Git Tutorials and Training | Atlassian Git Tutorial*. Atlassian. https://www.atlassian.com/git/tutorials (Accessed 2021, December 22).

Docker, *What is a Container? | Docker*. Docker. (2013). https://www.docker.com/resources/what-container

How to Containerize a Python Application. *Engineering Education (EngEd) Program | Section*. https://www.section.io/engineering-education/how-to-containerize-a-python-application/ (Accessed 2021, December 22).

Python Exceptions: *An Introduction—Real Python*. https://realpython.com/python-exceptions (Accessed 2021, December 22).

Python Security Practices You Should Maintain. *SecureCoding*. (2020, May 18). https://www.securecoding.com/blog/python-security-practices-you-should-maintain/

Python Tutorial. www.tutorialspoint.com. https://www.tutorialspoint.com/python. (Accessed 2021, December 22).

Welcome to Python.org. (2019, May 29). https://www.python.org/

Damith Herath is an Associate Professor in Robotics and Art at the University of Canberra. Damith is a multi-award winning entrepreneur and a roboticist with extensive experience leading multidisciplinary research teams on complex robotic integration, industrial and research projects for over two decades. He founded Australia's first collaborative robotics startup in 2011 and was named one of the most innovative young tech companies in Australia in 2014. Teams he led in 2015 and 2016 consecutively became finalists and, in 2016, a top-ten category winner in the coveted Amazon Robotics Challenge—an industry-focused competition amongst the robotics research elite. In addition, Damith has chaired several international workshops on Robots and Art and is the lead editor of the book "Robots and Art: Exploring an Unlikely Symbiosis"—the first significant work to feature leading roboticists and artists together in the field of Robotic Art.

Adam Haskard is a cyber security and technology professional with over 16 years' experience within the Department of Defence. Adam has led GRC and Security Engineering activities in Defence Gateway Operations, JP2047, AIR6000, 1771 and L4125 as the DIE ITSM. Adam possesses a strong understanding of information systems, cross domain solutions, the certification and accreditation process and the military and wider technology landscape. He has in-depth technical and GRC experience leading multi-disciplinary teams on sensitive and complex cyber security activities. Adam has gained significant work experience from his various roles in the ADF and Industry, which included Cyber Security Professional, Engineer and Network Security Administrator, that enabled him to develop his cyber security and ICT skills. He was a member of the ADF between 2006–2013 where he progressed through information system (CIS) and leadership-based trainings. Adam's expertise includes evaluating, designing, monitoring, administering and implementing cybersecurity systems, protections and capabilities.

Niranjan Shukla has 15 years of prior experience working as a Software Engineer, Team Lead and TechnoloArchitect with experience in Data-driven development, API Design, Frontend technologies, Data Visualization, Virtual Reality and Cloud. He practices Design thinking through digital-Art on-the-side.

Chapter 5
The Robot Operating System (ROS1&2): Programming Paradigms and Deployment

David St-Onge and Damith Herath

5.1 Learning Objectives

The objective at the end of this chapter is to be able to:

- to know how to use (run and launch) ROS nodes and packages;
- to understand the messaging structure, including topics and services;
- to know about some of the core modules of ROS, including the Gazebo simulator, ROSbags, MoveIt! and the navigation stack.

5.2 Introduction

We expect most readers of this book to aim at the development of a new robot or at adapting one for specific tasks. As we mentioned in the introduction, the content of this book covers all of the required grounds to know "what has to be done" with an overview of several ways to address "how can it be done". If you do not know it already, you will quickly understand through this book that robot design calls to many different disciplines. The amount of knowledge needed to deploy a robotic system can sometimes feel overwhelming. However, many individual problems were solved already, including software ecosystems to simulate and then deploy our robots seamlessly. Advanced toolset and libraries are certainly integrated in the proprietary solution stack of the main robotic system manufacturers (such as ABB RobotStudio and DJI UAV simulator), but can everybody benefit of the last decades of public research for their own robots? This is a recurrent issue in many fields, and several libraries have been created in specific domains, such as to gather vision algorithms

D. St-Onge (✉)
Department of Mechanical Engineering, ÉTS Montréal, Montreal, Canada
e-mail: david.st-onge@etsmtl.ca

D. Herath
Collaborative Robotics Lab, University of Canberra, Canberra, Australia
e-mail: Damith.Herath@Canberra.edu.au

© The Author(s) 2022
D. Herath and D. St-Onge (eds.), *Foundations of Robotics*,
https://doi.org/10.1007/978-981-19-1983-1_5

(OpenCV) and machine learning algorithms (TensorFlow). The Robot Operating System (ROS) is an open-source solution addressing this critical sharing need for robotic sensing, control, planning, simulation, and deployment. Not to be confused with a library, it is a software ecosystem (the concept of an operating system might be too strong) facilitating the integration, maintenance, and deployment of new functionalities and hardware from simulations to physical deployment. While ROS can run code the same from several popular languages, in order to use it you will need good knowledge of the infrastructure's underlying concepts (and honestly quite a bit of practice). ROS is renowned to have a steep learning curve and even more so for developers not familiar with software engineering. This chapter aims at giving you an overview of ROS and setting the bases to use it without being specific to any version (only few code examples are provided).

Since ROS is made to run predominantly on Linux operating system, we will end the chapter with a quick overview of Linux fundamental tools useful for roboticists and ROS developers.

An Industry Perspective

Alexandre Vannobel, Team Lead,
Kortex Applications Team

Kinova inc.

I have a bachelor's degree in biomedical engineering from Polytechnique Montréal. I was especially interested in software development through my studies, most especially newer technologies such as AI, robotics, and cloud computing. I had the chance to work as an intern for one summer at Kinova. Needless to say, I learned a lot about robots during those four months I never really learned the basics of robotics in a classroom. It was more of a learn-by-doing experience (and it still is).

Learning the details and intricacies of ROS, Gazebo, and MoveIt was certainly a challenge! I have also been responsible for interfacing our robots with this framework, and there were some development and integration issues, as the goals and objectives of people who create robots and those who use robots do sometimes differ. It is of importance in those cases to consider what users want and how they want to use the robot, but also to consider implementation costs and time of features.

I have witnessed the acceleration of ROS2's development in the last few months/years, and I think this is where the field is going. ROS1 is a centralized framework made to "unite" all of the robotics paradigms and tools in one big system, but it suffers from a lot of legacy design choices that make the industry really refractory from using it, starting with communication layers and the lack of real-time support. I think ROS2, which was designed with the same paradigms as ROS1 but with an emphasis on addressing those issues will bring the industrial and the research worlds closer.

5.3   Why ROS?

Before you dive into ROS usage, you must understand its roots, as they motivated several design decisions along the way, up to the need to redefine the whole ecosystem for industrial and decentralized applications in ROS2. ROS is a big part of the recent advances in robotics and its history is as important as any of the works presented in Chap. 1.

It all started with two PhDs students at Stanford: Eric Berger and Keenan Wyrobek. Early in their research, around 2005, they both needed a robotic platform to deploy and test their scientific contributions: the design of an intrinsically safe personal robot (Wyrobek et al., 2008). In their search for the best robotic platform, they ended up talking to several researchers, each developing their own hardware and software. The amount of duplicated work stunned them. They will later argue that 90% of the roboticists work involve re-writing code and building prototypes, as illustrated in Fig. 5.1. They made it their mission to change how things worked by developing a new common software stack and a versatile physical robot, the PR1. The fund-raising and marketing of their idea are out of scope here, but let us just mention that they had to work hard in order to gain some credibility (Wyrobek, 2017). While still at Stanford, they made the first PR1 prototype, alongside its modular software stack (inspired from *Switchyard* by Nate Koenig) and validated its versatility with a student coding competition and an in-house demonstration (a living room cleaning robot).

Berger and Wyrobek's vision of a universal operating system for robots definitely stroke right in the ambitious work of Scott Hassan (Silicon Valley billionaire). At the time, Scott Hassan was directing a research laboratory, Willow Garage, focused on autonomous vehicles. Over time, ROS (Willow Garage new name for Berger–Wyrobek–Koenig-inspired software stack) and PR became the main activity of Willow Garage, involving investments of several millions of dollars. These considerable resources clearly contributed to the rapid growth of ROS, namely by financially supporting a great team of engineers. However, there was already a hand-

Fig. 5.1 Comic commissioned at Willow Garage, from Jorge Cham, to illustrate the wasted time in robotics R&D

ful of open-source projects for robotics at the time, including Player/Stage (Gerkey et al., 2003), the Carnegie Mellon Navigation Toolkit (CARMEN) (Montemerlo et al., 2003), Microsoft Robotics Studio (Jackson, 2007), OROCOS (Bruyninckx, 2001), YARP (Metta et al., 2006), and more recently the Lightweight Communications and Marshalling (LCM) (Huang et al., 2010), as well as other systems (Kramer and Scheutz, 2007). These systems provide common interfaces that allow code sharing and reuse, but did not survive as strong as ROS did. Money itself could not ensure ROS success, they needed a community.

> In Silicon Valley, people are in a secret place working on something that may or may not ever see the light of day. They may or may not ever be able to talk about it. It's a very different experience to be able to - as we do here - all day, every day, just write code and put it out in the world—Brian Gerkey, chief executive officer at Open Robotics (Huet, 2017)

The ROS community is nowadays clearly what makes ROS unique,[1] powerful, and impossible to avoid when working in robotics. Following Berker testimony (Wyrobek, 2017), they built that community over three strategies:

1. They secured the support of the other major players in open-source robotics by involving them from the start in the definition of what ROS must be. These people became early ambassadors of ROS.
2. They started a wide internship program, hosting PhD students, postdoctoral fellows, professors, and industry engineers from all over the world, all contributing to ROS and then using it in their own work. Berker mentions that Willow Garage was hosting at some point more interns than employees, counting hundreds of them.
3. They gave away 11 of their first PR2 prototypes, running exclusively on ROS, to major research laboratories around the world as the result of a competitive call. The new owners had to commit to contribute significantly to the ROS code base and to provide a proof that their institution allows them to share their research publicly.

Unfortunately, after Willow Garage skyrocketed ROS popularity and usage worldwide, the company was dissolved in 2013. It was never meant to be the end of ROS and PR2: the hardware customer service was taken over by Clearpath Robotics and the open-source software development by a new entity, the Open Source Robotics Foundation (non-profit). Under OSRF, they developed the first set of ROS distributions (*Distro*), from Medusa Hydro (2013) to Melodic Morenia (2018), but the foundation was growing with more requests for commercial contracts. In 2017, it splits to create the Open Source Robotics Corporation (known as Open Robotics,[2]) while the foundation still maintains the ROS code base. Open robotics released the last version of ROS1, Noetic, the first to be based on Python 3 (all previous versions used Python 2) and a whole new version, ROS2.

5.4 What Is ROS?

Now you may wonder if ROS is not just a glorified library. . . What is so special about it? The minimal answer is twofold: 1. It provides mechanisms for code maintenance and extensibility (adding new features), and 2. it **connects** a large community. The ROS wiki provides a more complete answer:[3]

> ROS is an open-source, meta-operating system for your robot. It provides the services you would expect from an operating system, including hardware abstraction, low-level device control, implementation of commonly-used functionality, message-passing between processes, and package management. It also provides tools and libraries for obtaining, building, writing, and running code across multiple computers.

---

[1] ROS users' world map: http://metrorobots.com/rosmap.html.

[2] https://www.osrfoundation.org/welcome-to-open-robotics/.

[3] http://wiki.ros.org/ROS/Introduction.

Fig. 5.2 ROS workspace
folder structure from the
assignments detailed in
Chap. 18

We will stick to our two-item list and just scratch the surface of some core concepts of software engineering to understand a bit better how they unfold in ROS. Implementing software engineering best practices is at the core of ROS, from a modular architecture to a full code-building workflow. The concept of an *operating system* may be a bit stretched as ROS is closer to a *middleware*: the abstract interface to the hardware (*POSIX of robots*).

As users (i.e., not developers) of ROS, we usually do not need to know the details of the building structure, but it is mandatory to learn the basics in order to know how to properly use it. ROS provides a meta-builder, a uniform set of tools to build code in several languages for several different computer's environments and architecture. In ROS1, this is done by `catkin` (formerly by `rosbuild`); while in ROS2 `ament` takes over. In the end, they are really similar things, both just wrappers around *CMake* (Cross-platform Makefile system).[4] If you are a Python developer, you may think this kind of structure is unnecessary, but that is notwithstanding how it contributes to the portability and modularity of ROS. Portability here refers to the deployment of your code easily in different environments, as long as it follows the ROS building structure. Different environments can be for other users, new robots, but also in order to be seamlessly compatible with a testing environment. We will discuss more in details the simulation infrastructure provided by ROS in Sect. 5.6.3. Using the ROS build tools helps integrate your code with the rest of the ROS ecosystem. The meta-builder will generate the custom messages (topics), services, and actions your node requires (described in Sect. 5.5.1) and make them available to other executable (alike libraries). It will also add several paths and files to the environment in order to execute your code and quickly find your files (e.g., configuration files). The meta-builder will organize your work space over `build`, `devel` and `src` folders, as shown in Fig. 5.2.

Let us have a quick look at this ROS folder structure. In a glimpse, the `build` folder will host all the final files generated from the meta-builder while the `devel` folder keeps track of the files generated by the process for testing and debugging purpose. The `devel` folder will also include the essential `setup.bash` file, which, when sourced (`#source devel/setup.bash`), adds the location of the packages built to your ROS environment. Sourcing the system ROS (`#source /opt/<ROS Distro>/setup.bash`) and your local work space is mandatory to run any executable using ROS commands. This is usually part of any ROS installation procedures, both for maintained packages and third-party ones. The folder `src` is the one you will end up using the most. It contains a separated folder for each `package` of your work space. Software in ROS is organized in *packages*. A package might contain ROS nodes, a ROS-independent library, a dataset, configuration files, a third-party piece of software, or anything else that logically constitutes a useful module. The goal of these packages is to provide their intended functionality in an easy-to-consume manner so that software can be easily reused.

Each of the package's folders must respect a structure, as shown in Fig. 5.2, with the subfolders: `include`, `launch`, `src` as well as optional ones related to the use of Python code (`script`) and simulation (`models`, `urdf`, `worlds`). The `include` and `src` folders are part of common C/C++ code structure, the first for headers (declarations) and the second for content (definitions). `launch` contains the launched files discussed in Sect. 5.5.2. The `src` folder contains one or more `nodes`. The nodes are executable with dedicated functionalities and specific inputs and outputs (when applicable). The work space shown in Fig. 5.2 is extracted from the assignments in Project Chap. 18. It combines third-party packages from Intel for the cameras (`realsense-occupancy`), from

---

[4] https://cmake.org/.

Kinova for the Gen3 lite arm (`ros_kortex`), from Clearpath for the wheeled base (`dingo`, `dingo_robot`, `jackal`, `puma_motor_driver`) and packages specific to the assignments (`mobile_manip`, `realsense_simulator`, `common_gazebo_models`).

To deploy a ROS work space, you must follow the ROS installation instructions,[5] and then either copy a third-party node (clone a Git repository) in order to work on it, or make your own fresh work space.[6] In both cases, you will end up writing code inside the package folder, for instance inside `mobile_manip_ws/src/mobile_manip` shown in Fig. 5.2. Inside of your package, if a node (an executable file in `script` or C/C++ code in `src`) is new, you need to add it to the `CMakelist.txt` building configuration file at the root of your work space for your meta-builder to be aware of the node existence. When setting up a new ROS environment, be aware that there is a compatibility matrix to fit each ROS distribution with Linux distributions.[7]

Now that we have a better idea of how the meta-builder works to provide portability (dealing with different environments) and modularity (packages and nodes), we can look into how modularity help **connects** the ROS community. ROS developers can share their nodes on any online platform (e.g., GitHub), or make it official by including it to a ROS distribution (indexed). A ROS indexed package must follow perfectly the ROS structure standard as well as programming best practices (unit tests, well commented, etc.). After a bit of training, it becomes easy to download, build, and run nodes made by any contributor around the world. This helped strengthen a community, one so enthusiast that it creates its own annual event, entitled *ROSCon* (*ROSWorld* in 2021 for the Virtual version), gathering hundreds of developers and users. ROS community is growing pretty fast, with new groups emerging, such as ROS Industrial[8] focused on developing industry-relevant capability in ROS. Where the community can easily exchange, their software must also be able to communicate. A large part of the modularity of ROS is provided by its communication infrastructure. A library of message types, extendable, guarantees the data format is compatible between all users nodes. The messages, i.e., simple data structures, can then be called in the form of topics or as part of services, as will be explained in Sect. 5.5.1.

5.4.1   ROS1&2: ROSCore Versus DDS

ROS distributions are frequently released with major updates (enhancements). Since 2017, the core of ROS was revisited, leading to the release of a first stable ROS2 distribution in 2020, Foxy Fitzroy. The last distribution of ROS1, Noetic, will be

---

[5] https://wiki.ros.org/ROS/Installation.

[6] https://wiki.ros.org/ROS/Tutorials/InstallingandConfiguringROSEnvironment#Create_a_ROS_Workspace.

[7] https://www.ros.org/reps/rep-0003.html#platforms-by-distribution.

[8] https://rosindustrial.org/.

Fig. 5.3 ROS Core role: the librarian connecting the nodes' topics and services

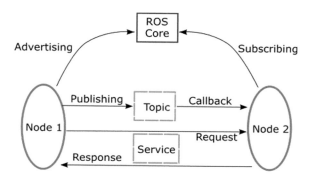

officially supported until May 2025 and may very well be active longer than that, but at some point all ROS users are expected to transit to ROS2. We quickly mentioned the new building mechanism of ROS2, `ament`, and we will discuss some format changes (e.g., launch files) in the upcoming sections, but the main difference is at the core, the `roscore`. In ROS1, `roscore` is a collection of nodes and programs that are prerequisites of a ROS-based system. You must have a roscore running in order for ROS nodes to communicate. Launching the `roscore` (either automatically with a launch file or manually with the `roscore` command) starts the ROS Core, i.e., the ROS1 librarian. As shown in Fig. 5.3, the ROS Master (i.e. ROS Core) is the one responsible for indexing all nodes running (the *slaves*) along with their communication modality. In other words, in ROS1, without the ROS Core, the nodes cannot be aware of the others, let alone start to communicate with one another. However, when all nodes are launched and aware of the others, theoretically the ROS Core could be killed without any node noticing.

At a glimpse: `roscore` is dead in ROS2, no more *master* and *slaves*. The communication infrastructure is fundamentally decentralized in ROS2, based on a peer-to-peer strategy, the Data Distribution Service (DDS). Where ROS1 had a critical single point of failure, no node can block the others from running in ROS2. DDS includes packet transport protocol and a distributed *discovery* service to grab information from the other running nodes.[9] This paves the way to facilitating the development and deployment of multi-robot systems, maybe even so-called robotic swarms.

Before getting into the ROS world, you need to pick your version. If you are looking for more existing packages and a more stable API, use ROS1. If you are

---

[9] For more information: https://design.ros2.org/articles/ros_on_dds.html.

looking for long-term stability, better performance, and newer algorithms, use ROS2. Just do not try to learn both from scratch! If you are still in doubt about which one to go for, ignore ROS1 and use ROS2, since ROS 1 will be going away in a couple of years.

5.4.2 ROS Industrial

While we will be limiting our discussions to ROS 1 & ROS 2 in this book, it is worth noting that another flavor of ROS exists called ROS Industrial or ROS-I for short. [10] As the name suggests, ROS-I is a concerted effort to bring the best of ROS to industrial-scale robotics. While, in general, research robotics systems such as the PR2 follow an open-source ethos, most commercial robotic systems use closed and proprietary software. This makes it extremely difficult to develop cross-platform projects using them or adapt existing commercial hardware systems outside their intended ecosystems. Frustrated by this situation, Shaun Edwards, in 2012, created the initial ROS-I repository in collaboration with Yaskawa Motoman Robotics company and Willow Garage while he was at Southwest Research Institute to facilitate the adoption of ROS in manufacturing and automation. Since then, many commercial robotic platforms have been integrated within ROS-I. Core developments of ROS-I are independently managed through several industrial consortia that require a paid membership to participate. A good understanding of ROS should set you up for a relatively easy transition to ROS-I if you eventually venture into commercial robotics.

5.5 Key Features from the Core

The following sections will give an overview of the main features included in ROS. While the focus is on ROS1 (the assignments presented in Project Chap. 18 run on Noetic), the concepts are shared with ROS2, but some format differences are discussed when applicable.

5.5.1 Communication Protocols

Whether it is decentralized (ROS2) or centralized (ROS1), the communication between nodes is structured in *messages*.[11] Figure 5.4 shows the `Odometry` mes-

---

[10] https://rosindustrial.org/.

[11] http://wiki.ros.org/Messages.

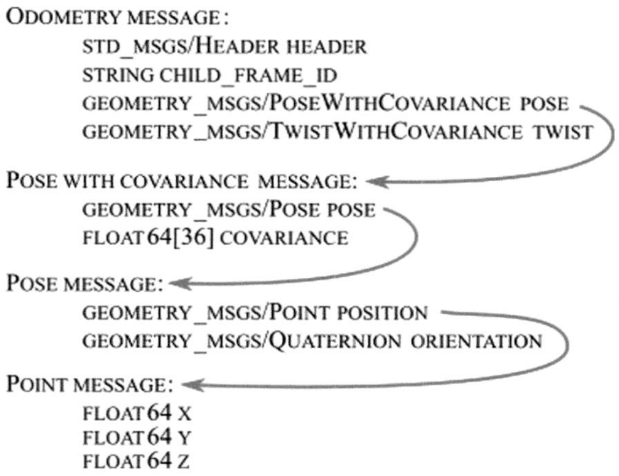

Fig. 5.4 Content of ROS topic Odometry

ODOMETRY MESSAGE:
    STD_MSGS/HEADER HEADER
    STRING CHILD_FRAME_ID
    GEOMETRY_MSGS/POSEWITHCOVARIANCE POSE
    GEOMETRY_MSGS/TWISTWITHCOVARIANCE TWIST

POSE WITH COVARIANCE MESSAGE:
    GEOMETRY_MSGS/POSE POSE
    FLOAT 64[36] COVARIANCE

POSE MESSAGE:
    GEOMETRY_MSGS/POINT POSITION
    GEOMETRY_MSGS/QUATERNION ORIENTATION

POINT MESSAGE:
    FLOAT 64 X
    FLOAT 64 Y
    FLOAT 64 Z

sage with some of the message types it contains. Several message libraries come along with a ROS installation, but developers can also generate custom messages for their node. At run time, the availability of these data structure can be advertised over *topics*. Topics are barely names, i.e., labels, put on a given data structure (message) from a given node. A node may publish data to any number of topics and simultaneously have subscriptions to any number of topics. Topics are one of the main ways in which data is exchanged between nodes and therefore between different parts of the system (between robots and with a monitoring ground station). In order to share information, a node needs to *advertise* a topic and then *publish* content (messages) into it. The first part is done in the initialization part of the node's code, while the latter is done each time new data must be shared, commonly inside the code's main loop at a fixed frequency. On the other side, the node(s) that needs a topic's content will *subscribe* to it. The subscriber will associate a callback function triggered for each new incoming message.

ROS comes with a really handy debugging tool for topics, the terminal command `rostopic` (`ros2 topic` in ROS2). It can be used to show all available topics from the nodes running: `rostopic list` (`ros2 topic list`), to print the content (message) of a given topic: `rostopic echo  odom` (`ros2 topic echo  odom`) and to show the publishing frequency of a topic: `rostopic hz odom` (`ros2 topic hz  odom`).

Topics are connectionless communication (classic publisher/subscriber system) in the sense that the publisher of the message does not know if any other node is listening. ROS also provides with a connection-oriented protocol (synchronous RPC calls), the *services*. Services have a client and a server, and both will acknowledge the information received by the other at each transaction. Topics and services use the same containers (message types) for information, but are better suited to different applications. For instance, topics are useful to stream the reading from a sensor, while services are better suited to share the configuration of a node or change a

```
1  <launch>
2    <arg name="use_sim_time" default="true" />
3    <arg name="gui" default="false" />
4    <arg name="headless" default="true" />
5    <arg name="world_name" default="$(find robot_manip)/worlds/jackal_maze.world" />
6
7    <!-- Launch Gazebo with the specified world -->
8    <include file="$(find gazebo_ros)/launch/empty_world.launch">
9      <arg name="debug" value="0" />
10     <arg name="gui" value="$(arg gui)" />
11     <arg name="use_sim_time" value="$(arg use_sim_time)" />
12     <arg name="headless" value="$(arg headless)" />
13     <arg name="world_name" value="$(arg world_name)" />
14   </include>
15
16   <!-- Spawn Jackal -->
17   <include file="$(find robot_manip)/launch/include/spawn_jackal.launch" >
18     <arg name="init_pose" value="-x 2.0 -y -4.0 -z 1.0" />
19     <arg name="use_state_estimation" value="false" />
20     <arg name="control_config" value="$(find robot_manip)/config/jackal_control_tf.yaml" />
21   </include>
22   }
23 </launch>
```

```python
from launch import LaunchDescription
from launch_ros.actions import Node

def generate_launch_description():
    return LaunchDescription([
        Node(
            package='turtlesim',
            namespace='turtlesim1',
            executable='turtlesim_node',
            name='sim'
        ),
        Node(
            package='turtlesim',
            namespace='turtlesim2',
            executable='turtlesim_node',
            name='sim'
        ),
        Node(
            package='turtlesim',
            executable='mimic',
            name='mimic',
            remappings=[
                ('/input/pose', '/turtlesim1/turtle1/pose'),
                ('/output/cmd_vel', '/turtlesim2/turtle1/cmd_vel'),
            ]
        )
    ])
```

Fig. 5.5 Example of a launch file for: left is the XML format for ROS1 and right, the Python format new to ROS2

node's state. Finally, ROS provides the *actions* protocol (asynchronous RPC calls), combining topics and services. A basic action includes a goal service, a result service, and a feedback topic. Its format is well suited to interface with mission planners, such as QGroundControl.[12]

5.5.2 Launch and Run

To deploy a ROS system means to start several executable files, i.e., *nodes*. The most basic command to do so is rosrun <package name> <node name> (ros2 run <package name> <node name>), which is most often run in a different terminal for each node. However, in ROS1 you need a roscore before any node can be run, so you must use the command roscore beforehand. Using this strategy to start the nodes individually will lead to numerous terminal tabs that must be monitored simultaneously. ROS provides another way to launch several nodes altogether: the launch files. In ROS1, using a launch file will also automatically start the roscore.

The format of the launch file differs between ROS1 and ROS2, as shown in Fig. 5.5. ROS1 uses an XML file while ROS2 encourages the use of Python scripts (ROS2 still supports XML format). Nevertheless, both serve to call nodes with parameters and can nest other launch files. Calling several nodes simultaneously is great, but what happens if you need twice the same node, for instance to process images from two cameras? You can always use the rosrun command to launch nodes afterward that are not in the launch file; they will connect to the same ecosystem automatically. However, a powerful feature of launch files is the group tag to force nodes into a given namespace: the same node can then be launched several times in different parallel namespaces without interfering with one another. This is essential to simulate multi-robot systems.

---

[12] http://qgroundcontrol.com/.

5.5.3 ROS Bags

Now say you developed a new collision avoidance algorithm, based on the data of several sensors. You deploy it on your robot and go for a run with it. No matter how well it goes, you will want to extract performance metrics and assess afterward the issues you faced. This calls for a logging system, luckily ROS provides a robust and versatile one out-of-the-box! The ROS bag format is a logging format for storing ROS messages in files. Files using this format are called bags and have the file extension .bag. Bags are recorded, played back, and generally manipulated by tools in the rosbag (ros2 bag) and rqt_bag (no counterpart yet available in ROS2) packages. You can replay your field experiments: republish all sensor data at their real frequency (or simulate different publishing rates), including packet loss or any disturbance from the experiment. rosbags also has an API that provides features to quickly parse and analyze or export your data. For instance in Python, it may look like:

```
import rosbag bag = rosbag.Bag('test.bag') for topic, msg, t
in bag.read_messages(topics='odom'):
    print("Odometry is x={}, y={} and z={} at time {} sec".
format(
        msg.pose.pose.position.x, msg.pose.pose.position.y,
msg.pose.pose.position.z,
        t.toSec())
bag.close()
```

Rosbags are key to tuning your algorithms and sharing your time-consuming experimental data with your peers.

5.5.4 Transforms and Visualization

Can you imagine a useful, physical robot that does not move or watch something else move? Any useful application in ROS will inevitably have some component that needs to monitor the position of a part, a robot link, or a tool. The ROS way of dealing with relative motion is encompassed in TF (transforms). TF allows seeking the geometrical transformation between any connected frames, even back through time. It allows you to ask questions like "What was the transform between A and B 10 seconds ago?"

One possible example is a camera, mounted on a robot, tracking markers in the scene, as shown in Fig. 5.6. This example shows the robot odometry frame (the mobile base motion—camera_odom_frame), the camera pose (fixed to the base—camera_fisheye2_frame), and the frame of the tag detected (ETS_target). The tag is detected in the camera_fisheye2_frame, and its pose is extracted and transformed directly in camera_odom_frame to visualize all frames together.

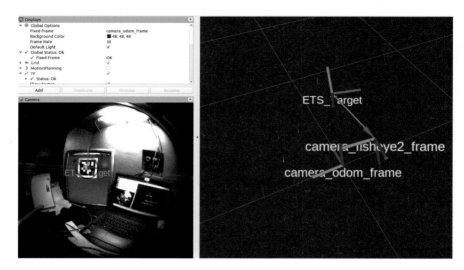

Fig. 5.6 Visualization in RViz of a fish-eye camera feed and the reference frames resulting from a fiducial marker detection

As long as you have the position and orientation of an object (six degrees of freedom), you can broadcast its `TF` in ROS. For instance in Python, it may look like:

```
import tf2_ros # import the TF library br =
tf2_ros.TransformBroadcaster() # create de broadcaster t =
geometry_msgs.msg.TransformStamped() # create the message
container
# fill the message: t.header.stamp = rospy.Time.now()
t.header.frame_id = "world"
t.child_frame_id = "myrobotframe"
t.transform.translation.x = x
t.transform.translation.y = y
t.transform.translation.z = z
q = tf_conversions.transformations.quaternion_from_euler
(psi, phi, theta)
t.transform.rotation.x = q[0]
t.transform.rotation.y = q[1]
t.transform.rotation.z = q[2]
t.transform.rotation.w = q[3]
br.sendTransform(t) # broadcast the transform
```

Notice in the snippet above the format of the orientation (rotation): ROS, by default, requires to use quaternions. `tf_conversions` library provides the tool to convert rotation matrices and Euler angles to quaternions and back, but for more information about the mathematical representation of the quaternions, read Chap. 6. Often `TF` are used to define the fixed geometrical relations between a robot's parts. You can then rather easily use the pose of an object detected by a camera mounted somewhere on your robot to feed the wheels motors with appropriate commands, such as "my camera sees the door 2 m ahead, but is positioned 50cm from the wheel axis, so let's go forward by only 1.5 m".

The viewer shown in Figs. 5.6, 5.7, and 5.8 is RViz, short for ROS Visualization. It is a 3D viewer supporting almost all types of topics, namely 2D and 3D LiDAR point clouds, camera stream, and dynamic reference frames motion. The viewer is launched simply with `rosrun rviz rviz` (or simply `rviz`). Then using the graphical interface *Add* button, you can select the topic you want to monitor. While RViz was made to monitor your robot's topics, it can also host *interactive markers* that can be moved in the visualization window and will broadcast their updated position out in ROS. An example used to command a robotic arm is shown in Fig. 5.8.

5.6   Additional Useful Features

Several community contributions went into the essential toolset of ROS and greatly contribute to its popularity. This section covers a handful of what we consider to be the most important for mobile robots and manipulators. All of these packages are leveraged in at least one of the assignments of Project Chap. 18.

5.6.1   ROS Perception and Hardware Drivers

When dealing with your hardware integration, the same logic applies as for the software parts discussed previously: you do not want to waste time in reproducing what was done already to interface with each component. Manufacturers have their counterpart to this logic: it can be really expensive to develop drivers for several different operating systems and software solutions to accommodate potential clients. ROS acts here again as a standard, connecting the manufacturers to a large community. Hundreds of hardware manufacturers deliver ROS nodes with their products, namely SICK, Clearpath, Kinova, Velodyne, Bosch, and Intel. The driver node made by the manufacturer most often deals only with low-level communication into ROS compatible topics and services. From that point, the meta-package ROS perception helps with filtering, synchronizing, and visualization of the data. For instance, ROS perception includes `pcl_ros` to manage point clouds. It includes filters such as voxel grid filter and pass-through filter, but also geometrical segmentation of the data to extract planes or polygons from the point cloud. An example point cloud published as a ROS topic is shown in Fig. 5.7. For dealing with images (cameras), `cv_bridge` and several other packages bring the powerful features of the open library OpenCV to process images within ROS code. This provides the classic algorithms for contours detection, images filtering (blur, etc.), and histogram generation. From there, many

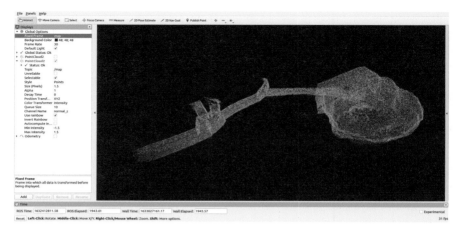

Fig. 5.7 DARPA subterranean 2021 spot-1 finals map made by CTU-CRAS-NORLAB team

machine learning algorithms have ROS wrappers, such as the powerful You Only Look Once (YOLO)[13] for object recognition.

Finally, ROS perception also contains a package integrating several of the most up-to-date algorithms for simultaneously localization and mapping (SLAM), `gmapping`. Based on either on 2D LiDAR, 3D LiDAR, stereo camera, or a single camera, the package outputs a rough map of the environment explored by the robot without any a priori knowledge of the robot position in the map. These powerful algorithms are nowadays essential to any mobile robot deployment in GPS-denied environment. Several other, and more recent, SLAM solutions are also available on GitHub from research laboratories around the world, but `gmapping` is maintained by OSRF. When the environment is known (a 2D map is available), you may prefer to use the ROS package for adaptive Monte Carlo localization (AMCL). This one uses a particle filter to find the best candidates in position when simulating your laser scan from the map provided. This is the strategy deployed in the assessments 4 and 5 of Project Chap. 18.

5.6.2 ROS Navigation and MoveIt!

Let us assume that the perception stack grants us with the position of the robot and a map of its environment. In order to fulfill any mission, the robot will need to move in this environment, either by finding an optimal trajectory (mobile robot) or by computing an optimal posture for a manipulator to reach a given pose with its tool (i.e., gripper). For mobile robots, conventional methods of indoor path planning

---

[13] https://github.com/leggedrobotics/darknet_ros.

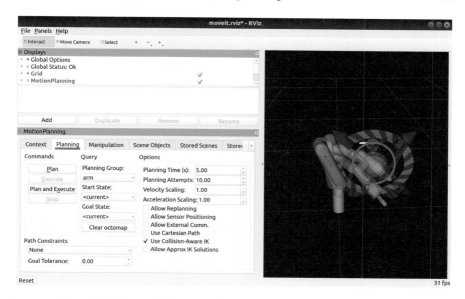

Fig. 5.8   Kinova Gen3 lite manipulator controlled by interactive markers and MoveIt! planner from RViz

often refer to the optimal path as the shortest path that can be obtained from various algorithms such as A*, Dijkstra's (Palacz et al., 2019) or rapid-exploring random trees (RRT). These algorithms, and a lot more, are available out of the box from public ROS packages.

For manipulators, many numerical solvers for multibody dynamics have been proposed over the past decades and along with them path planners that either use sampling-based algorithms or optimization-based algorithms. These algorithms and several others were integrated in the Open Motion Planning Library,[14] itself integrated in the MoveIt! ROS planning package.[15]

Figure 5.8 shows MoveIt! in action through RViz using interactive markers. These markers can simply be dragged to the desired goal and then the left menu grants the user access to different planners and their configurable parameters. MoveIt! can also consider static objects in the scene to plan a solution considering collision avoidance. These objects can be added manually or imported from the Gazebo simulator.

5.6.3   Gazebo Simulator

Robot simulation is an essential tool in every roboticist's toolbox. A well-designed simulator makes it possible to rapidly test algorithms, design robots, perform regression testing, and train artificial intelligence systems using realistic scenarios. Gazebo

---

[14] https://ompl.kavrakilab.org/.

[15] https://moveit.ros.org/.

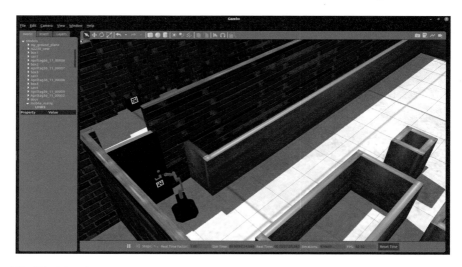

Fig. 5.9 View from Gazebo simulator with the mobile manipulator of the assignment in Project Chap. 18

offers the ability to accurately and efficiently simulate robots in complex indoor and outdoor environments. It encompasses a robust physics engine, with convenient programmatic and graphical interfaces. Best of all, alike ROS, Gazebo is free, open source, and has a vibrant community.

Gazebo simulator can load any mesh in `obj` or `dae` format and then use it with realistic dynamics to simulate robot motion and collisions. Alike ROS, Gazebo is modular, so the simulation plugins for dynamics, can be customized as well as any sensor data. Several manufacturers provide plugins (e.g., Intel cameras) and models (e.g., Kinova robots) to simulate their hardware within Gazebo. Figure 5.9 shows a simulation environment from the Project Chap. 18, including Intel cameras, the fully actuated Kinova Gen3 lite manipulator, the differential drive Clearpath Dingo mobile base, and a world made out of walls, furniture, and functional doors.

Gazebo is by far the most popular simulator for ROS users, but it lacks realistic rendering and can be pretty heavy to run for a large number of robots (swarms). To address the first limitation, Gazebo is being phased out in favor of Ignition.[16] Nevertheless, developers in vision-based machine learning will prefer more realistic environments such as Unreal[17] and Unity[18] (which has a ROS plugin[19]). For the latter, swarm roboticists will use dedicated simulators, such as ARGoS[20] (which also has a ROS plugin[21]).

---

[16] https://ignitionrobotics.org/.

[17] https://www.unrealengine.com/.

[18] https://unity.com/.

[19] https://resources.unity.com/unitenow/onlinesessions/simulating-robots-with-ros-and-unity/.

[20] https://www.argos-sim.info/

[21] https://github.com/BOTSlab/argos_bridge/.

5.7   Linux for Robotics

We mentioned previously that ROS is not exactly an operating system, but rather a middleware. Still, many people are referring to it as the *Linux for robotics* (Wyrobek, 2017). There is some truth in this name, as ROS is extending the Linux operating system to robotic applications. Until ROS2, it was only able to run properly on Linux. It means that the majority of ROS users must know their way around in a Linux environment.

We will take for granted that you start on a computer already set up with Linux (Dell sells certified computers preloaded with Linux[22]) or that you know how to launch a Linux virtual machine in Windows or OSX (although virtual machines are not recommended for hardware experiments and computer-intense simulations).

As we mentioned earlier, when installing ROS on a Linux system, look into the ROS-Linux compatibility matrix first.[23] In all of Linux distributions, you will need to input some terminal commands to get things done. Knowing the basic commands in a Linux terminal is also rather essential for embedded development, as the most popular on-board computers (e.g., Raspberry Pi, NVidia) will run a version of Linux and can be accessed through a remote terminal session (e.g., ssh). The most essentials terminal commands are as follows:

- cd: Change Directory. cd .. is used to get to the parent directory.
- ls: List Files. ls -la: will list all files (hidden ones too) along with the properties (permissions and size).
- mv: MoVe file.
- cp: CoPy file.
- rm: ReMove file.
- df: Disk Filesystem (disk usage). df -h allows to see the memory usage on all disks in human readable format.
- reset: to remove all output from the terminal screen and remove any local environment variables changes.

To edit and compile your ROS code, you want an integrated development environment (IDE) that can help you find the right names and definitions of functions, as well as compile and even debug your code. IDEs are like glasses: you need to try them to find the one that fits you best. A lot of IDEs are available for Python (Atom, Eclipse, PyCharm, etc.) and C/C++ (Visual Code, CLion). Linux experts sometimes prefer the highly configurable text editors such as Sublime, Emacs, and Vim, for which plugins and tutorials are available for ROS. However, the majority of the ROS developers seems to prefer Eclipse, for its user-friendly interface, its support for several programming languages, and its ROS plugin seamlessly integrated. Other more recent options are drawing attention: Microsoft Visual Code, or its open-source ROS version, Roboware, and the web-based ROS Development Studio (RDS). While they

---

[22] https://www.dell.com/en-us/work/shop/overview/cp/linuxsystems/.

[23] https://www.ros.org/reps/rep-0003.html#platforms-by-distribution.

all have pros and cons, they also all do essentially the same thing. If you are looking for an IDE, we suggest VS Code. If you just want a code editor, we like Sublime Text.

5.8 Chapter Summary

This chapter introduced the Robotic Operating System, ROS. We first discussed the motivation for its conception by going through its origin and then we gave an overview of its core advantages, leading to its current popularity. The chapter covered both ROS1 and ROS2, with a short stopover on the centralized versus decentralized differences between them. We then covered the essential features from the ROS Core and third-party additions. Finally, we gave essential hints to new Linux users, as this operating system is still the best suited one for ROS development.

5.9 Revision Questions

Question #1
In ROS1, what is the result of the command `rosrun robot_manip dingo_control`?

1. It launches the `robot_manip` node of the `dingo_control` package, but a `roscore` must have been started beforehand.
2. It launches the `dingo_control` node of the `robot_manip` package, but a `roscore` must have been started beforehand.
3. It launches the `robot_manip` node of the `dingo_control` package and a `roscore` if none is present.
4. It launches the `dingo_control` node of the `robot_manip` package and a `roscore` if none is prese[]nt.

Question #2
Associate the following ROS concepts:

1. Topic
2. Service
3. Message

with their definition:

A A link created by a node to post information to those who *subscribe* to it.
B A standardized container for the exchange of information between nodes.
C A blocking communication that awaits the response of the called node.

Question #3
Is ROS1 a completely decentralized software ecosystem? Explain why.

Question #4
Give the relative path in the ROS workspace to a C++ node source file (`doit.cpp`)
of a package named `realsense_occupancy`.

5.10   Further Reading

The best way to learn ROS is to play with it. ROS wiki[24] is a great place to start
learning more about the core packages. ROS wiki also contains several basic tutorials
to practice with topics, services, actions, and launch file either in C++ or in Python.
If you are looking for an extension to this chapter, including more explanations on
the functionalities of ROS, the open access online book of Jason M. O'Kane, *A
Gentle Introduction to ROS*[25] is a perfect resource. For the one that prefers physical
books, going in depth in all of the ROS components, along with detailed example,
look into the book of Quigley, Gerkey, and Smart, *Programming Robots with ROS*.
Unfortunately, there is still a lack of good books specific to ROS2, but the online
official documentation is always a great resource.[26]

References

Bruyninckx, H. (2001). Open robot control software: The orocos project. In *Proceedings 2001 ICRA.
IEEE International Conference on Robotics and Automation (Cat. No.01CH37164)*, vol. 3, pp.
2523–2528, vol. 3. https://doi.org/10.1109/ROBOT.2001.933002
Gerkey, B. P., Vaughan, R. T., & Howard, A. (2003). The player/stage project: Tools for multi-robot
and distributed sensor systems. In *Proceedings of the 11th International Conference on Advanced
Robotics*, pp. 317–323.
Huang, A. S., Olson, E., & Moore, D. C. (2010). LCM: lightweight communications and mar-
shalling. In *2010 IEEE/RSJ International Conference on Intelligent Robots and Systems*, pp.
4057–4062. https://doi.org/10.1109/IROS.2010.5649358
Huet, E. (2017). *The not-so-secret code that powers robots around the globe*. Bloomberg The Quint.
Jackson, J. (2007). Microsoft robotics studio: A technical introduction. *IEEE Robotics Automation
Magazine, 14*(4), 82–87. https://doi.org/10.1109/M-RA.2007.905745
Kramer, J., & Scheutz, M. (2007). Development environments for autonomous mobile robots: A
survey. *Autonomous Robots, 22*(2), 101–132. https://doi.org/10.1007/s10514-006-9013-8
Metta, G., Fitzpatrick, P., & Natale, L. (2006). Yarp: Yet another robot platform. *International
Journal of Advanced Robotic Systems, 3*(1), 8.
Montemerlo, M., Roy, N., & Thrun, S. (2003). Perspectives on standardization in mobile robot
programming: The carnegie mellon navigation (carmen) toolkit. In *Proceedings 2003 IEEE/RSJ
International Conference on Intelligent Robots and Systems (IROS 2003) (Cat. No.03CH37453)*,
vol. 3, pp. 2436–2441. https://doi.org/10.1109/IROS.2003.1249235

---

[24] https://docs.ros.org/.

[25] https://www.cse.sc.edu/~jokane/agitr.

[26] such as https://docs.ros.org/en/rolling/.

Palacz, W., Ślusarczyk, G., Strug, B., & Grabska, E. (2019). Indoor robot navigation using graph models based on bim/ifc. In *International Conference on Artificial Intelligence and Soft Computing*, Springer, pp 654–665.

Wyrobek, K. (2017). The origin story of ros, the linux of robotics. In *IEEE Spectrum*.

Wyrobek, K. A., Berger, E. H., Van der Loos, H. M., & Salisbury, J. K. (2008). Towards a personal robotics development platform: Rationale and design of an intrinsically safe personal robot. In *2008 IEEE International Conference on Robotics and Automation*, pp. 2165–2170. https://doi.org/10.1109/ROBOT.2008.4543527

David St-Onge (Ph.D., Mech. Eng.) is an Associate Professor in the Mechanical Engineering Department at the École de technologie supérieure and director of the INIT Robots Lab (initrobots.ca). David's research focuses on human-swarm collaboration more specifically with respect to operators' cognitive load and motion-based interactions. He has over 10 years' experience in the field of interactive media (structure, automatization and sensing) as workshop production director and as R&D engineer. He is an active member of national clusters centered on human-robot interaction (REPARTI) and art-science collaborations (Hexagram). He participates in national training programs for highly qualified personnel for drone services (UTILI), as well as for the deployment of industrial cobots (CoRoM). He led the team effort to present the first large-scale symbiotic integration of robotic art at the IEEE International Conference on Robotics and Automation (ICRA 2019).

Damith Herath is an associate professor in Robotics and Art at the University of Canberra. He is a multi-award winning entrepreneur and a roboticist with extensive experience leading multidisciplinary research teams on complex robotic integration, industrial, and research projects for over two decades. He founded Australia's first collaborative robotics start-up in 2011 and was named one of the most innovative young tech companies in Australia in 2014. Teams he led in 2015 and 2016 consecutively became finalists and, in 2016, a top-ten category winner in the coveted Amazon Robotics Challenge—an industry-focused competition among the robotics research elite. In addition, he has chaired several international workshops on Robots and Art and is the lead editor of the book "Robots and Art: Exploring an Unlikely Symbiosis"—the first significant work to feature leading roboticists and artists together in the field of Robotic Art.

Chapter 6
Mathematical Building Blocks: From Geometry to Quaternions to Bayesian

Rebecca Stower⑩, Bruno Belzile⑩, and David St-Onge⑩

6.1 Learning Objectives

The objective at the end of this chapter is to be able to:

- use vector and matrix operations;
- represent translation, scaling, and symmetry in matrix operations;
- understand the use and limitation of Euler's angles and quaternions;
- use homogeneous transformations;
- use derivatives to find a function optimums and linearize a function;
- understand the importance and the definition of a Gaussian distribution;
- use t-tests and ANOVAs to validate statistical hypothesis.

6.2 Introduction

Several of the bodies of knowledge related to robotics are grounded in physics and statistics. While this book tries to cover each topic in an accessible manner, the large majority of these book chapters expect a minimal background in mathematics. The following pages summarize a wide range of mathematical concepts from geometry to statistics. Throughout this chapter, relevant Python functions are included.

R. Stower (✉)
Department of Psychology, Université Vincennes-Paris 8, Saint-Denis, France
e-mail: becstower@gmail.com

B. Belzile · D. St-Onge
Department of Mechanical Engineering, ÉTS Montréal, Montreal, Canada
e-mail: bruno.belzile.1@ens.etsmtl.ca

D. St-Onge
e-mail: david.st-onge@etsmtl.ca

© The Author(s) 2022
D. Herath and D. St-Onge (eds.), *Foundations of Robotics*,
https://doi.org/10.1007/978-981-19-1983-1_6

Fig. 6.1 Different
coordinate systems in 3D
space

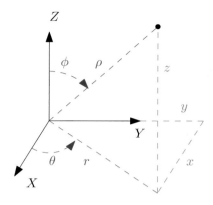

6.3 Basic Geometry and Linear Algebra

In this section, a brief non-exhaustive summary of basic concepts in Euclidean geometry is given. Moreover, some linear algebra operations, useful for the manipulations of components in different arrays, are recalled.

6.3.1 Coordinate Systems

A coordinate system is a "system for specifying points using coordinates measured in some specified way."[1] The most common, which you have most probably used in the past is the *Cartesian* coordinate system, is shown in Fig. 6.1. In this case, more precisely in 3D space, we have an origin, i.e., the point from where the coordinates are measured, and three independent and orthogonal axes, X, Y, and Z. Three axes are needed and they must be independent, but they do not need to be orthogonal. However, for practical reasons in most (but not all) applications, orthogonal axes are preferred (Hassenpflug, 1995).

You may encounter some common alternatives to Cartesian coordinates that can be more appropriate for some applications, such as spherical and cylindrical coordinates. In the former, the coordinates are defined by a distance ρ from the origin and two angles, i.e., θ and ϕ. In the latter, which is an extension of polar coordinates in 2D, a radial distance r, an azimuth (angle) θ, and an axial coordinate (height) z are needed. While a point is uniquely defined with Cartesian coordinates, it is not totally the case with spherical and cylindrical coordinates; more precisely, the origin is defined by an infinite set of coordinates with those two systems, as the angles are not defined at the origin. Moreover, you can add/subtract multiples of 360° to every angle and you will end up with the same point, but different coordinates. Moreover, you should be

---

[1] https://mathworld.wolfram.com/CoordinateSystem.html.

careful with cylindrical and spherical coordinates, as the variables used to define the individual coordinates may be switched, depending on the convention used, which usually differs if you are talking to a physicist, a mathematician, or an engineer.[2]

6.3.2 Vector/Matrix Representation

In mathematics, a vector is "a quantity that has magnitude and direction and that is commonly represented by a directed line segment whose length represents the magnitude and whose orientation in space represents the direction."[3] As you may wonder, this definition does not refer to components and reference frames, which we often come across when vectors are involved. This is because there is a common confusion between the physical quantity represented by a vector and the representation of that same quantity in a coordinate system with one-dimensional arrays. The same word, vector, is used to refer to these arrays, but you should be careful to distinguish the two. Commonly, an arrow over a lower case letter defines a vector, the physical quantity, for example $\vec{a}$, and a lower case bold letter represents a vector defined in a coordinate system, i.e., with components, for example, $\mathbf{a}$. You should note, however, that authors sometimes use different conventions. In this book, the coordinate system used to represent a vector is denoted by a superscript. For example, the variable $\mathbf{b}^S$ is the embodiment of $\vec{b}$ in frame S, while $\mathbf{b}^T$ is the embodiment of $\vec{b}$ in frame T. They do not have the same components, but they remain the same vector.

Vectors $\vec{a}$ and $\vec{b}$ in a n-dimensional Euclidean space can be displayed with their components as

$$\mathbf{a} = \begin{bmatrix} a_1 \\ a_2 \\ a_3 \\ \vdots \\ a_{n-1} \\ a_n \end{bmatrix}, \quad \mathbf{b} = \begin{bmatrix} b_1 \\ b_2 \\ b_3 \\ \vdots \\ b_{n-1} \\ b_n \end{bmatrix} \tag{6.1}$$

For example, vectors $\vec{c}$ and $\vec{d}$ are shown in Fig. 6.2. As can be seen, two reference frames are also displayed. Their components in these frames are

$$\mathbf{c}^S = \begin{bmatrix} 1 \\ 1 \end{bmatrix}, \quad \mathbf{c}^T = \begin{bmatrix} 0 \\ 1.4142 \end{bmatrix}, \quad \mathbf{d}^S = \begin{bmatrix} 1 \\ 3 \end{bmatrix}, \quad \mathbf{d}^T = \begin{bmatrix} -1.4142 \\ 2.8284 \end{bmatrix} \tag{6.2}$$

---

[2] See https://mathworld.wolfram.com/SphericalCoordinates.html.

[3] https://www.merriam-webster.com/dictionary/vector.

Fig. 6.2 Planar vectors and their components in different frames

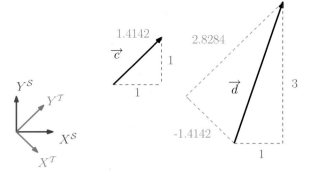

```
import numpy as np # Import library
# arrays
a = np.array([1,1]) # vector
A = np.array([1,2],
             [3,4]) # matrix
```

Similarly, tensors are used to represent physical properties of a body (and many other things). More formally, tensors are algebraic objects defining multilinear relationships between other objects in a vector space. Do not focus to much on the mathematical definition, but instead on what you already know. You have already encountered some tensors in this chapter, since scalars and vectors (the physical quantity, not the array) are, respectively, rank-0 and rank-1 tensors.[4] Therefore, tensors can be seen as their generalization. One example of rank-2 tensors is the inertia tensor of a rigid body, which basically represents how the mass is distributed in a rigid body (which does not depend on a reference frame). For the sake of numerical computation, the representation of a rank-2 tensor in a coordinate system can be done with what we call a matrix. You should be careful, however, not to confuse matrices and rank-2 tensors. Indeed, all rank-2 tensors can be represented by a matrix, but not all matrices are rank-2 tensors. In other words, matrices are just boxes (arrays) with numbers inside (components) that can be used to represent different objects, rank-2 tensors among them. Matrices are generally represented by upper case bold letters, eg. **A**. Matrices, which have components, can also be defined in specific reference frames. Therefore, the superscript to denote the reference frame also applies to matrices in the book, e.g., $\mathbf{H}^S$ is a homogeneous transformation matrix (will be seen in Sect. 6.4.4) defined in S.

Other common matrices with typical characteristics include:

- the square matrix, which is a matrix with an equal number of rows and columns;
- the diagonal matrix, which only has nonzero components on its diagonal, i.e., components $(1, 1), (2, 2), \ldots, (n, n)$;
- the identity matrix **1**, which is a $(n \times n)$ matrix with only 1 on the diagonal, the other components all being equal to 0.

---

[4] For more information on tensors and their rank: https://mathworld.wolfram.com/Tensor.html.

6.3.3 Basic Vector/Matrix Operations

Vectors and matrices are powerful and versatile mathematical tools with several handful properties and operations. We will recall the most useful in robotics in the following.

Dot Product

The addition and the multiplication with a scalar operations with vectors are simply distributed over the components. Otherwise, two most relevant operations in robotics are the dot and cross products. The dot product is also known as the scalar product, as the result of the dot product of two arbitrary vectors is a scalar. Let $\vec{a}$ and $\vec{b}$ be two arbitrary vectors and their corresponding magnitude[5] be $\|\vec{a}\|$ and $\|\vec{b}\|$, then the dot product of these two vectors is

$$\vec{a} \cdot \vec{b} = \|\vec{a}\|\|\vec{b}\| \cos\theta \tag{6.3}$$

where θ is the angle between those two vectors. If the two vectors are orthogonal, by definition, the result will be zero. If components are used, then we have

$$\mathbf{a} \cdot \mathbf{b} = a_1 b_1 + a_2 b_2 + a_3 b_3 + \cdots + a_{n-1} b_{n-1} + a_n b_n \tag{6.4}$$

```
import numpy as np # Import library
# dot product
np.dot(a,b)                    # dot product of two array-like inputs
np.linalg.multi_dot(a,b,c) # dot product of two or more arrays in a single call
# magnitude of a vector
np.linalg.norm(a)
```

Using the numerical values previously given in (6.2), the dot product of $\vec{a}$ and $\vec{b}$ is:

$$\vec{a} \cdot \vec{b} = 1.4142 \cdot 3.1623 \cos(0.4636) = 4 \tag{6.5}$$

$$\mathbf{a}^S \cdot \mathbf{b}^S = 1 \cdot 1 + 1 \cdot 3 = 4 \tag{6.6}$$

$$\mathbf{a}^T \cdot \mathbf{b}^T = 0 \cdot -1.4142 + 1.4142 \cdot 2.8284 = 4 \tag{6.7}$$

As you can see from this example, both the geometric and algebraic definitions of the dot product are equivalent.

Cross Product

The other type of multiplication with vectors is the cross product. Contrary to the dot product, the cross product of two vectors results in another vector, not a scalar. Again, both vectors must have the same dimension. With $\vec{a}$ and $\vec{b}$ used above, the cross product is defined as

---

[5] Length, always positive.

$$\vec{a} \times \vec{b} = \|\vec{a}\| \|\vec{b}\| \sin\theta \, \vec{e} \tag{6.8}$$

where, as with the dot product, θ is the angle between $\vec{a}$ and $\vec{b}$, and $\vec{e}$ is a unit vector[6] orthogonal to the first two. Its direction is established with the right-hand rule. In 3D space, the components of the resulting vector can be computed with the following formula:

$$\mathbf{a} \times \mathbf{b} = \begin{bmatrix} a_2b_3 - a_3b_2 \\ a_3b_1 - a_1b_3 \\ a_1b_2 - a_2b_1 \end{bmatrix} \tag{6.9}$$

where $\mathbf{a} = [a_1 \ a_2 \ a_3]^T$ and $\mathbf{b} = [b_1 \ b_2 \ b_3]^T$.

> The right-hand rule is used to easily determine the direction of a vector resulting from the cross product of two others. First, you point in the direction of the first vector with your remaining fingers, then curl them to point in the direction of the second vector. According to this rule, the thumb of the right hand will point along the direction of the resulting vector, which is normal to the plane formed by the two initial vectors.

```
import numpy as np # Import library
# cross product
np.cross(a,b)
```

Again, using the numerical values used above in (6.2), we can compute the cross product. Of course, since these two vectors are planar and the cross product is defined over 3D space, the third component in Z is assumed equal to zero. The result is given below:

$$\vec{a} \times \vec{b} = 1.4142 \cdot 3.1623 \sin(0.4636) \, \vec{k} = 2\,\vec{k} \tag{6.10}$$

$$\mathbf{a}^S \times \mathbf{b}^S = \begin{bmatrix} 1 \cdot 0 - 0 \cdot 3 \\ 0 \cdot 1 - 1 \cdot 0 \\ 1 \cdot 3 - 1 \cdot 1 \end{bmatrix} = \begin{bmatrix} 0 \\ 0 \\ 2 \end{bmatrix} \tag{6.11}$$

$$\mathbf{a}^T \times \mathbf{b}^T = \begin{bmatrix} 1.4142 \cdot 0 - 0 \cdot 2.8284 \\ 0 \cdot -1.41421356 - 1.4142 \cdot 0 \\ 0 \cdot 2.8284 - 1.4142 \cdot -1.4142 \end{bmatrix} = \begin{bmatrix} 0 \\ 0 \\ 2 \end{bmatrix} \tag{6.12}$$

where $\vec{k}$ is the unit vector parallel to the Z-axis. By this definition, you can observe that the unit vector defining the Z-axis of a Cartesian coordinate frame is simply the cross product of the unit vectors defining the X- and Y-axes, following the order given by the right-hand rule. These three unit vectors are commonly labeled $\vec{i}$, $\vec{j}$ and $\vec{k}$, as shown in Fig. 6.3. You should note that the cross product of unit vector $\vec{a}$ with $\vec{j}$ also results in $\vec{k}$, since $\vec{a}$ is also in the XY-plane. Moreover, as you

---

[6] With a magnitude of 1.

Fig. 6.3 Unit vectors
defining a Cartesian frame

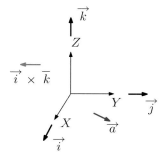

can see with the cross product of $\vec{i}$ and $\vec{k}$ illustrated in the same figure, a vector
is not attached to a particular point in space. As mentioned before, it is defined by
a direction and a magnitude, thus the location where it is represented does not have
any impact on the cross product result.

Matrix Multiplication

Similarly to vectors, the addition and multiplication by a scalar are also distributed
over the components for matrices. On the other hand, the matrix multiplication is
a little more complicated. Let matrix **A** be defined by row vectors and matrix **B** be
defined by column vectors, i.e.,

$$
\mathbf{A} = \begin{bmatrix} \mathbf{a}_1 \\ \mathbf{a}_2 \\ \mathbf{a}_3 \\ \vdots \\ \mathbf{a}_{n-1} \\ \mathbf{a}_n \end{bmatrix}, \quad \mathbf{B} = \begin{bmatrix} \mathbf{b}_1 & \mathbf{b}_2 & \mathbf{b}_3 & \dots & \mathbf{b}_{n-1} & \mathbf{b}_n \end{bmatrix} \tag{6.13}
$$

Then, the matrix multiplication is defined as

$$
\mathbf{AB} = \begin{bmatrix}
\mathbf{a}_1 \cdot \mathbf{b}_1 & \mathbf{a}_1 \cdot \mathbf{b}_2 & \mathbf{a}_1 \cdot \mathbf{b}_3 & \dots & \mathbf{a}_1 \cdot \mathbf{b}_{n-1} & \mathbf{a}_1 \cdot \mathbf{b}_n \\
\mathbf{a}_2 \cdot \mathbf{b}_1 & \mathbf{a}_2 \cdot \mathbf{b}_2 & \mathbf{a}_2 \cdot \mathbf{b}_3 & \dots & \mathbf{a}_2 \cdot \mathbf{b}_{n-1} & \mathbf{a}_2 \cdot \mathbf{b}_n \\
\mathbf{a}_3 \cdot \mathbf{b}_1 & \mathbf{a}_3 \cdot \mathbf{b}_2 & \mathbf{a}_3 \cdot \mathbf{b}_3 & \dots & \mathbf{a}_3 \cdot \mathbf{b}_{n-1} & \mathbf{a}_3 \cdot \mathbf{b}_n \\
\vdots & \vdots & \vdots & \ddots & \vdots & \\
\mathbf{a}_{n-1} \cdot \mathbf{b}_1 & \mathbf{a}_{n-1} \cdot \mathbf{b}_2 & \mathbf{a}_{n-1} \cdot \mathbf{b}_3 & \dots & \mathbf{a}_{n-1} \cdot \mathbf{b}_{n-1} & \mathbf{a}_{n-1} \cdot \mathbf{b}_n \\
\mathbf{a}_n \cdot \mathbf{b}_1 & \mathbf{a}_n \cdot \mathbf{b}_2 & \mathbf{a}_n \cdot \mathbf{b}_3 & \dots & \mathbf{a}_n \cdot \mathbf{b}_{n-1} & \mathbf{a}_n \cdot \mathbf{b}_n
\end{bmatrix} \tag{6.14}
$$

While this result may seem scary at first, you can see that the (i, j) component[7] is
simply the dot product of the ith row of the first matrix and the jth column of the

---

[7] The (i, j) component is the component on the ith row and jth column.

second matrix. The number of columns of the first matrix (**A**) must be equal to the number of rows of the second matrix (**B**).

```
import numpy as np # Import library
# matrix multiplication
np.matmul(A,B)  # for array-like inputs
A @ B           # for ndarray inputs
```

To illustrate this operation, let **A** and **B** be (2×2) matrices, i.e.,

$$\mathbf{A} = \begin{bmatrix} 1 & 2 \\ 3 & 4 \end{bmatrix}, \quad \mathbf{B} = \begin{bmatrix} 1 & 0 \\ -1 & 2 \end{bmatrix} \tag{6.15}$$

then, the result of the matrix multiplication is

$$\mathbf{AB} = \begin{bmatrix} 1 \cdot 1 - 2 \cdot 1 & 1 \cdot 0 + 2 \cdot 2 \\ 3 \cdot 1 - 4 \cdot 1 & 3 \cdot 0 + 4 \cdot 2 \end{bmatrix} = \begin{bmatrix} -1 & 4 \\ -1 & 8 \end{bmatrix} \tag{6.16}$$

It is critical that you understand that matrix multiplication is **not commutative**, which means the order matters, as you can see in the following example with matrices **A** and **B** used above:

$$\mathbf{AB} = \begin{bmatrix} -1 & 4 \\ -1 & 8 \end{bmatrix}, \text{ but } \mathbf{BA} = \begin{bmatrix} 1 & 2 \\ 5 & 6 \end{bmatrix} \tag{6.17}$$

Transpose of a Matrix

Another common operation on a matrix is the computation of its transpose, namely an operation which flips a matrix over its diagonal. The generated matrix, denoted $\mathbf{A}^T$ has the row and column indices switched with respect to **A**. For instance, with a (3×3) matrix **C**, its transpose is defined as

$$\mathbf{C}^T = \begin{bmatrix} c_{1,1} & c_{1,2} & c_{1,3} \\ c_{2,1} & c_{2,2} & c_{2,3} \\ c_{3,1} & c_{3,2} & c_{3,3} \end{bmatrix}^T = \begin{bmatrix} c_{1,1} & c_{2,1} & c_{3,1} \\ c_{1,2} & c_{2,2} & c_{3,2} \\ c_{1,3} & c_{2,3} & c_{3,3} \end{bmatrix} \tag{6.18}$$

```
import numpy as np # Import library
# matrix transpose
np.transpose(A) # function for array-like input
A.transpose()   # method for ndarray
A.T             # attribute for ndarray
```

Since vectors (array of components) are basically ($1 \times n$) matrices, the transpose can be used to compute the dot product of two vectors with a matrix multiplication, i.e.,

$$\mathbf{a} \cdot \mathbf{b} = \mathbf{a}^T \mathbf{b} = a_1 b_1 + a_2 b_2 + \cdots + a_n b_n \tag{6.19}$$

Determinant and Inverse of a Matrix

Finally, a brief introduction to the inverse of a matrix is necessary, as it is quite common in robotics, from the mechanics to control to optimization. Let $\mathbf{A}$ be a $(n \times n)$ square matrix;[8] this matrix is invertible if

$$\mathbf{AB} = 1, \quad \text{and} \quad \mathbf{BA} = 1 \tag{6.20}$$

Then, matrix $\mathbf{B}$ is the inverse of $\mathbf{A}$ and therefore can be written as $\mathbf{A}^{-1}$. The components of $\mathbf{A}^{-1}$ can be computed formally with the following formula:

$$\mathbf{A}^{-1} = \frac{1}{\det(\mathbf{A})} \mathbf{C}^T \tag{6.21}$$

where $\det(\mathbf{A})$ is called the *determinant* of $\mathbf{A}$ and $\mathbf{C}$ is the cofactor matrix[9] of $\mathbf{A}$. The determinant of a matrix, a scalar sometimes labeled $\|\mathbf{A}\|$, is equal to, in the case of a (2×2) matrix,

$$\det(\mathbf{A}) = ad - bc, \quad \text{where} \quad \mathbf{A} = \begin{bmatrix} a & b \\ c & d \end{bmatrix} \tag{6.22}$$

Similarly, for a 3×3 matrix, we have

$$\det(\mathbf{A}) = a(ei - fh) - b(di - fg) + c(dh - eg), \quad \text{where} \quad \mathbf{A} = \begin{bmatrix} a & b & c \\ d & e & f \\ g & h & i \end{bmatrix} \tag{6.23}$$

The determinant of a matrix is critical when it comes to the computation of its inverse, as a determinant of 0 corresponds to a *singular* matrix, which does not have an inverse. The inverse of a (2×2) matrix can be computed with the following formula

$$\mathbf{A}^{-1} = \frac{1}{ad - bc} \begin{bmatrix} d & -b \\ -c & a \end{bmatrix}, \quad \text{where} \quad \mathbf{A} = \begin{bmatrix} a & b \\ c & d \end{bmatrix} \tag{6.24}$$

Similarly, for a 3×3 matrix, we have

$$\mathbf{A}^{-1} = \frac{1}{\det(\mathbf{A})} \begin{bmatrix} (ei - fg) & -(bi - ch) & (bf - ce) \\ -(di - fg) & (ai - cg) & -(af - cd) \\ (dh - eg) & -(ah - bg) & (ae - bd) \end{bmatrix}, \quad \text{where} \quad \mathbf{A} = \begin{bmatrix} a & b & c \\ d & e & f \\ g & h & i \end{bmatrix} \tag{6.25}$$

---

[8] Same number of rows and columns.

[9] The cofactor matrix will not be introduced here for the sake brevity, but its definition can be found in any linear algebra textbook.

```
import numpy as np # Import library
# matrix determinant
np.linalg.det(A)
# matrix inverse
np.linalg.inv(A)
```

As you can see from Eq. (6.25), you cannot inverse a matrix with a determinant equal to zero, since it would result in a division by zero. The inverse of a matrix is a useful tool to solve a system of linear equations. Indeed, a system of n equations with n unknowns can be casted in matrix form as

$$\mathbf{Ax} = \mathbf{b} \tag{6.26}$$

where the unknowns are the components of $\mathbf{x}$, the constants are the components of $\mathbf{b}$ and the factors in front of each unknowns are the components of matrix $\mathbf{A}$. Therefore, we can find the solution of this system, namely the values of the unknown variables, as

$$\mathbf{x} = \mathbf{A}^{-1}\mathbf{b} \tag{6.27}$$

Generalized Inverses

However, if we have more equations (m) than the number of unknowns (n), the system is overdetermined, and thus $\mathbf{A}$ is no longer a square matrix. Its dimensions are ($m \times n$). An exact solution to this system of equations cannot generally be found. In this case, we use a generalized inverse; a strategy to find an optimal solution. Several generalized inverse, or *pseudo-inverse*, can be found in the literature (Ben-Israel and Greville, 2003), each with different optimization criterion. For the sake of this book, only one type is presented here, the Moore–Penrose generalized inverse (MPGI). In the case of overdetermined systems, the MPGI is used to find the approximate solution that minimized the Euclidean norm of the error, which is defined as

$$\mathbf{e}_0 = \mathbf{b} - \mathbf{Ax}_0 \tag{6.28}$$

where $\mathbf{x}_0$ and $\mathbf{e}_0$ are the approximate solution and the residual error, respectively. The approximate solution is computed with

$$\mathbf{x}_0 = \mathbf{A}^L\mathbf{b}, \quad \mathbf{A}^L = (\mathbf{A}^T\mathbf{A})^{-1}\mathbf{A}^T \tag{6.29}$$

where $\mathbf{A}^L$ is named the *left Moore–Penrose generalized inverse* (LMPGI), since $\mathbf{A}^I\mathbf{A} = \mathbf{1}$. As an exercise, you can try to prove this equation.

There is another MPGI that can be useful in robotics, but not quite as common as the LMPGI, the *right Moore–Penrose generalized inverse* (RMPGI). The right generalized inverse is defined as

$$\mathbf{A}^R \equiv \mathbf{A}^T(\mathbf{AA}^T)^{-1}, \quad \mathbf{AA}^R = \mathbf{1} \tag{6.30}$$

where $\mathbf{A}$ is a $m \times n$ matrix with $m < n$, i.e., representing a system of linear equations with more unknowns than equations. In this case, this system admits infinitely many solutions. Therefore, we are not looking for the best approximate solution, but one solution with the minimum-(Euclidean) norm. For example, in robotics, when there is an infinite set of joint configurations possible to perfectly reach an arbitrary position with a manipulator, the RMPGI can give you the one minimizing the joint rotations.

With both generalized inverses presented here, we assume that $\mathbf{A}$ is full rank, which means that its individual columns are independent if $m > n$, or its individual rows are independent if $m < n$. In the case of a square matrix ($m = n$), a full rank matrix is simply non-singular.

6.4 Geometric Transformations

It is crucial in robotics to be able to describe geometric relations in a clear and unambiguous way. This is done with coordinate systems and reference frames as mentioned above. You may have studied already four kinds of geometric transformation: translation, scaling, symmetry (mirror), and rotation. We will quickly go over each of them, as they all are useful for computer-assisted design. However, keep in mind that transformations used to map coordinates in one frame into another use only translation and rotation.

For clarity, we will present all geometric transformations in matrix form, to leverage the powerful operations and properties as well as their condensed format. Using the vector introduction above (Sect. 6.3.2), the simplest geometric element will be used to introduce the transformation, the point:

$$P_{2D}(x, y) = \begin{bmatrix} x \\ y \end{bmatrix}, \quad P_{3D}(x, y, z) = \begin{bmatrix} x \\ y \\ z \end{bmatrix} \tag{6.31}$$

In fact, you only need to apply transformations to point entities in order to transform any 2D and 3D geometry. From a set of points, you can define connected pairs, i.e., edges or lines, and from a set of lines you can define loops, i.e., surfaces. Finally, a set of selected surfaces can define a solid (Fig. 6.4).

6.4.1 Basic Transformations

Let's start with a numerical example: given a point in $x = 1$ and $y = 2$ that we intend to move by 2 units toward x positive and by 4 units toward y positive. The algebraic form of this operation is simply $x' = x + 2$ and $y' = y + 4$, which can be written in matrix form:

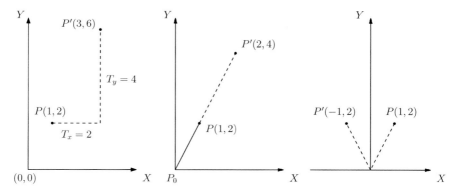

Fig. 6.4 Basic geometrical transformations, from left to right: translation, scaling and mirror (symmetry)

$$P' = P + T = \begin{bmatrix} 1 \\ 2 \end{bmatrix} + \begin{bmatrix} 2 \\ 4 \end{bmatrix} \tag{6.32}$$

Similar reasoning applies in three dimensions. Now imagine we use point P to define a line with the origin $P_0 = \begin{bmatrix} 0 \\ 0 \end{bmatrix}$ and that we want to stretch this line with a scaling factor of 2. The algebraic form of this operation is $x' = x \times 2$ and $y' = y \times 2$, which can be written in matrix form:

$$P' = SP = \begin{bmatrix} 2 & 0 \\ 0 & 2 \end{bmatrix} \begin{bmatrix} 1 \\ 2 \end{bmatrix} \tag{6.33}$$

This scaling operation is referred to as *proportional*, since both axes have the same scaling factor. Using different scaling factors will deform the geometry. If, instead of scaling the geometry, we use a similar diagonal matrix to change the sign of one or more of its components, it will generate a symmetry. For instance, a symmetry with respect to y is written:

$$P' = SP = \begin{bmatrix} -1 & 0 \\ 0 & 1 \end{bmatrix} \begin{bmatrix} 1 \\ 2 \end{bmatrix} \tag{6.34}$$

These operations are simple and do not change with increasing the dimensions from two to three. The rotations, however, are not as such.

6.4.2   2D/3D Rotations

A rotation is a geometric transformation that is more easily introduced with polar coordinates (see Fig. 6.5):

Fig. 6.5 Planar rotation and polar coordinates

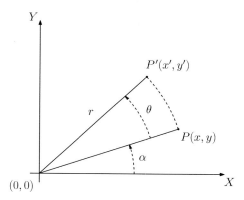

$$P = \begin{bmatrix} x \\ y \end{bmatrix} = \begin{bmatrix} r\cos(\alpha) \\ r\sin(\alpha) \end{bmatrix}. \tag{6.35}$$

Then a rotation θ applied to this vector consists in:

$$P' = \begin{pmatrix} r\cos(\alpha + \theta) \\ r\sin(\alpha + \theta) \end{pmatrix}, \tag{6.36}$$

which can be split with respect to the angles using common trigonometric identities leading to

$$P' = \begin{bmatrix} x\cos(\theta) - y\sin(\theta) \\ x\sin(\theta) + y\cos(\theta) \end{bmatrix} = \begin{bmatrix} \cos(\theta) & -\sin(\theta) \\ \sin(\theta) & \cos(\theta) \end{bmatrix} \begin{bmatrix} x \\ y \end{bmatrix}. \tag{6.37}$$

The resulting 2×2 matrix is referred to as the rotation matrix, and its format is unique in 2D. Any rotation in the plane can be represented by this matrix, using the right-hand rule for the sign of θ. This matrix is unique because a single rotation axis exists for planar geometry: the perpendicular to the plane (often set as the z-axis). For geometry in three-dimensional space, there is an infinite number of potential rotation axis; just visualize the rotational motions you can apply to an object in your hand. One approach to this challenge consists in defining a direction vector in space and a rotation angle around it, since Leonhard Euler taught us that "*in three-dimensional space, any displacement of a rigid body such that a point on the rigid body remains fixed, is equivalent to a single rotation about some axis that runs through the fixed point.*" While this representation is appealing to humans fond of geometry, it is not practical to implement in computer programs for generalized rotations. Instead, we can decompose any three-dimensional rotation into a sequence of three rotations around principal axis. This approach is called the *Euler's Angles* and is the most common representation of three-dimensional rotation. We only need to define three matrices:

$$\mathbf{R}_x = \begin{bmatrix} 1 & 0 & 0 \\ 0 & \cos(\psi) & -\sin(\psi) \\ 0 & \sin(\psi) & \cos(\psi) \end{bmatrix}, \tag{6.38}$$

$$\mathbf{R}_y = \begin{bmatrix} \cos(\phi) & 0 & \sin(\phi) \\ 0 & 1 & 0 \\ -\sin(\phi) & 0 & \cos(\phi) \end{bmatrix}, \tag{6.39}$$

$$\mathbf{R}_z = \begin{bmatrix} \cos(\theta) & -\sin(\theta) & 0 \\ \sin(\theta) & \cos(\theta) & 0 \\ 0 & 0 & 1 \end{bmatrix}. \tag{6.40}$$

If these matrices are the only ones required to represent any rotation, they still leave two arbitrary definitions: 1. the orientation of the principal axes $(x - y - z)$ in space, 2. the order of the rotations. Rotation matrices are multiplication operations over geometry features, and, as mentioned above, these operations are not commutative. The solution is to agree over a universal set of *conventions*:

$$XYX, \ XYZ, \ XZX, \ XZY, \ YXY, \ YXZ, \ YZX,$$
$$YZY, \ ZXY, \ ZXZ, \ ZYX, \text{ and } ZYZ. \tag{6.41}$$

These twelves conventions still need their axes orientation to be defined: Each axis can either be fixed to the inertial frame (often referred to as *extrinsic* rotations) or attached to the body rotating (often referred to as *intrinsic* rotations). For instance, the fixed rotation matrix for the XYZ convention is:

$$\mathbf{R}_z\mathbf{R}_y\mathbf{R}_x = \begin{bmatrix} \cos_\theta \cos_\phi & \cos_\theta \sin_\phi \sin_\psi - \sin_\theta \cos_\psi & \cos_\theta \sin_\phi \cos_\psi + \sin_\theta \sin_\psi \\ \sin_\theta \cos_\phi & \sin_\theta \sin_\phi \sin_\psi - \cos_\theta \cos_\psi & \sin_\theta \sin_\phi \cos_\psi - \cos_\theta \sin_\psi \\ -\sin_\phi & \cos_\phi \sin_\psi & \cos_\phi \cos_\psi \end{bmatrix}. \tag{6.42}$$

While using a fixed frame may seem easier to visualize, most embedded controllers require their rotational motion to be expressed in the body frame; one attached to the object and moving with it. The same convention XYZ, but in mobile frame is:

$$\mathbf{R}'_x\mathbf{R}'_y\mathbf{R}'_z = \begin{bmatrix} \cos_\phi \cos_\theta & -\cos\phi \sin_\theta & \sin_\phi \\ \cos_\psi \sin_\theta + \sin_\psi \sin_\phi \cos_\theta & \cos_\psi \cos_\theta - \sin_\psi \sin_\phi \sin_\theta & -\sin_\psi \cos_\phi \\ \sin_\psi \sin_\theta - \cos_\psi \sin_\phi \cos_\theta & \sin_\psi \cos_\theta + \cos_\psi \sin_\phi \sin_\theta & \cos_\psi \cos_\phi \end{bmatrix}. \tag{6.43}$$

In aviation, the most common convention is the ZYX (roll–pitch–yaw) also called the *Tait–Bryan* variant. In robotics, each manufacturer and software developer decides on the convention they prefer to use, for instance, FANUC and KUKA use the fixed XYZ Euler angle convention, while ABB uses the mobile ZYX Euler angle convention. As for computer-assisted design, the Euler angles used in CATIA and SolidWorks are described by the mobile ZYZ Euler angles convention.

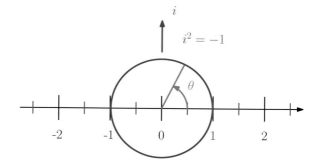

Fig. 6.6 Vector representation of planar rotation using the imaginary axis i

Euler's angle representation is known to have a significant limitation: gimbal lock. In a glimpse, each convention suffers from singular orientation(s), i.e., orientation at which two axes are overlaid, thus both having the same effect on rotation. With two axes generating the same rotation, our three-dimensional space is no longer fully reachable; i.e., one rotation is not possible anymore. Gimbal lock has become a rather popular issue in spacecraft control since Apollo's mission suffered from it (Jones and Fjeld, 2006). Nevertheless, Euler's angles stay the most common and intuitive representation of three-dimensional rotation and orientation, but others, often more complex, representation were introduced to cope with this limitation.

6.4.3 Quaternion

One such gimbal-lock-free representation is the quaternion. Quaternion is a rather complex mathematical concept with respect to the level required for this textbook. We will not try to define exactly the quaternion in terms of their mathematical construction, and we will not detail all of their properties and operations. Instead, you should be able to grasp the concept thanks to a comparison with the imaginary numbers, a more common mathematical concept.

We recall that the imaginary axis (i) is orthogonal to the real numbers one (see Fig. 6.6), with the unique property $i^2 = -1$. Together they create a planar reference frame that can be used to express rotations:

$$R(\theta) = \cos(\theta) + \sin(\theta)i. \tag{6.44}$$

In other words, we can write a rotation in the plane as a vector with an imaginary part. Now, imagine adding two more rotations as defined above with Euler's angles: we will need two more "imaginary" orthogonal axes to represent these rotations. Equation 6.44 becomes:

$$R(\theta) = \cos(\theta) + \sin(\theta)(xi + yj + zk). \tag{6.45}$$

While this can be easily confused with a vector-angle representation, remember that $i - j - k$ define "imaginary" axes; not coordinates in the Cartesian space. These axes hold similar properties as the more common i imaginary axis:

$$\|i, j, k\| = 1, \ ji = -k, \ ij = k, \ i^2 = -1. \tag{6.46}$$

For most people, quaternions are not easy to visualize compared to Euler angles, but they provide a singularity-free representation and several computing advantages. This is why ROS (see Chap. 5) developers selected this representation as their standard.

In Python, the **scipy** library contains a set of functions to easily change from one representation to another:

```python
# Import the library
from scipy.spatial.transform import Rotation as R
# Create a rotation with Euler angles
mat = R.from_euler('yxz', [45, 0, 30], degrees=True)
print("Euler: ", mat.as_euler('yxz', degrees=True))
# Print the resulting quaternion
print("Quaternion: ", mat.as_quat())
```

6.4.4 Homogeneous Transformation Matrices

A standardized way to apply a transformation from one coordinate system to another, i.e., to map a vector from one reference frame to another, is to use homogeneous transformation matrices. Indeed, a homogeneous transformation matrix can be used to describe both the position and orientation of an object.

The (4×4) homogeneous transformation matrix is defined as

$$\mathbf{H}_{\mathcal{S}}^{\mathcal{T}} \equiv \begin{bmatrix} \mathbf{Q} & \mathbf{p} \\ \mathbf{0}^T & 1 \end{bmatrix} \tag{6.47}$$

where $\mathbf{Q}$ is the (3×3) rotation (orientation) matrix, $\mathbf{p}$ is the three-dimensional vector defining the Cartesian position $[x, \ y, \ z]$ of the origin and $\mathbf{0}$ is the three-dimensional null vector. As can be seen with the superscript and subscript of $\mathbf{H}$, the matrix defines the reference frame $\mathcal{T}$ in the reference frame $\mathcal{S}$. While being composed of 9 components, there are not all independent, since the position and orientation in the Cartesian space add up to 6 degrees-of-freedom (DoF). Whereas the translation introduced above were defined as additions, the homogeneous matrix merges it with rotation and makes it possible to use only multiplications.

6.5 Basic Probability

6.5.1 Likelihood

When we talk about probability, we are typically interested in predicting the likelihood of some event occurring, expressed as $P(event)$. On the most basic level, this can be conceptualized as a proportion representing the number of event(s) we are interested in (i.e., that fulfill some particular criteria), divided by the total number of equally likely events.

Below is a summary of the notation for describing the different kinds and combinations of probability events which will be used throughout the rest of this section (Table 6.1).

As an example, imagine we have a typical (non-loaded) 6-sided die. Each of the six sides has an equal likelihood of occurring each time we roll the die. So, the total number of possible outcomes on a single dice roll, each with equal probability of occurring is 6. Thus, we can represent the probability of any specific number occurring on a roll as a proportion over 6.

For example, the probability of rolling a 3 is expressed as:

$$P(3) = \frac{1}{6} \qquad (6.48)$$

The probability of an event *not* occurring is always the inverse of the probability of it occurring, or $1 - P(event)$. This is known as the **rule of subtraction**.

$$P(A) = 1 - P(A') \qquad (6.49)$$

So in the aforementioned example, the probability of *not* rolling a 3 is:

$$P(3') = 1 - \frac{1}{6} = \frac{5}{6} \qquad (6.50)$$

We could also change our criteria to be more general, for example to calculate the probability of rolling an even number. In this case, we can now count 3 possible outcomes which match our criteria (rolling a 2, 4, or 6), but the total number of possible events remains at 6. So, the probability of rolling an even number is:

Table 6.1 Common probability notations

$P(A)$	Probability of A occurring
$P(A')$	Probability of A *not* occurring
$P(A \cap B)$	Probability of both A and B occurring
$P(A \cup B)$	Probability of either A or B occurring
$P(A|B)$	Probability of A occurring given B occurs

$$P(\text{even}) = \frac{3}{6} = \frac{1}{2} \tag{6.51}$$

Now, imagine we expanded on this criterion of rolling even numbers, to calculate the probability of rolling either an even number OR a number greater than 3. We now have two different criteria which we are interested in (being an even number or being greater than 3) and want to calculate the probability that a single dice roll results in either of these outcomes.

To begin with, we could try simply adding the probability of each individual outcome together:

$$P(\text{even} \cup > 3) = \frac{3}{6} + \frac{3}{6} = \frac{6}{6} = 1 \tag{6.52}$$

We have ended up with a probability of 1, or in other words, a 100% chance of rolling a number which is either even or greater than 3. Since we already know there are numbers on the die which do not meet either of the criteria, we can deduce that this conclusion is incorrect.

The miscalculation stems from the fact that there are numbers which are both even numbers AND greater than 3 (namely 4 and 6). By just adding the probabilities together, we have "double-counted" their likelihood of occurring. In Fig. 6.7, we can see that if we create a Venn diagram of even numbers and numbers > 3, they overlap in the middle with the values of 4 and 6. If we think of probability as calculating the total area of these circles, then we only need to count the overlap once.

So to overcome this double-counting, we subtract the probability of both events occurring simultaneously (in this example, the probability of rolling a number which is both an even number AND greater than 3) from the summed probability of the individual events occurring;

$$P(\text{even} \cup > 3) = \frac{3}{6} + \frac{3}{6} - \frac{2}{6} = \frac{4}{6} = \frac{2}{3} \tag{6.53}$$

More generally, this is known as the **rule of addition** and takes the general form:

$$P(A \cup B) = P(A) + P(B) - P(A \cap B) \tag{6.54}$$

Fig. 6.7 Venn diagram of even numbers and numbers greater than 3

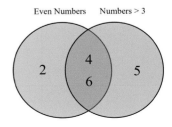

In the case where two outcomes cannot happen simultaneously (i.e., there is no overlap in the venn diagram), then $P(A \cup B) = P(A) + P(B)$, as $P(A \cap B) = 0$. This is known as *mutually exclusive events*.

Finally, imagine we slightly changed our criteria again, so that we are now interested in the probability of rolling both an even number AND a number greater than 3. You might have noticed we actually already used the probability of both an even number and a number greater than three occurring in the previous equation to calculate the probability of either of the two events occurring, $P(\text{even} \cap > 3) = \frac{2}{6} = \frac{1}{3}$. This is because in this example we have a small number of outcomes, meaning it is relatively easy to just count the number of outcomes which match our criteria. However, in more complicated scenarios the calculation is not as straightforward.

So, to begin thinking about the question of how to calculate the probability of two events happening simultaneously, we can first ask what is the probability of one of the events occurring, *given the other event has already occurred*. In this example, we could calculate the probability of rolling a number greater than 3, given that the number rolled is already even. That is, if we have already rolled the die and know that the outcome is an even number, what is the likelihood that it is *also* greater than 3?

We already know that there are three sides of the die which have even numbers (2, 4, or 6). This means our number of possible outcomes, if we know the outcome is even, is reduced from 6 to 3. We can then count the number of outcomes from this set which are greater than 3. This gives us two outcomes (4 and 6). Thus, the probability of rolling a number greater than 3, given that it is also even is:

$$P(> 3|\text{even}) = \frac{2}{3} \tag{6.55}$$

However, this calculation still overestimates the probability of both events occurring simultaneously, as we have reduced our scenario to one where we are 100% sure one of the outcomes has occurred (we have already assumed that the outcome of the roll is an even number). So, to overcome this, we can then multiply this equation by the overall probability of rolling an even number, which we know from before is $P = \frac{3}{6}$.

$$P(\text{even} \cap > 3) = \frac{3}{6} \times \frac{2}{3} = \frac{6}{18} = \frac{1}{3} \tag{6.56}$$

This gives us the same value, $P(A \cap B) = \frac{1}{3}$ that we saw in our previous equation. This is also called the **rule of multiplication**, with the general form:

$$P(A \cap B) = P(A)P(B|A) \tag{6.57}$$

One additional factor to consider when calculating probability is whether events are dependent or independent. In the dice example, these events are dependent, as one event happening (rolling an even number) affects the probability of the other event happening (rolling a number greater than 3). The overall probability of rolling a number greater than 3 is $\frac{1}{2}$, but increases to $\frac{2}{3}$ if we already know that the number rolled is even.

If events are independent, i.e., do not affect each other's probability of occurring, the rule of multiplication reduces to:

$$P(A \cap B) = P(A) \times P(B) \qquad (6.58)$$

The rule of multiplication also forms the basis for Bayes' theorem, to be discussed in the next section.

6.5.2 Bayes' Theorem

Bayes' rule is a prominent principle used in artificial intelligence to calculate the probability of a robot's next steps given the steps the robot has already executed. Bayes' theorem is defined as:

$$P(A \cap B) = \frac{P(A)P(B|A)}{P(B)} \qquad (6.59)$$

Robots (and sometimes humans) are equipped with noisy sensors and have limited information on their environment. Imagine a mobile robot using vision to detect objects and its own location. If it detects an oven it can use that information to infer where it is. What you know is that the probability of seeing an oven in a bathroom is pretty low, whereas it is high in a kitchen. You are not 100% sure about this, because you might have just bought it and left it in the living room, or your eyes are "wrong" (your vision sensors are noisy and erroneous), but it is probabilistically more likely. Then, it seems reasonable to guess that, given you have seen an oven, you are "more likely" to be in a kitchen than in bathroom. Bayes' theorem provides one (not the only one) mechanism to perform this reasoning.

$P(room)$ is the "prior" belief before you've seen the oven, $P(oven|room)$ provides the likelihood of seeing an oven in some room, and $P(room|oven)$ is your new belief after seeing the oven. This is also called the "posterior" probability, the conditional probability that results after considering the available evidence (in this case an observation of the oven).

6.5.3 Gaussian Distribution

Moving away from our dice example, we know that in real-life things do not always have an equal probability of occurring. When different outcomes have different probabilities of occurring, we can think about these probabilities in terms of frequencies. That is, in a given number of repetitions of an event, how frequently is a specific

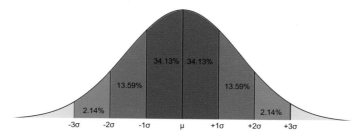

Fig. 6.8 Normal distribution

outcome likely to occur? We can plot these frequencies on a *frequency histogram*, which counts the number of times each event has occurred. This logic forms the basic of *frequentist statistics*, which we discuss more of in Sect. 6.7.

The Gaussian, or normal, distribution (aka the "Bell Curve") refers to a frequency distribution or histogram of data where the data points are symmetrically distributed—that is, there is a "peak" in the distribution (representing the mean) under which most values in the dataset occur, which then decreases symmetrically on either side as the values become less frequent (see Fig. 6.8). Many naturally occurring datasets follow a normal distribution, for example, average height of the population, test scores on many exams, and the weight of lubber grasshoppers. In robotics, we can see a normal distribution on the output of several sensors. In fact, the *central limit theorem* suggests that, with a big enough sample size, many variables will come to approximate a normal distribution (even if they were not necessarily normally distributed to begin with), making it a useful starting point for many statistical analyses.

We can use the normal distribution to predict the likelihood of a data point falling within a certain area under the curve. Specifically, we know that if our data is normally distributed, 68.27% of data points will fall within 1 standard deviation of the mean, 95.45% will fall within 2 standard deviations, and 99.73% will fall within 3 standard deviations. In probability terms, we could phrase this as "there is a 68.27% likelihood that a value picked at random will be within one standard deviation of the mean." The further away from the mean (the peak of the curve) a value is, the lower its probability of occurring. The total probability of all values in the normal distribution (i.e., the total area under the curve) is equal to 1.

Mathematically, the area under the curve is represented by a *probability density function*, where the probability of falling within a given interval is equal to the area under the curve for this interval. In other words, we can use the normal distribution to calculate the probability density of seeing a value, x, given the mean, μ, and standard deviation, σ^2.

$$p(x|\mu, \sigma^2) = \frac{1}{\sqrt{2\pi\sigma^2}} e^{-\frac{1}{2}\frac{(x-\mu)^2}{2\sigma^2}} \tag{6.60}$$

Fig. 6.9 Derivative of a
function gives the
instantaneous slope of that
function. Locations with null
derivative are in green: the
optimums

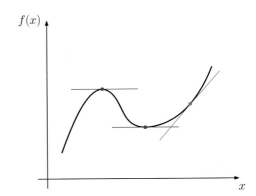

We can see that there are actually only two parameters which need to be input, μ, and σ^2. The simplicity of this representation is also relevant to computer science and robotics applications.

In a classic normal distribution, the mean is equal to 0, and the standard deviation is 1. The mean and standard deviation of any normally distributed dataset can then be transformed to fit these parameters using the following formula:

$$z = \frac{x - \mu}{\sigma} \tag{6.61}$$

These transformed values are known as *z-scores*. Thus, if we have the mean and standard deviation of any normally distributed dataset, we can convert it into *z*-scores. This process is called *standardization*, and it is useful because it means we can then use the aforementioned properties of the normal distribution to work out the likelihood of a specific value occurring in any dataset which is normally distributed, independent of its actual mean and standard deviation. This is because each *z*-score is associated with a specific probability of occurring (we already know the probabilities for z-scores at exactly 1, 2, and 3 standard deviations above/below the mean). You can check all *z*-score probabilities using *z-tables*.[10] From these, we can calculate the percentage of the population which falls either above or below a certain *z*-score. A *z-score* can then be considered a *test statistic* representing the likelihood of a specific result occurring in a (normally distributed) dataset. This becomes important when conducting inferential statistics, to be discussed later in this chapter.

6.6 Derivatives

Differential calculus is an essential tool for most of the mathematical concepts in robotics: from finding optimal gains to the linearization of complex dynamic systems.

---

[10] https://www.ztable.net/.

The derivative of a function $f(x)$ is the rate at which its value changes. It can be approximated by $f'(x) = \frac{\Delta f(x)}{\Delta x}$. However, several algebraic functions have known exact derivatives, such as $\dot{v}x^n x = nx^{n-1}$. In robotics, we manipulate derivatives for physical variables such as the velocity $(\dot{x})$, the derivative of the position (x), and the acceleration $(\ddot{x})$, the derivative of the velocity. On top of this, derivative can be helpful to find a function optimum: when the derivative of a function is equal to zero we are either at a (local) minimum or a (local) maximum (see Fig. 6.9). Several properties are useful to remember, such as the derivative operator can be distributed over addition:

$$[f(x) + g(x)]' = f'(x) + g'(x), \tag{6.62}$$

and distributed over nested functions:

$$f(g(x))' = f'(g(x))g'(x). \tag{6.63}$$

Finally, derivative operators can be distributed over a multivariate function, using *partial derivatives*, i.e., derivatives with respect to each variable independently. For instance:

$$\partial[Ax_1 + Bx_2]x_1 = A. \tag{6.64}$$

6.6.1 Taylor Series

Robotics is all about trying to control complex dynamic systems in complex dynamic environments. Most often these systems and models present nonlinear dynamics. For instance, airplane and submarines drag forces impact the vehicle acceleration with regard to the (square of) its velocity. One way to cope with this complexity is to simplify the equation using polynomial (an addition of various powers) approximation. The most popular is certainly the Taylor series:

$$f(x)|_a = f(a) + \frac{f'(a)}{1!}(x-a) + \frac{f''(a)}{2!}(x-a)^2 + \frac{f'''(a)}{3!}(x-a)^3 + \cdots \tag{6.65}$$

which approximate $f(x)$ around the point $x = a$ using a combination of its derivatives. If we want our approximation to linearize the function, we will keep only the first two terms:

$$f(x) \approx f(a) + f'(a)(x-a) \tag{6.66}$$

6.6.2 Jacobian

Now instead of a single function depending of a single variable, you will often find yourself with a set of equations each depending of several variables. For instance,

$$f_1 = Axy, \quad f_2 = Cy^2 + Dz, \quad \text{and} \quad f_3 = E/x + Fy + Gz \qquad (6.67)$$

which can be written as a vector:

$$\mathbf{F} = \begin{bmatrix} f_1 \\ f_2 \\ f_3 \end{bmatrix}. \qquad (6.68)$$

You can linearize this system of equations using Taylor's series:

$$\mathbf{F} \approx \mathbf{F}(a) + \mathbf{J} \begin{bmatrix} x - x_a \\ y - y_a \\ z - z_a \end{bmatrix}, \qquad (6.69)$$

where $\mathbf{J}$ is the matrix of partial derivatives of the functions, often referred to as the Jacobian, in this case:

$$\mathbf{J} = \begin{bmatrix} \partial f_1 x & \partial f_1 y & \partial f_1 z \\ \partial f_2 x & \partial f_2 y & \partial f_3 z \\ \partial f_3 x & \partial f_3 y & \partial f_3 z \end{bmatrix} = \begin{bmatrix} Ay & Ax & 0 \\ 0 & 2Cy & D \\ -E/x^2 & F & G \end{bmatrix}. \qquad (6.70)$$

In Chap. 10, the Jacobian is leveraged as a matrix to relate the task space (end effector velocities) to the joint space (actuator velocities). A Jacobian matrix derived for a single function, i.e., a single row matrix, is called a gradient, noted (for a geometric function in Cartesian space):

$$\nabla f = \begin{bmatrix} \partial f x & \partial f y & \partial f z \end{bmatrix}. \qquad (6.71)$$

The gradient is a useful tool to find the optimum of a function by *traveling* on it; a stochastic approach very useful in machine learning (see Chap. 15).

6.7   Basic Statistics

When conducting research in robotics, and especially user studies, you will often have data you have collected in pursuit of answering a specific research question. Typically, such research questions are framed around the relationship between an *independent variable* and a *dependent variable*. For example, you might ask how the number of drones (independent variable) in a mission affects the operator's cognitive workload (dependent variable). Being able to analyze the data you have collected is then necessary to communicate the outcomes from your research. Chapter 13 gives more detail on how to design and conduct user studies, for now we will begin explaining some of the analyses you can perform once you have obtained some data!

Table 6.2 Common parameter notations for samples versus populations

Parameter	Sample	Population
Mean	$\bar{x}$	μ
Standard deviation	s	σ
Variance	s^2	σ^2
Number of data points	n	N

The first step of analyzing any dataset is usually to describe its properties in a way that is meaningful to your audience (**descriptive statistics**). This involves taking the raw data and transforming it (e.g., into visualizations or summary statistics). The second step is then to determine how you can use your data to answer a specific research question and/or generalize the results to a broader population (**inferential statistics**). Here, it is important to distinguish between a *sample* of data collected, and the *population* the data is intended to generalize to (see also Chap. 13). Critically, descriptive statistics only relate to the actual sample of data you have collected, whereas inferential statistics try to make generalizations to the population. Typically, formulas relating to calculating values of a sample use Greek letters, whereas formulas relating to a population use Roman letters. Below is a table with some of the most common notations for both samples and populations (Table 6.2).

When we collect data our samples can either be independent (the data is from two different groups of people) or repeated (from the same group). For example, imagine we wanted to test robotics students' knowledge of basic geometry and linear algebra. We could either take a single sample of students, and test their knowledge before and after reading this chapter—this would be a *within-groups* study, as the same students were tested each time. Alternatively, we could take a sample of students who have read this book chapter and compare them against a sample who have not read this chapter. There is no overlap between these two groups; thus, it is a *between-groups* study design.

You can first begin describing the properties of your sample using three different **measures of central tendency**; the mean, the median, and the mode. The mode represents the most common response value in your data . That is, if you took all of the individual values from your dataset and counted how many times each occurred, the mode is the value which occurred the most number of times. For example, imagine we asked 10 robotics professors how many robots they have in their laboratory (see Table 6.3).

We can see that the most common value reported is 12 robots—this is the mode. The mode can be most easily identified by creating a frequency distribution of the values in your dataset (see Fig. 6.10).

The median is the value which is in the middle of your range of values. Using the aforementioned example, if we ranked the number of robots in each laboratory from smallest to largest, the median is the value which falls exactly in the middle (or, if there is an even number of data points, the sum of the two middle values divided by

Table 6.3 Sample data of robots per professor

Professor ID	Number of Robots
1	1
2	5
3	7
4	10
5	10
6	12
7	12
8	12
9	15
10	20

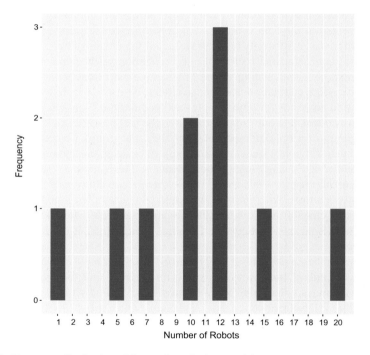

Fig. 6.10 Frequency distribution of the number of robots per laboratory

2). In this case, we have 10 values, so the median is the average of the 5th and 6th values, $\frac{10+12}{2} = 11$.

However, the median and the mode both rely on single values, and thus, ignore much of the information which is available in a dataset. The final measure of central tendency is then the mean, which takes into account all of the values in the data by summing the total of all the values, and dividing them by the total number of observations. The formula to calculate the mean of a sample is expressed as:

$$\bar{x} = \frac{\sum_{i=1}^{N} x_i}{n} \tag{6.72}$$

where $\bar{x}$ represents the mean of the sample, x represents an individual value, and n represents the number of values in the dataset.

In our example, this would be:

$$\bar{x}_{\text{robots}} = \frac{1 + 5 + 7 + 10 + 10 + 12 + 12 + 12 + 15 + 20}{10} = 10.4 \tag{6.73}$$

Conversely to the median and the mode, this value does not actually have to exist in the dataset (e.g., if the average number of robots in the laboratory is actually 10.4, some students probably have some questions to answer . . .)

Many basic statistics can be computed in Python using the **numpy** library:

```
import numpy as np # Import the library
mu = np.mean(data) # Mean of the sample ''data''
mod = np.mode(data) # Mode of the sample ''data''
med = np.median(data) # Mode of the sample  ''data''
```

In the classic normal distribution, the mean, the median, and the mode are equal to each other. However, in real life, data often does not conform perfectly to this distribution, thus, these measures can differ from each other. In particular, while the median and the mode are relatively robust to extreme values (outliers), the value of the mean can change substantially. For example, imagine our sample included one professor who works with microrobots who reported having hundreds of robots in their lab. This would obviously skew the mean by a lot while not being representative of the majority of the sample.

Let's say we asked another 90 robotics professors about the number of robots they have, so we now have sampled a total of 100 robotics professors. Our results now show the frequency distribution shown in Fig. 6.11.

We can see that although the mean is still 10.4, the mode is now 8 robots, and the median is 10. These values, although similar to each other, are not identical, although the data is normally distributed. We can check this using the probability density function. Again, this is not perfectly represented by the normal distribution, but it makes a very good approximation of the data.

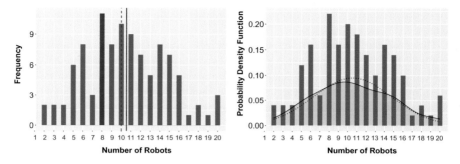

Fig. 6.11 Frequency histogram and probability density function for a normally distributed dataset. The graph on the left shows the measures of central tendency, with the dark purple bar representing the model, the dashed purple line representing the median, and the solid black line representing the mean. The graph on the right shows the actual probability distribution of the data contrasted with the normal distribution

6.7.1 Variance

The sensitivity of our dataset descriptive metrics to new data points can be grasped in terms of its *variability*. We can measure the amount of variance in any given sample, as well as detect outliers, in multiple different ways. The first one is the *standard deviation*. This represents on average, how far away values are from the mean. The smaller the standard deviation, the closer the values in the sample are on average to the mean, and the more accurate the mean is at representing the sample. We can also use the standard deviation to create a cutoff for extreme values—any value which falls above or below 3 standard deviations from the mean is likely to be an outlier (i.e., not representative of the population) and can often be excluded.

To calculate the standard deviation of a variable, we first take each individual value and subtract the mean from it, resulting in a range of values representing the deviances from the mean. The total magnitude of these deviances is equal to the total variance in the sample. However, given that some individual values will be above the mean, and some below, we need to square these values so that they are all positive, to avoid positive and negative values canceling each other out. We then sum the squared deviances to get a total value of the error in the sample data (called the *sum of squares*). Next, we divide by the number of data points in the sample (n), minus one. Because we are calculating the sample mean, and not the population mean, $n - 1$ represents the *degrees of freedom* in the sample. This is because we know both, the sample mean and the number of data points. Thus, if we have the values of all the data points bar one, the last data point can only be whatever value is needed to get that specific mean. For example, if we go back to our first sample of 10 robotics professors, and took the values of the first 9, knowing that the mean is 10.4 and that we sampled 10 robotics professors total, the number of robots in the laboratory of the last professor must have a fixed value.

$$10.4 = \frac{1 + 5 + 7 + 10 + 10 + 12 + 12 + 12 + 15 + x}{10}$$

$$x = 20 \qquad (6.74)$$

That is, this value of x is not free to vary. So, the degrees of freedom are always one less than the number of data points in the sample.

Finally, since we initially squared our deviance values, we then take the square root of the whole equation so that the standard deviation is still expressed in the same units as the mean.

The full formula for calculating the standard deviation of a sample is described below. Note that if we were to calculate the standard deviation of the *population* mean instead, the first part would be replaced with $\frac{1}{N}$, rather than $\frac{1}{n-1}$.

$$s = \sqrt{\frac{1}{n-1} \sum_{i=1}^{n} (x_i - \bar{x})^2} \qquad (6.75)$$

In Python, we can compute this using:

```
import numpy as np # Import the library
stddev = np.std(data) # Standard deviation of the sample ''data''
```

If we don't take the square root of the equation, and instead leave it as is, this is known as the *variance*, denoted by s^2.

In statistical testing, we are interested in explaining what causes this variance around the mean. In this context, the mean can be considered as a very basic model of the data, with the variance acting as an indicator of how well this model describes our data. Means with a very large variance are a poor representation of the data, whereas means with a very small variance are likely to be a good representation.

The variance for any given variable is made up of two different sources; *systematic variance*, which is variance that can be explained (potentially by another variable), and *unsystematic variance*, which is due to error in our measurements.

We therefore often in our experiments have more than one variable, and we might be interested in describing the relationship between these variables—that is, as the values in one variable change, do the values for the other variable *also* change? This is known as *covariance*.

The total variance of a sample with two variables is then made up of the variance attributed to variable x, the variance attributed to variable y, *and* the variance attributed to both. Remembering that variance is simply the square of the formula for the standard deviation, or s^2, we can frame the sum of the total variance for two variables as:

$$(s_x + s_y)^2 = s_x^2 + s_y^2 + 2s_{xy} \qquad (6.76)$$

It is this last term, $2s_{xy}$ that we are interested in, as this represents the covariance between the two variables. To calculate this, we take the equation for variance, but rather than squaring the deviance of x, $(x - \bar{x})$, we multiply it by the deviance of the other variable, $y - \bar{y}$. This ensures we still avoid positive and negative deviations canceling each other out. These combined deviances are called the *cross product deviation*.

$$\mathrm{cov}(x, y) = \frac{1}{n - 1} \sum_{i=1}^{n} (x_i - \bar{x})(y_i - \bar{y}) \tag{6.77}$$

To get the covariance between two variables in Python, we can use:

```
import numpy as np # Import the library
cov = np.cov(data,ddof=0) #compute the covariance matrix
```

6.7.2 General Population and Samples

In the aforementioned example, we have a specific population that we are interested in robotics professors. However, as it would be difficult to test every single robotics professor in the world, we took only a subset of robotics professors and asked them about the number of robots they have in their laboratories. In this case, the mean number of robots is an *estimation* of the general population mean. This is different from the true population mean, which is the mean we would get if we actually were able to ask every single robotics professor how many robots they have. In an ideal world, the sample you have collected would be perfectly representative of the entire population, and thus, the sample mean would be equal to the true mean. However, as there is always some error associated with the data, the sample mean will likely always vary slightly from the true mean.

If we were to take several different samples of different robotics professors, these samples would each have their own mean and standard deviation, some of which might over or underestimate the true population mean. If we were to plot the means of each of our samples in a frequency distribution, the center of this distribution would also be representative of the population mean. Importantly, if the population is normally distributed, the distribution of samples will also be normally distributed. Thus, knowing the variance in a distribution of samples would allow us to know how likely it is that any one specific sample is close to the true population mean, exactly the same as the standard deviation of individual values around a sample mean allows to estimate the error in that sample. The standard deviation of a distribution of samples around the population mean is then known as the *standard error* and is expressed as.

$$\sigma = \frac{s}{\sqrt{n}} \tag{6.78}$$

The standard error allows us to determine how far away, on average, the mean of a sample taken at random from the population is likely to be from the population mean. Of course, when we conduct experiments, we cannot actually repeatedly take different samples from the population—normally we only have one sample. However, the concept of the standard error is theoretically important to understand how we can generalize our results from our sample to a population. Going back to the central limit theorem, if you have a large enough sample size, the sample mean will become more similar to the true population mean. Similarly, the standard error will also come to approximate the standard error of the population. For this reason, the standard error is often used in place of the standard deviation when using inferential statistics.

6.7.3 The Null Hypothesis

Hypothesis testing involves generating some prediction about our data and then testing whether this prediction is correct. Normally, these predictions relate to the presence or absence of an effect (i.e., there is some relationship between variables, or not).

The *null hypothesis*, typically denoted as H_0, is the assumption that there will be *no effect* present in the data. For example, if you are comparing two different robots on some feature (e.g., appearance) and how much appearance affects the robots' likability, H_0 would state that there is no difference between the two robots. Relating this back to our normal distribution, H_0 is the assumption that the data from the two groups come from the same population (i.e., are represented by the same distribution, with the same mean). That is, do we happen to have two samples that vary in their mean and standard deviation by chance, but are actually from the same population, or, is their a systematic difference between the two (see Fig 6.12)?

In contrast, the *alternative hypothesis*, or H_1, relates to the *presence* of some effect (in the aforementioned example, H_1 would be that there is an effect of robot appearance on likeability). Again putting this in context of the normal distribution, H_1 is the idea that the data comes from two different population distributions, with different means and standard deviations. In this context the "populations" can also refer to an experimental manipulation—e.g., is a population of people who saw a robot with glowing red buttons and aggressive beeping more likely, on average, to rank this robot as less likeable than a population of people who saw a robot with colorful lights and calm beeping?

In inferential testing, we work on the basis that H_0 is true by default. Thus, the goal is not to prove that H_1 is true, but rather to try and demonstrate that H_0 is false. That is, we want to show that it is very unlikely that the two (or more) groups come from the same population distribution.

So, when we have two sample means of different values, we can test whether the differences in these values are due to chance (i.e., random variation), or, if they actually come from different populations. The likelihood that we would have obtained these data, given the null hypothesis is true, is represented by the *p*-value,

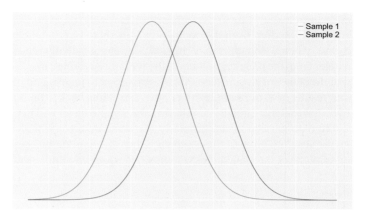

Fig. 6.12 Two overlapping bell curves from different samples

Fig. 6.13 *p*-values in relation to the normal distribution

see Fig. 6.13. Typically, the threshold for this likelihood is set at 95%. That is, if we assume the null hypothesis is true, and the results from our model indicate that the likelihood of observing these results is 5% or less, then the null hypothesis is likely not the correct explanation for the data. In this case, we would reject H_0 and accept H_1. In other words, we call the result *statistically significant*. The smaller the *p*-value, the lower the probability that H_0 is true. Although $p < .05$ is the minimum threshold that is typically accepted, $p < .01$ and $p < .001$ may also be used.

Note that all these thresholds still leave some margin for error—it is possible that we could observe these results even if H_0 is true, just unlikely. That is, by chance we have picked two samples that differ substantially from each other (remember that our distribution of samples from the general population also follows a normal distribution, thus, there is always the chance to have a sample that is not actually representative of the population). This is called a *Type-I* error, or a false positive—we have incorrectly deduced that the samples come from different populations, when in fact they come from the same one. The inverse of this, if we incorrectly conclude that the samples come from the same population, when in reality they come from different ones, is called a *Type-II* error; see Table 6.4.

Table 6.4 Type I and II errors

	H_0 is true	H_0 is false
Reject H_0	Type I error α	Correct $1 - \beta$
Accept H_0	Correct $1 - \alpha$	Type-II error β

An additional factor to consider when setting the p-value threshold is the directionality of our test. If we predict that there will be a significant difference between our two sample means, we could choose to test precisely whether one of the two samples, specifically, will have a higher mean than the other. For example, we could test whether an older versus newer model of a robot have different levels of battery performance, *or* we could test specifically whether the newer model has a better battery performance than the older model. In the former scenario, we would use *two-tailed hypothesis testing*. That is, we don't know which side of the normal distribution our test statistic (e.g., the *z-value*) will fall, so we consider both. In the latter scenario, we are specifically saying that the mean for the newer robot model will be higher than the mean of the old model, thus, we only look at the probabilities for that side of the distribution with a test statistic in that direction, called *one-tailed hypothesis testing*. However, one-tailed hypothesis testing is generally used sparingly, and usually only in contexts where it is logistically impossible or irrelevant to have results in both directions. That is, even if we have a directional hypothesis (e.g., that the newer model has a better battery performance), if it is theoretically possible that the older model has a better battery performance, we need to test both sides of the probability distribution. In this example, if we used a one-tailed hypothesis test assuming that the newer model is better, and in fact it is actually worse than the older model, we would likely get a non-significant result and incorrectly conclude that there is no difference in battery performance between the two models. For this reason, most hypothesis testing in robotics is two-sided.

6.7.4 The General Linear Model

So far, we have discussed measures of central tendency and different measures of variance as ways of describing variables. However, as mentioned at the beginning of this section, we are usually interested in not only describing our data, but using it to *predict* some outcome. That is, we want to create a model of our data so that we can accurately predict the outcome for any given set of parameters. We can then conceptualize any outcome or variable we are trying to predict as a function of both the true value of the model and the error, such that:

$$\text{outcome}_i = \text{model}_i + \text{error}_i \qquad (6.79)$$

Where $model_i$ can be replaced with any number of predictor variables. This forms the basis for the *general linear model*. Mathematically, this model can be expressed as:

$$Y_i = b + w X_i + \epsilon_i \tag{6.80}$$

Where Y_i represents the outcome variable, b is where all predictors are 0, and w represents the *strength* and *direction* of an effect.

As mentioned before, this can then be expanded to any number of predictor variables:

$$Y_i = b_0 + b_1 X_i + b_2 X_i + \cdots + b_n X_i + \epsilon_i \tag{6.81}$$

Once we have defined a model, we want to test how well it actually predicts the data that we have. We can do this by comparing the amount of variance in the data that is explained by our model, divided by the unexplained variance (error) to get different test statistics. We can then use the normal distribution to check the likelihood that we would have obtained a specific test statistic, given the null hypothesis is true.

$$\text{test statistic} = \frac{\text{variance explained by model}}{\text{unexplained variance (error)}} \tag{6.82}$$

To get the ratio of explained to unexplained variance, we start by calculating the total variance in our sample. To do this, we need to go back to the formula for the sum of squares, which is simply:

$$SS_{\text{total}} = \sum_{i=1}^{n} (x_i - \bar{x}_{\text{grand}})^2 \tag{6.83}$$

Where x_i is an individual data point, $\bar{x}_{\text{grand}}$ is the *grand mean*, or the mean of the total dataset, and n is the number of datapoints.

We also know that variance is equal to the sum of squares divided by the degrees of freedom, so, the sum of squares can be rearranged as:

$$
\begin{aligned}
s^2 &= \frac{1}{n-1} \sum_{i=1}^{n} (x_i - \bar{x}_{\text{grand}})^2 \\
s^2 &= \frac{1}{n-1} SS_{\text{total}} \\
SS_{\text{total}} &= s^2 (n-1)
\end{aligned}
\tag{6.84}
$$

This gives us the total amount of variation in the data (the sum of the deviation of each individual data point from the grand mean). We are interested in how much of this variation can be explained by our model (remembering that total variation = explained variation + unexplained variation).

To get the amount of variation explained by our model, we then need to look at our group means, rather than the grand mean. In this case, our model predicts that an individual from Group A will have a value equal to the mean of Group A, an individual from Group B will have a value equal to the mean of Group B, etc.

We can then take the deviance of each group mean from the grand mean, and square it (exactly the same as calculating normal sums of squares). We then multiply each value by the number of participants in that group. Finally, we add all of these values together.

So, if we have three groups, this would look like:

$$SS_{model} = n_a(\bar{x}_a - \bar{x}_{grand})^2 + n_b(\bar{x}_b - \bar{x}_{grand})^2 + n_c(\bar{x}_c - \bar{x}_{grand})^2 \qquad (6.85)$$

Where n_a represents the number of datapoints in group A, $\bar{x}_a$ is the mean of group A, and $\bar{x}_{grand}$ is the grand mean.

This can be expanded to k number of groups with the general form:

$$SS_{model} = \sum_{n=1}^{k} n_k(\bar{x}_k - \bar{x}_{grand})^2 \qquad (6.86)$$

Where k is the number of groups, n_k is the number of datapoints in group k, $\bar{x}_k$ is the mean of group k, and $\bar{x}_{grand}$ is the grand mean.

So, now we have the total variance, and the variance explained by our model. Intuitively, the variance that is left must be the error variance, or variance not explained by the model. This *residual variance* is the difference between what our model predicted (based on the group means) and our actual data. Although in theory we can get this value by subtracting the model variance from the total variance, we can also calculate it independently.

Remember that our model predicts that an individual from Group A will have a score equal to the mean of Group A. So, to get the residual variance we first calculate the deviance of each individual in Group A from the mean of Group A, and the same for Group B and so on and so forth. This can be expressed as:

$$SS_{residual} = \sum_{i=1}^{n} (x_{ik} - \bar{x}_k)^2 \qquad (6.87)$$

Where n is the total number of data points, i is an individual datapoint, x_{ik} is the value of an individual, i in group k, and $\bar{x}_k$ is the mean of that group.

This takes the deviance of each individual datapoint from its associated group mean and sums them together. However, we could also conceptualize residual variance as the sum of the variance of Group A, plus the variance of Group B and so on for k number of groups. We also saw before how the sum of squares can be expressed in terms of the variance (see Eq. 6.84). The same logic can be applied here for adding the group variances together to give us:

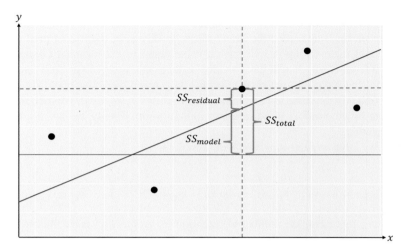

Fig. 6.14 Illustration of total sum of squares, the model sum of squares, and the residual sum of squares. The solid red line represents the grand mean, the dashed red lines indicate a specific data point (or group mean), and the solid blue line represents the predicted value of y for a given value of x

$$SS_{\text{residual}} = \sum s_k^2 (n_k - 1) \tag{6.88}$$

Where s_k^2 is the variance of group k, and n_k is the number of data points for that group.

Visually, the total sum of squares (SS_{total}), the model sum of squares (SS_{model}), and the residual sum of squares (SS_{residual}) can be represented by the three values illustrated in Fig. 6.14.

However, right now these are biased by the number of data points used to calculate them—the model sum of squares is based on the number of groups (e.g., 3), whereas the total and residual sum of squares are based on individual data points (which could be 5, or 15, or 50, or 500). To rectify this, we can divide each sum of squares by the degrees of freedom to get the *mean squares (MS)*. For MS_{model} the degrees of freedom are equal to the number of groups minus one, whereas for the MS_{residual} they are calculated by the number of total data points minus the number of groups.

$$
\begin{aligned}
MS_{\text{model}} &= \frac{SS_{\text{model}}}{k - 1} \\
MS_{\text{residual}} &= \frac{SS_{\text{residual}}}{n - k}
\end{aligned}
\tag{6.89}
$$

Where k is the total number of groups and n is the total number of data points.

From here, we are able to compare the variance explained by our model to the residual, or error variance and test whether this ratio is significant. Although there are many different kinds of test statistics that we can use to see whether our model is significant or not, we will focus on only two of them: the t-test and the ANOVA.

Fig. 6.15 Comparison of t and z distributions

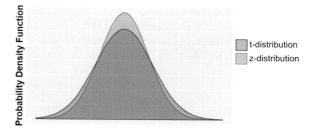

6.7.5   T-test

The t-test is used to compare the means of two different samples, to test whether there is a statistically significant difference between the two (i.e., a less than 5% chance of observing this difference in means, given the null hypothesis is true).

As we discussed before in Sect. 6.7.2, when sample sizes are sufficiently large, the sampling distribution of a population will approximate the normal distribution, and we can use z-scores to calculate probabilities (using p-values associated with specific z-scores). However, if we have small sample sizes (which can often be the case in user studies), then we cannot reliably estimate the variance of the population. In this case, we use a t-distribution, which is a more conservative estimate of the normal distribution. It is the t-distribution that we use to calculate our p-values for the t-test. See Fig. 6.15 for a comparison between the z and t distributions.

The value of the t-test is then a function of the mean and the standard error of the two samples we are comparing. If we have a difference between two means, then intuitively the larger this difference is, the more likely it is there is an actual difference between the samples. However, if the standard error is *also* very large, and the difference in means is equal to or smaller than this value, then it is unlikely that it represents a true difference between the samples—the difference between means could simply be accounted for by a large variance in a single population.

So, to perform a t-test, we want to compare the difference in means we actually saw, to the difference in means we would expect if they come from the same population (which is typically 0). Going back to the previous section, we also saw that in general the test statistic (which in this case, is the t-test) can be calculated by dividing variance explained by the model by the error variance (see Eq. 6.82). In this case, the model we are testing is the difference between the actual and expected means. So, we take this value (which, as the expected difference between means is 0, is actually just the value of the observed difference) and divide it by the standard error of the differences between the means.

$$t = \frac{(\bar{x}_1 - \bar{x}_2) - (\mu_1 - \mu_2)}{\text{standard error of the difference}} = \frac{(\bar{x}_1 - \bar{x}_2)}{\text{standard error of the difference}} \qquad (6.90)$$

To get the standard error of the differences between means, we first start by summing together the variance for each sample, divided by the sample size of each. This is based on the *variance sum law*, which states that the variance of the difference between two samples is equal to the sum of their individual variances.

$$\frac{s_1{}^2}{n_1} + \frac{s_2{}^2}{n_2} \tag{6.91}$$

We then take the square root of this value to get the standard error of the difference.

$$\sqrt{\frac{s_1{}^2}{n_1} + \frac{s_2{}^2}{n_2}} \tag{6.92}$$

So, Eq. 6.90 becomes:

$$t = \frac{(\bar{x}_1 - \bar{x}_2)}{\sqrt{\frac{s_1{}^2}{n_1} + \frac{s_2{}^2}{n_2}}} \tag{6.93}$$

However, this assumes that the sample sizes of each group are equal. In the case that they are not (which is often), we replace s_1 and s_2 with an estimate of the pooled variance, which weights each sample's variance by its sample size.

$$s_{\text{pooled}}^2 = \frac{(n_1 - 1)s_1^2 + (n_2 - 1)s_2^2}{n_1 + n_2 - 2} \tag{6.94}$$

In turn, the t-test statistic becomes:

$$t = \frac{(\bar{x}_1 - \bar{x}_2)}{\sqrt{\frac{s_{\text{pooled}}{}^2}{n_1} + \frac{s_{\text{pooled}}{}^2}{n_2}}} \tag{6.95}$$

Note that we are also assuming the data comes from two different groups (i.e., an independent groups t-test). When we have a within-groups design, we instead use a *dependent t-test*.

In Python, t-tests for both within- and between-groups samples can be computed with:

```
from scipy import stats # Import library
res = stats.ttest_rel(x1, x2) # Run test for dependent sample
res = stats.ttest_ind(x1, x2) # Run test for independent sample
print(res[1])
```

We can then calculate the probability that we would have seen this *t*-value if the samples actually did come from the same population, which gives us the *p*-value. We can do this using t-distribution tables, or, since you are likely using some form of statistical software, read this value from the output. The important thing to know is that, because our data is normally distributed, the *p*-values for each *t*-value remain

Table 6.5 Independent groups t-test

Mean (SD)		Estimate	t-value	p-value
Own algorithm	Competing algorithm			
1055 (408)	4042 (605)	−2986.9	−28.94	<.001

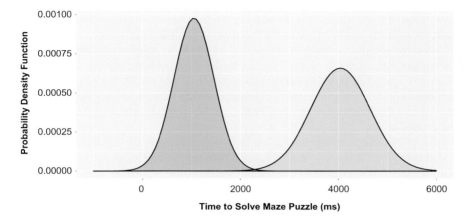

Fig. 6.16 Probability density function for each algorithm

consistent. That is, if we conducted two completely different experiments and ended up with the same or similar t-values, the p-values would also be the same.

As a practical example, imagine we want to compare two different navigation algorithms, one we have developed and one a competing laboratory has developed, in terms of how fast they can solve a maze puzzle (in milliseconds). We run an independent t-test comparing the two algorithms and find the following results in Table 6.5:

From this, we can see that our algorithm solves the puzzle significantly faster. We can also compare the probability distributions of the two groups; see Fig. 6.16. This also confirms that there is only a very small overlap in the values that occur in both samples and that these values have a very low probability of occurring.

For more t-test examples, a set of Python examples based on a public dataset of task load surveys is available online.[11]

6.7.6 ANOVA

ANOVA stands for "Analysis of Variance" and is an extension of the t-test when we have more than two groups. That is, we are again interested in comparing the means of

---

[11] https://github.com/Foundations-of-Robotics/Stats-examples.

different samples to determine if there is a statistically significant difference between them (i.e., whether they come from the same or different population distributions). To do this, we use the F-test. This is simply another test statistic, which, as we have seen before, is a measure of the total variance explained by our model divided by the amount of error in the model.

To explain more about the difference between a t-test, and an F-test, imagine we have three different groups (A, B, and C). We then have multiple different possible outcomes for the results: First, there could be no significant difference between any of the three groups. Second, A, B, and C could all be significantly different from each other. Alternatively, A and B could be different from each other, but not from C, and so on for all possible combinations of A, B, and C. So, we can already see that this is quite a few more options compared to the t-test where we have only two groups and the outcome is binary—there is either a significant difference between the groups or not.

An ANOVA is therefore conducted in two stages: First, we conduct an *omnibus* test to determine if there is any difference between the means at all. However, this does not tell us which groups, specifically, might be different from each other. Thus, if the result of this test is significant, then we conduct a series of t-tests for each of the possible two-way combinations of the groups. The reason that we do not start straight away with t-tests is because this inflates our chance of making a type I error (incorrectly stating that the samples come from different populations, when in fact they come from the same one). This is because if we set our significance threshold to 95%, then we are still allowing for a 5% chance of incorrectly rejecting the null hypothesis. If we perform three t-tests independently of each other, each with a significance threshold of 95%, then we can see how this error compounds: $0.95^3 = 0.8571$ and $1 - 0.857 = 14.3\%$. So, instead of having a 5% chance of incorrectly rejecting the null hypothesis, we now have a 14.3% chance. This is known as the *familywise error rate* and increases with the more comparisons we make. By starting our analysis with an omnibus test, we are trying to mitigate this error.

To get the omnibus F-statistic, we need to go back to Eq. 6.89, where we can see that we actually can calculate values for our model variance and residual variance! Thus, the F-statistic can be expressed as:

$$F = \frac{MS_{\text{model}}}{MS_{\text{residual}}} \tag{6.96}$$

Any value greater than 1 means that our model explains more variance than random individual differences (which is a good thing!) However, it still does not tell us whether this value is significant. To check this, we again go back to our p-values to determine, with a given F-statistic and associated degrees of freedom, what the likelihood of obtaining this value is, if the null hypothesis is true. To get the degrees of freedom, we need to consider the two parts that make up our ratio: our model variance and the residual variance. We also mentioned before about how the sum of squares for each of these was calculated using their respective number of data points—for the model variance this is equal to the number of groups, and for the

Table 6.6 Mean and SD for perceived task difficulty in each task environment

Environment	Mean (SD)
Land	3.12 (1.00)
Water	5.16 (0.95)
Air	5.38 (1.28)

residual variance this is equal to the number of data points. It is these values which we use to get the degrees of freedom.

$$\mathrm{d}f = \frac{k-1}{n-k} \qquad (6.97)$$

Where k is equal to the number of groups and n is equal to the sample size. This is also why $k-1$ is sometimes called the *numerator degrees of freedom*, whereas $n-k$ is called the *denominator degrees of freedom*.

Having determined our degrees of freedom and our F-statistic, we then need to use F-distribution tables (or our statistics software) to look up the corresponding p-value. Again, the p-value for all F-*values* with specific degrees of freedom will always be the same.

Following a significant ANOVA, the next step is to conduct individual t-tests between each pair of groups, called *pairwise comparisons*. Again, however, we have to be a bit cautious of inflating our type I error. When the number of comparisons is less than 5, it is generally considered okay to use the p-values as-is. Anything above this however, and it is recommended to use an adjustment method. This typically involves applying a correction to the p-values to make their estimates more conservative, see (Bender and Lange, 2001) for an explanation of the different types of corrections.

Now that we have covered some of the logic underpinning the ANOVA, we can consider what this looks like in practice. Imagine we have a sample of robot operators, and we are interested in understanding the difficulty of using unmanned robots to explore different types of environments. So, we design an experiment into how the type of environment affects the perceived task difficulty. Here, our independent variable is task environment (land, water, air) and our dependent variable is the perceived task difficulty, measured on a 7-point scale from 1 (very easy) to 7 (very difficult). We test a total of 150 robot operators, 50 in each environment.

In this case, the null hypothesis (H_0) is that there will be no difference between the three task environments. The alternative hypothesis (H_1) is that the perceived task difficulty will change according to the task environment.

After running our descriptive statistics, we observe the following means and standard deviations for each group; see Table 6.6.

As our first step of the ANOVA, we conduct the omnibus F-test, to determine if there is any overall difference between the groups, Table 6.7.

Table 6.7 Results of Omnibus F-test for one-way ANOVA

	DF	Sums of squares	Mean squares	F-value	p
Environment	2	155.3	77.65	65.68	<.001
Residuals	147	173.8	1.18		

Table 6.8 Post-hoc pairwise comparisons for each task environment with no correction

	Estimate	SE	t-value	p-value
Land versus Water	−2.04	0.22	−9.38	<.001
Land versus Air	−2.26	0.22	−10.39	<.001
Water versus Air	−0.22	0.22	−1.01	.313

What you might be able to see from these tables is that the values for each column match exactly the formulas we discussed for the general linear model. That is, the mean squares are equal to the sums of squares divided by the DF for each row, and the F-value is the ratio of the mean squares. So, in case you are ever stuck in a room with only your experimental data and no Internet access or statistics software downloaded, you can still calculate your ANOVAs by hand!

From looking at the p-value, we can see that the overall F-test is significant ($p <$ .001). However, we don't yet know where this difference lies (i.e., we don't know which of the environments are perceived as significantly more or less difficult). So, the next step is to compare the groups, using our pairwise comparisons; see Table 6.8.

From these results, we can now determine that both air and water environments are perceived as more difficult to explore than land environments, but that there is no difference between these two. We could then write up the results from this test as follows:

> The results from a one-way between-groups ANOVA revealed a significant effect of task environment on perceived task difficulty, $F(2, 147) =$ 65.68, $p <$ .001. Post-hoc pairwise comparisons with no correction indicate that the land environment was perceived as significantly less difficult than both the water ($t = -9.38$, $p <$ .001) and air ($t = -10.39$, $p <$ .001) environments, respectively. However, there was no difference in perceived task difficulty between the water and air environments ($t = -1.01$, $p = .313$).

In the aforementioned example, we only had one independent variable, task environment, and thus, it is a *one-way ANOVA*. Now, imagine we expanded our experimental design to include not only the task environment, but also the type of robot being used for exploration, unmanned aerial vehicles (UAVs) versus unmanned ground vehicles (UGVs). Now we have two independent variables, task environment (again with three levels, land, water, and air) and robot type (UAV versus UGV). Our dependent variable, perceived task difficulty, remains the same. This is called a *two-way ANOVA*.

In this case, we now have two different *main effects* we are interested in; the effect of robot type on task difficulty, and the effect of task environment. However, there is also a third effect—the *interaction* between the two variables. An interaction effect indicates that at different levels of one variable, the effect of the other variable on the dependent variable changes. To keep things simple, these kinds of effects are called . . . *simple effects*. The directionality of the interaction hypotheses is normally theoretically driven and specified before conducting the analysis. However, usually we only conduct one set (i.e., either the effect of variable A at different levels of variable B, or the effect of variable B at different levels of variable A, but not both). This again has to do with minimizing our chances of making a type I error—remember that every analysis we run comes with a small chance of incorrectly rejecting the null, so the more analyses we run, the more this chance compounds.

In this case, we will look at the simple effects of robot type over the levels of task environment. That is, at each level of task environment (land, water, air) we will run an analysis of the effect of robot type on perceived task difficulty. However, we could just as equally say that depending on the type of robot, the effect of task environment on perceived task difficulty changes.

The syntax to compute this analysis in python looks like:

```
# Import libraries
from statsmodels.formula.api import ols
from statsmodels.stats.anova import anova_lm
# Create the model (two factors - last term is interaction)
formula = 'task_difficulty ~ C(environment) + C(robot) + C(environment):C(robot)'
# Test the model against the data (must have column headers as in the model)
model = ols(formula, data).fit()
# Run a two-way ANOVA
aov_table = anova_lm(model, typ=2)
print(aov_table.round(4))
```

We can see the means and standard deviations for our new dataset below in Table 6.9:

So to recap, we now have two *main* effects which we are looking at, and an *interaction effect*. Each main effect has an F-value associated with it, as does the interaction. We can see these and their associated significance's in the table below (Table 6.10).

Table 6.9 Means and SDs for task environment and Robot type

Environment	Robot type	Mean (SD)
Land	UAV	5.12 (1.47)
Land	UGV	3.20 (1.34)
Water	UAV	6.16 (0.79)
Water	UGV	6.00 (0.70)
Air	UAV	2.92 (1.45)
Air	UGV	5.08 (1.42)

Table 6.10 Results of Omnibus F-test for two-way ANOVA

	DF	Sums of squares	Mean squares	F-value	p
Environment	2	133.97	66.99	43.91	<.001
Robot type	1	0.03	0.03	0.017	.900
Environment * Robot type	2	104.69	52.35	34.31	<.001
Residuals	144	219.68	1.53		

Table 6.11 Post-hoc pairwise comparisons for two-way ANOVA with no correction

	Estimate	t-value	p-value
Land	1.92	4.78	<.001
Air	−2.16	−5.41	<.001
Water	0.16	0.749	.457

We can see, based on this table, that there is still the main effect of task environment, but no main effect of robot type. However, the interaction between task environment and robot type is significant.

As with the previous one-way ANOVA, we can follow up the significant F-test for the interaction with pairwise comparisons. In this case, however, we take each level of environment (land, water, air) and look at the effect of robot type on task difficulty *within* each of these conditions. As we only have two levels of robot type, we can go straight to t-tests comparing UAVs and UGVs within each task environment. However, if we had more than two levels (e.g., if we had also tested unmanned underwater vehicles), we would need to conduct another one-way ANOVA for each environment type, *then* conduct the pairwise comparisons between the robot types depending on which environment was significant.

The results of the pairwise comparisons for the effect of robot type within each task environment are in Table 6.11.

Now things are starting to get a little bit interesting. From this table, we can see that, in the water environment, there is no difference between UGVs and UAVs. In fact, the mean perceived task difficulty for both of these groups is quite high (probably because neither UAVs nor UGVs are suited for underwater exploration). Conversely, in the land environment, the UGV is rated as having a significantly lower task difficulty than the UAV, and vice versa for the air environment, where the UAV has a lower task difficulty.

We can plot a graph of this interaction as seen in Fig. 6.17.

Looking at this graph, we can begin to get an idea of why the main effect for robot type was non-significant. Because the means for the UAV and UGV were flipped for the land and air environments, and similar for the water environment, when averaged all together, they cancel each other out. So when we look at the aggregated means for the two robot types (see Fig .6.18), ignoring whether they were in a land, water, or

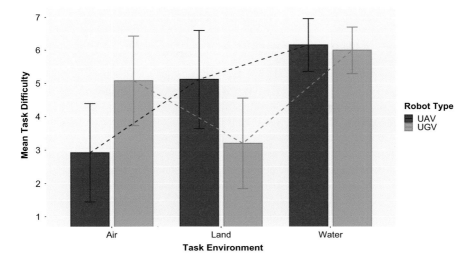

Fig. 6.17 Two-way interaction between task environment and robot type

Fig. 6.18 Two-way interaction between task environment and robot type

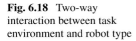

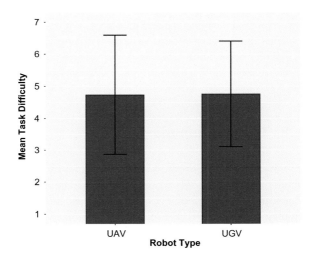

air environment, there does not appear to be a big difference between them. We can also see, if we plot some lines connecting the means (the dashed purple and green lines), that they intersect. This usually indicates the presence of an interaction.

Thus, when we find an interaction, the results from this interaction supersede the results of the main effects. That is, we can say that the main effects were *qualified* by the presence of an interaction. If we have no significant interaction, then we can follow up any significant main effects exactly the same way as for the one-way ANOVA.

The write-up for this analysis would look something like:

> The results of the two-way analysis of variance revealed a significant main effect of task environment, $F(2, 144) = 43.91$, $p < .001$, but no significant main effect of robot type $F(1, 144) = 0.017$, $p = .900$. However, these effects were qualified by the presence of an interaction between task environment and robot type, $F(2, 144) = 34.31$, $p < .001$. Follow up tests for the simple effect of robot type at each level of environment indicates that in land environments, UGVs were rated significantly lower for perceived task difficulty than UAVs $t = 4.78$, $p < .001$, whereas for air environments the opposite is true, with UAVs being rated significantly lower for perceived task difficulty $t = -5.41$, $p < .001$. However, in underwater environments, there was no difference between UAVs and UGVs—each of them was rated equally as difficult for exploration ($t = 0.75$. $p = .457$).

In sum, ANOVAs follow the same logic for test statistics that we have consistently seen throughout this chapter; that is, they rely on the ratio of explained to unexplained variance. This logic can be extended to more complex analyses, for example if you have three independent variables (three-way ANOVA), or a within-groups experimental design (repeated measures ANOVA), or a design which combines both within- and between-groups variables (mixed ANOVA). The math to compute these is slightly more complicated, but they all stem from the same basic principles of the general linear model. Thus, if you understand the content from this chapter, you will be well placed to conduct other more advanced statistical analyses in the future.

6.8 Chapter Summary

In this chapter, we covered a lot of ground on various mathematical tools essential to modern roboticists. We expect most of it to be merely a reminder for most readers, but with a twist toward how we need and use these tools in robotics. From geometry to matrix calculus to quaternions and inference statistics, this chapter is meant to be a reference you will come back to when reading the rest of this book.

6.9 Revision Questions

Question #1

Consider the following system of equations:

$$2x + 3y = 12 \tag{6.98}$$
$$y - 2z = 0 \tag{6.99}$$
$$x - y + 2z = 3 \tag{6.100}$$

Write this system in matrix form ($\mathbf{Ax} = \mathbf{b}$), compute the determinant of $\mathbf{A}$, its inverse and finally, find the values of x, y, and z.

Question #2

Demonstrate the equality in Eq. 6.37.

Question #3

Define what a p-value is and explain how it is related to the normal distribution.

Question #4

State the ratio needed to compute a test statistic and why.
     More examples and exercises on statistical tests are available online.[12]

6.10   Further Reading

While the theory behind basic linear algebra was presented in this chapter, some practical limitations must be known before solving a numerical problem. For instance, even if the determinant of a square matrix is not equal to zero, it may not be a good idea to inverse it to solve a system of linear equation. This is where you must consider the conditioning of a matrix, quantified by the condition number, which should not be close to 1. If it is, numerical approximations during the computation will be amplified and this will result in significant errors on the obtained solution. Moreover, to solve a numerical system of equations, the inverse (and generalized inverse) of a matrix is generally only of theoretical value, as algorithms such as the LU-decomposition, the Gram–Schmidt orthogonalization procedure, and the Householder reflections are used to avoid numerical errors. For further information, you can refer to a textbook on numerical analysis (Gilat and Subramaniam, 2008; Kong et al., 2020).[13]

     The statistics covered in this chapter are only a starting point for many other techniques for analysing experimental data. If you are interested in learning more about the theory behind different statistical methods, you can read *Discovering Statistics Using R* by Field et al. (2012), also available online.[14]

References

Ben-Israel, A., & Greville, T. N. (2003). *Generalized inverses: Theory and applications* (Vol. 15). Springer Science & Business Media.

Bender, R., & Lange, S. (2001). Adjusting for multiple testing-when and how? *Journal of Clinical Epidemiology, 54*(4), 343–349.

Field, A. P., Miles, J., & Field, Z. C. (2012). *Discovering statistics using R*. SAGE Publications.

Gilat, A., & Subramaniam, V. (2008). *Numerical methods for engineers and scientists: An introduction with applications using MATLAB*. Wiley.

---

[12] https://github.com/Foundations-of-Robotics/Stats-examples.

[13] *Python Programming and Numerical Methods* also available online: https://pythonnumericalmethods.berkeley.edu/notebooks/Index.html.

[14] https://www.discoveringstatistics.com/.

Hassenpflug, W. (1995). Matrix tensor notation part ii. skew and curved coordinates. *Computers & Mathematics with Applications, 29*(11), 1–103. https://doi.org/10.1016/0898-1221(95)00050-9, https://www.sciencedirect.com/science/article/pii/0898122195000509

Jones, E. M., & Fjeld, P. (2006). Gimbal angles, gimbal lock and a fourth gimbal for Christmas. *Apollo Lunar Surface Journal.* http://history.nasa.gov/alsj/gimbals.html

Kong, Q., Siauw, T., & Bayen, A. (2020). *Python programming and numerical methods: A guide for engineers and scientists.* Academic Press.

Rebecca Stower is a postdoctoral research fellow at the CHArt Laboratory at Paris 8 in collaboration with the INIT lab at ETS, Montreal. She holds a PhD in Psychology and a Bachelor of Psychological Science (Honours First Class). Her PhD centred on the occurrence of robot errors during child-robot-interactions and how this impacts children's attitudes and behaviours towards robots. Separately, she is also interested in the conceptualisation of social intelligence in robots and the design and measurement of social robot behaviour. More generally, she is passionate about the intersection of psychology and technology and how psychological research methods can be applied to robotics. She is also highly involved with open science and has contributed to the organisation of multiple interdisciplinary and cross-industry events.

Bruno Belzile is a postdoctoral fellow at the INIT Robots Lab. of ÉTS Montréal in Canada. He holds a B.Eng. degree and Ph.D. in mechanical engineering from Polytechnique Montréal. His thesis focused on underactuated robotic grippers and proprioceptive tactile sensing. He then worked at the Center for Intelligent Machines at McGill University, where his main areas of research were kinematics, dynamics and control of parallel robots. At ÉTS Montréal, he aims at creating spherical mobile robots for planetary exploration, from the conceptual design to the prototype.

David St-Onge (Ph.D., Mech. Eng.) is an Associate Professor in the Mechanical Engineering Department at the École de technologie supérieure and director of the INIT Robots Lab (initrobots.ca). David's research focuses on human-swarm collaboration more specifically with respect to operators' cognitive load and motion-based interactions. He has over 10 years' experience in the field of interactive media (structure, automatization and sensing) as workshop production director and as R&D engineer. He is an active member of national clusters centered on human-robot interaction (REPARTI) and art-science collaborations (Hexagram). He participates in national training programs for highly qualified personnel for drone services (UTILI), as well as for the deployment of industrial cobots (CoRoM). He led the team effort to present the first large-scale symbiotic integration of robotic art at the IEEE International Conference on Robotics and Automation (ICRA 2019).

Part II
Embedded Design

Chapter 7
What Makes Robots? Sensors, Actuators, and Algorithms

Jiefei Wang and Damith Herath

7.1 Learning Objectives

This chapter explores a framework and some of the main building blocks in developing robots. You will learn about:

- The Sense, Think, Act loop.
- Different types of sensors that make robots 'feel' the world and find suitable sensors for use in specific scenarios.
- Algorithms that make the robots' 'intelligent'.
- Actuators that make robots move.
- Commonly used computer vision algorithms that make robots 'see'.

7.2 Introduction

In Chap. 4, we discussed that programming could be thought of as input, process, and output. *Sense, Think, Act* is a similar paradigm used in robotics. A robot could be thought of rudimentarily as analogous to how a human or an animal responds to environmental stimuli. For example, we humans perceive the environment through the five senses (e.g. sight). We might then 'decide' the following action based on these incoming signals and, finally, execute the action through our limbs. For example, you

J. Wang (✉)
The School of Engineering and Information Technology, University of New South Wales, Canberra, Australia
e-mail: Jiefei.wang@adfa.edu.au

D. Herath
Collaborative Robotics Lab, University of Canberra, Canberra, Australia
e-mail: Damith.Herath@Canberra.edu.au

© The Author(s) 2022
D. Herath and D. St-Onge (eds.), *Foundations of Robotics*,
https://doi.org/10.1007/978-981-19-1983-1_7

Fig. 7.1 Sense, Think, Act loop in robotics

might see (sense) a familiar face in the crowd and think it would be good to grab their attention and then act on this thought by waving your hand.

Similarly, a robot may have several sensors through which it could sense the environment. An algorithm could then be used to interpret and decide on an action based on the incoming sensory information. This computational process could be thought of as analogous to the thinking process in humans. Finally, the algorithm sends out a set of instructions to the robot's actuators to carry out the actions based on the sensor information and goals (Fig. 7.1).

The current configuration of the robot is called the robot's *state*. The robot *state space* is all possible states a robot could be in. Observable states are the set of fully visible states to the robot, while other states might be hidden or partly visible to the robot. Such states are called *partially observable* states. Some states are discrete (e.g. motor on or off), and others could be continuous (e.g. rotational speed of the motor). In the above paradigm, the sense element observes the state, and the Act element proceeds to alter the state.

An Industry Perspective

Vitaliy Khomko, Vision Application Developer

Kinova Inc.

My journey into robotics started when I joined JCA Technologies, Manitoba, in 2015. At that time, the maturity of sensors and controllers technology allowed innovators to create smart agricultural and construction equipment capable of performing many complex operations autonomously with very little input from a machine operator. Frankly speaking, I did not get to choose robotics. I simply got sucked into the technological vortex because the industry was screaming for innovation as well as researchers and developers to drive it. In 2018, I was welcomed by the team of very passionate roboticists at Kinova, Quebec, in the position of Vision Technology Developer. The creative atmosphere fuelled by Kinova's employees kept driving me for many months. I worked hard during the day trying my best to fit in and kept learning new stuff in the evening to fill in the blanks. All this hard work paid off well in the end. I must admit, now, I can wield magic with a vision-enabled robot.

Continuous learning and keeping up with all the industry trends is by far the most challenging and time-consuming. The theoretical knowledge alone, though, can only help with being on the right track. In reality, when working on delivering a real consumer product, an enormous effort goes into research and evaluation, work planning, development/coding, and testing. Maintaining a good relationship with your coworkers is essential. At the end of the day, it is your teammates who give you a hand when you get stuck, who share your passion and excitement, who appreciate your effort, and who let you feel connected. Nothing really compares with the satisfaction of joint accomplishment when you can pop a beer by the end of a long day with your colleagues after delivering the next milestone, watching the robot finally doing its thing over and over again.

With regard to evolution, I certainly noticed a shift from simple automated equipment controlled by human experts into very efficient autonomous machines capable of making decisions. Sensors have been around for a long time. By strategically placing them into a machine, one can achieve an unprecedented amount of feedback from a machine to allow better control and operation precision. The amount of information and real-time constraints, though, can be too much even for an expert human operator to process through. What really made a difference now is the availability of algorithms and computational devices to enable a certain degree of machine autonomy. For example, camera technology is widely available these days. But it is not the camera alone that enables vision-guided robotics. Its robot–camera calibration, 2D/3D object matching and localisation grasping clearance validation, etc. Some can argue that recent advancements in artificial intelligence mainly contributed to that evolution. I think AI is just another tool. And by no means an ultimate solution to every problem.

7.3 Sense: Sensing the World with Sensors

Everything changes in the real world. Some changes are notable while others are subtle, some are induced, and some are provoked. However, these changes always reveal information hidden from the initial perception. Sensing the changes in the environment are particularly meaningful and allow perception and interaction with the surrounding world. For example, humans perceive the displacements of the colour patterns on the retina and compute those displacements to understand the changes. Some animals, such as bats, can use echolocation to estimate their environment changes and localise themselves. Unlike humans or animals, robots do not have naturally occurring senses. Therefore, robots need extra sensors to help them sense the environment and use algorithms to process and understand the information. For example, a typical sensor such as a video camera can be considered the robot's 'eyes'. A sonar sensor could be thought of as equivalent to echolocation in a bat. By having different sensors integrated with robots, they can achieve various tasks like the one human being can do.

7.3.1 Typical Sensor Characteristics

Sensors could be characterised in various ways. Let us look at some of the common characteristics and their definitions first.

7.3.1.1 Proprioceptive and Exteroceptive

As humans perceive aches and pains internal to their body, so could robots sense various internal states of the robot, such as the speed of its wheels/motors or the current drawn by its internal power circuitry. Such sensors are called *proprioceptive* sensors. On the other hand, sensors that provide information about the robot's external environment are called *exteroceptive* sensors.

7.3.1.2 Passive and Active Sensors

A sensor that only has a detector to observe or measure the physical properties of the environment is categorised as a *passive* sensor. A light sensor is an example. In contrast, *active* sensors emit their own signal or energy to the environment and employ a detector to observe the reaction resulting from the emitted signal. A sonar sensor is a typical example.

7.3.1.3 Sensor Errors and Noise

However well made a sensor is, they are susceptible to various manufacturing errors and environmental noise. However, some of these errors could be anticipated and understood. Such errors that are deterministic and reproducible are called *systematic errors*. Systematic errors could be modelled and integrated as part of the sensor characteristics. Other errors are difficult to pinpoint. These could be due to environmental effects or other random processes. Such errors are called random errors. Understanding these errors is crucial to deploying a successful robotics system. When this information is not readily available for the sensor selected, you will need to conduct a thorough error analysis to isolate and quantify the systematic errors and figure out how to capture the random errors.

7.3.1.4 Other Common Sensor Characteristics

You may encounter the following terms describing various other characteristics of a sensor. It is important to understand what they mean in a given context to use the appropriate sensor for the job.

Resolution The minimum difference between two values that the sensor can measure.

Accuracy The uncertainty in a sensor measurement with respect to an absolute standard.

Sensitivity The smallest absolute change that a sensor can measure.

Linearity Whether the output produced by a sensor depends linearly on the input.

Precision The reproducibility of the sensor measurement.

Bandwidth The speed at which a sensor can provide measurements. Usually expressed in Hertz (Hz)readings per second.

Dynamic range Under normal operation, this is the ratio between the limits of the lower and upper sensor inputs. This is usually expressed in decibels (dB):

$$\text{Dynamic Range} = 10 \log \log_{10} \left(\frac{\text{upper limit}}{\text{lower limit}} \right)$$

7.3.2 Common Sensors in Robotics

7.3.2.1 Light Sensors

Light sensors are used to detect light that creates a difference in voltage signal to feedback to the robot's system. The two common light sensors that are widely used in the field of robotics are photoresistors and photovoltaic cells. The change in incident light intensity changes the photoresistor's resistance in a photoresistor. More light leads to less resistance, vice versa. Photovoltaic cells, on the other hand, convert solar radiation into electricity. This is especially helpful when planning a solar robot. While the photovoltaic cell is considered an energy source, a smart implementation combined with transistors and capacitors can convert this into a sensor. Other light sensors, such as phototransistors, phototubes, and charge-coupled devices (CCD), are also available (Fig. 7.2).

Fig. 7.2 A common light
sensor (a photoresistor)

7.3.2.2 Sonar (Ultrasonic) Sensors

Sonar sensors (also called ultrasonic sensors) utilise acoustic energy to detect objects and measure distances from the sensor to the target objects. Sonar sensors are composed of two main parts, a transmitter and receiver.

The transmitter sends a short ultrasonic pulse, and the receiver receives what comes back of the signal after it has reflected from the surface of nearby objects. The sensor measures the time from signal transmission to reception, namely the time-of-flight (TOF).

Knowing the transmission rate of an ultrasonic signal, the distance to the target that reflects the signal can be calculated using the following equation.

$$\text{Distance} = (\text{Time} \times \text{Speed Of Sound})/2$$

where '2' means the sound has to travel back and forth.

Sonar sensors can be used for mobile robot localisation through model matching or triangulation by computing the pose change between the inputs acquired at two different poses (Jiménez & Seco, 2005). Sonar sensors could also be used in detecting obstacles (see Fig. 7.3).

One of the challenges of using these sensors is that they are sensitive to noise from the surrounding and other sonar sensors with the same frequency. Moreover, they are highly dependent on the material and orientation of the object surface as these sensors make use of the reflection of the signal waves (Kreczmer, 2010). New techniques such

Fig. 7.3 Four sonar sensors are embedded in the chest of this NAO robot to help detect any obstacles in front of it. A tactile sensor is embedded on its head

as compressed high-intensity radar pulse (CHIRP) have been developed to improve sonar performance.

Sonar signals have a characteristic 3D beam pattern. This makes them suitable for detecting obstacles in a wide area when the exact geometric location is not needed. However, laser sensors provide a better solution for situations where precise geometry needs to be inferred.

7.3.2.3   Laser and LIDAR

Laser sensors can be utilised in several applications related to positioning. It is a remote sensing technology for distance measurement that involves transmitting a laser beam towards the target and analysing the reflected light. Laser-based range measurements depend on either TOF or phase-shift techniques. Like the sonar sensor, a short laser pulse is sent out in a TOF system, and the time, until it returns, is measured. A low-cost laser range finder popular in robotics is shown in Fig. 7.4. Also see Fig. 7.10.

LIDAR Light Detection And Ranging (LIDAR) has found many applications in robotics, including object detection, obstacle avoidance, mapping, and 3D motion capture. LIDAR can be integrated with GPS and INS to enhance the performance and accuracy of outdoor positioning applications (Aboelmagd et al., 2013).

One of the disadvantages of using LIDAR is that it requires high computational ability to process the data, which may affect mobile robot applications' real-time performance. Moreover, scanning can fail when the object's material appears transparent, such as glass, as the reflections on these surfaces can bring misleading and unreliable data (Takahashi, 2007).

Fig. 7.4 Hokuyo URG-04LX range finder

7.3.2.4 Visual Sensors

Compared with proximity sensors we mentioned above, optical cameras are low-cost sensors that provide a large amount of meaningful information.

The images captured by a camera can provide rich information about the robot's environment once processed using appropriate image processing algorithms. Some examples include localisation, visual odometry, object detection, and identification. There are different types of cameras, such as stereo, monocular, omnidirectional, and fisheye, that suit all manner of robotic applications.

Monocular cameras (Fig. 7.5) are especially suitable for applications where compactness and minimum weight are critical. Moreover, low cost and easy deployment are the primary motivations for using monocular cameras for mobile robots. However, monocular cameras can only obtain visual information and are not able to obtain depth information. On the other hand, a stereo camera is a pair of identical monocular cameras mounted on a rig. It provides everything that a single camera can offer and extra information that benefits from two views. Based on the parallax principle, the stereo camera can estimate the *depth map* (a 2D image that depicts the depth relationship between the objects in the scene and the camera's viewpoint)

Fig. 7.5 A popular monocular camera

by utilising the two views of the same scene slightly shifted. *Fisheye* cameras are a variant of monocular cameras that provide wide viewing angles and are attractive for obstacle avoidance in complex environments, such as narrow and cluttered environments.

7.3.2.5 RGB-D Sensors

RGB-D sensors are unconventional visual sensors that can simultaneously obtain a visible image (RGB image) and depth map of the same scene. They have been very popular in the robotics community for real-time image processing, robot localisation, obstacle avoidance. However, due to the limited range and sensitivity to noise, they are mostly used in indoor environments.

The Kinect sensor is one of the most well-known RGB-D sensors (Yes! The same sensor you use when playing video games on the Xbox), introduced to the market in November 2010 and has gained great popularity since then. The computer vision community quickly discovered that this depth-sensing technology could be used for other purposes while costing much less than some traditional three-dimensional (3D)

Fig. 7.6 A newer version of the Microsoft® Kinect sensor

cameras, such as time-of-flight-based ones. In June 2011, Microsoft released an SDK for the Kinect to be used as a tool for non-commercial products (Fig. 7.6).

The basic principle behind the Kinect depth sensor is the emission of an IR speckle pattern (invisible to the naked eye) and the simultaneous capture of an IR image by a CMOS camera fitted with an IR-pass filter. An image processing algorithm embedded inside the Kinect uses the relative positions of the dots in the speckle pattern (see Fig. 7.7) to calculate the depth displacement at each pixel position in the imagethe technique is called structured light. Hence, the depth sensor can provide the x-, y-, and z-coordinates of the surface of 3D objects.

The Kinect sensor consists of an IR laser emitter and IR and RGB cameras. It simultaneously captures depth and colour images at frame rates of up to 30 Hz. The RGB colour camera delivers images at 640 × 480 pixels and 24 bits at the highest frame rate. In contrast, the 640 × 480 and 11 bits per pixel IR camera provides 2048 levels of sensitivity with a field-of-view of 50° horizontal and 45° vertical. The operational range of the Kinect sensor is from 50 to 400 cm.

Fig. 7.7 A view from an RGB-D camera (from left to right—RGB image, depth image, IR image showing the projected pattern)

7.3.2.6   Inertial Measurement Units

An inertial measurement unit (IMU) utilises gyroscopes and accelerometers (and optionally magnetometers and barometers) to sense motion and orientation. An accelerometer is a device for measuring acceleration and tilt. Two types of forces affect an accelerometer: gravity which helps determine how much the robot tilts. This measurement helps balance the robot or determine whether a robot is driving on a flat or uphill surface—the other is the dynamic force which is the acceleration required to move an object. These sensors are useful in inferring incremental changes in motion and orientation. However, they suffer from bias, drift, and noise. This requires regular calibration of the system before use or sophisticated sensor fusion and filter techniques (such as the EKF described in Chap. 9). You will often see IMU units used with computer vision systems or combined with Global Navigation Satellite System (GNSS) information. Such systems are commonly called INS/GNSS systems (Intertial Navigation Systems/GNSS).

7.3.2.7   Encoders

Simply put, encoders record movement metrics in some form. There are three types of encoders: linear encoders, rotary encoders, and angle encoders.

Linear encoders measure straight-line motion. Sensor heads that attach to the moving piece of machinery run along guideways. Those sensors are linked to a scale inside the encoder that sends digital or analog signals to the control system. *Rotary encoders* measure rotational movement. They typically surround a rotating shaft, sensing and communicating changes in its angular motion. Traditionally, rotary encoders are classified as having accuracies above $\pm 10''$ (arcseconds). Rotary encoders are also available, equipped with important functional safety capabilities. Similar to their rotary counterpart, angle encoders measure rotation. These, however, are most often used in applications when a precise measurement is required.

Mobile robots often use encoders to calculate their *odometry*. Odometry is the use of motion sensors to determine the robot's temporal change in position relative to some known position. A simple example of using a rotary incremental encoder to calculate the robot's travel distance could be illustrated using Fig. 7.8. A light is shone through a slotted disc (usually made of metal or glass). As the disc rotates, the light passing through the slots is picked up by a light sensor mounted on the other side of the disc. This signal could be converted into a sinusoidal or square wave using electronic circuitry. If this encoder is attached to the axis of the robot's wheel, we can use the output signal to calculate the velocity at which the robot is moving.

To calculate the length travelled L (cm) using the output from an incremental encoder, we start by calculating the number of pulses per cm (PPCM):

$$\text{PPCM} = \frac{\text{PPR}}{2\pi r}$$

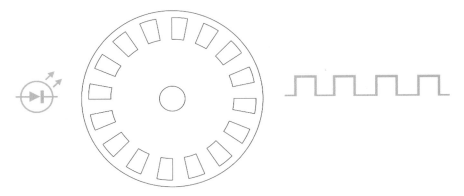

Fig. 7.8 A simplified rotary incremental encoder with 16 slots

where PPR is the pulses per revolutionwhich in the example in Fig. 7.8 is 16.
  Then the length L is given by:

$$L = \frac{\text{Pulses}}{\text{PPCM}}$$

The speed (S) is then calculated as:

$$S = \frac{L}{\text{Time Taken}}$$

  It is worth noting that the need to have these sensors closer to the motors often results in them being subject to electromagnetic noise. Therefore to improve the encoder's performance as well as to decipher the direction of rotation, a second set of light and sensor pair is included with a 90° a phase shift (Fig. 7.9).

7.3.2.8   Force and Tactile Sensors

Both these types of sensors measure physical interactions between the robot and the external environment. A typical force sensor is usually used to measure external mechanical force input, such as in the form of a load, pressure, or tension. Sensors such as strain gauges and pressure gauges fall into this category. On the other hand, tactile sensors are generally used to mimic the sense of touch. Usually, tactile sensors are expected to measure small variations in force or pressure with high sensitivity. Robots designed to be interactive integrate many tactile sensors so they can respond to touch (e.g. Fig. 7.3). Sophisticated sensors are emerging that could mimic skin-like sensitivity to touch. A more primitive version could be seen in most vacuum cleaning robots, where the front bumper acts as a collision detector (Fig. 7.10).

Fig. 7.9   A popular hobby rotary incremental encoder with two outputs (quadrature encoder)

7.3.2.9   Other Common Sensors in Robotics

Many other sensors are used in robotics, and new ones are developed in various research laboratories and commercialised regularly. These include microphones (auditory signals), compasses, temperature sensors (thermal and infrared sensors), chemical sensors, and many more. Therefore, it is prudent to research suitable sensors for your next project as new and more capable sensors may better suit your needs. Can you think of all the sensors that may be used in the robot shown in Fig. 7.10?

7.4   Think: Algorithms

A critical component of a robotic system is its ability to make control decisions based on the available sensory information and to realise the tasks and goals allocated to it. If the brains of a robot are the computers embedded in the robotic system, the *algorithms* are the software components that enable a robot to '*think*' and make decisions. Algorithms interpret the environment based on sensory input and decide

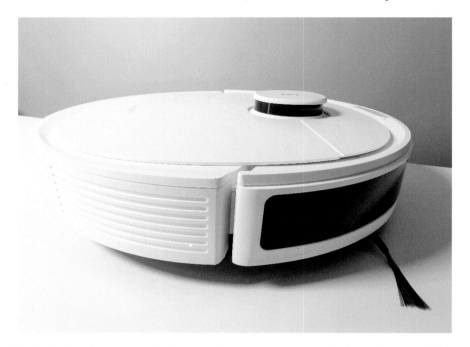

Fig. 7.10 A modern vacuum cleaning robot integrates many sensors. On the top is a time-of-flight laser scanner. The front bumper includes several tactile sensors to detect any frontal collisions. What other sensors do you think this robot may have?

what needs to be done at what given time and what is happening in the environment based on the allocated tasks.

In the most general sense, an algorithm is a finite list of instructions used to solve problems or perform tasks. To get a feel for the concept of algorithms, think about baking a sponge cake. How would you write down your whole process to make a sponge cake to a person who does not know baking at all? Answering these questions in a detailed and ordered way makes an algorithm. One of the attributes of an algorithm is that there is a systematic process that occurs in a specific order. The wrong order of the steps can result in a big difference. For example, if we change the order of steps in making sponge cake, for instance, put eggs and flour in the oven for half an hour before preheating the oven. That would not make any sense!

For a robotic system, algorithms are the specific recipes that help them 'think'. They are precise sequences of instructions implemented using programming languages. The essential elements of an algorithm are input, sequence, selection, iteration, and output.

- InputData, information or signals collected from the sensors or a command from a human operator.
- Sequence—The order in which behaviours and commands are combined to produce the desired result.

- Selection—Is the use of conditional statements in a process. For example, conditional statements such as [If then] or [If then else] can affect the process.
- Iteration—Algorithms can use repetition to execute steps a certain number of times or until a specific condition is reached. It is also known as 'looping'.
- OutputDesired result or expected outcome, such as the robot reaching the targeted location or avoiding the collision with certain obstacles.

Robotics is rife with all kinds of algorithms, from simple obstacle avoidance to complex scene understanding using multiple sensors. Among these, computer vision algorithms play a significant role in their ability to infer the rich information generated through various optical camera systems discussed earlier. Therefore, we discuss some common vision algorithms found in robotics next.

7.5 Act: Moving About with Actuators

We identify robots as things that move around or with moving parts. In the Sense, Think, Act paradigm, the Act refers to this dynamic aspect of robots. The robot acts on the environment by manipulating it using various appendages called manipulators (arm-type robots) or traversing it (mobile robots). In order to act, a robot needs actuators. An actuator is a device that requires energy, such as electric, hydraulic, pneumatic, and external signal input, then convert them to a form of motion that can be controlled as desired.

7.5.1 Common Actuators in Robotics

7.5.1.1 Motors

The electric motor is a typical example of an electrically driven actuator. As they can be made in different sizes, types, and capacities, they are suitable for use in a wide range of robotic applications. There are various electric motors, such as servo motors, stepper motors, and linear motors.

Servo motors

A servo motor is controlled with an electric signal, either analog or digital, which determines the amount of movement. It provides control of position, speed, and torque. Servo motors are classified into different types based on their application, such as the AC servo motor and DC servo motor.

The speed of a DC motor is directly proportional to the supply voltage with a constant load, whereas, in an AC motor, speed is determined by the frequency of the applied voltage and the number of magnetic poles. AC motors are commonly used

Fig. 7.11 Hobby DC servo motors (left) and a high-end actuator (right) used in an industrial robot arm (courtesy of Kinova Robotics)

in servo applications in robotics and in, in-line manufacturing, and other industrial applications where high repetitions and high precision are required.

DC servo motors are commutated mechanically with brushes, using a commutator, or electronically without brushes. Brushed motors are generally less expensive and simpler to operate, while brushless motors are more reliable, have higher efficiency, and are less noisy (Fig. 7.11).

Stepper motors

A stepper motor is a brushless synchronous DC motor that features precise discrete angular motions. A stepper motor is designed to break up a single complete rotation into a number of much smaller and essentially equal part rotations. For practical purposes, these can be used to instruct the stepper motor to move through set degrees or angles of rotation. The end result is that a stepper motor can be used to transfer accurate movements to mechanical parts that require a high degree of precision. Stepper motors are very versatile, reliable, cost-effective and provide precise motor movements, allowing users to increase the dexterity and efficiency of programmed movements across a huge variety of applications and industries. Most 3D printers, for example, use multiple stepper motors to precisely control the 3D print head.

Linear motors

A linear motor operates on the same principle as an electric motor but provides linear motion. Unlike a rotary machine, a linear motor moves the object in a straight line or along a curved track. Linear motors can reach very high acceleration, up to 6 g, and travel speeds of up to 13 m/s. Due to this character, they are especially suitable for use in machine tools, positioning and handling systems, and machining centres.

7.5.1.2   Hydraulic Actuators

Hydraulic actuators are driven by the pressure of the hydraulic fluid. It consists of a cylinder, piston, spring, hydraulic supply and return line, and stem. They can deliver large amounts of power. As such, they can be used in construction machinery and other heavy-duty equipment.

There are some advantages to using hydraulic actuators. A hydraulic actuator can hold force and torque constant without the pump supplying more fluid or pressure due to the incompressibility of fluids. Hydraulic actuators can have their pumps and motors located a considerable distance away with minimal loss of power. Comparing the pneumatic cylinder of equal size, the forces generated by hydraulic actuators are 25 times greater, ensuring they operate well in heavy-duty settings. One of the disadvantages of using hydraulic actuators is that they may leak fluid, leading to reduced efficiency and, in extreme cases, damage to nearby equipment due to spillage. Hydraulic actuators require many complementary parts, including a fluid reservoir, motor, pump, release valves, and heat exchangers, along with noise reduction equipment.

7.5.1.3   Pneumatic Actuators

Pneumatic actuators have been known for being highly reliable, efficient, and safe sources of motion control. These actuators are driven by pressurised air that can convert energy in the form of compressed air into linear or rotary mechanical motion. They feature both simple mechanical design and flexible operation. They are widely used in combustible automobile engines, railway applications, and aviation. Most of the benefits of choosing pneumatic actuators over alternative actuators, such as electric ones, boil down to the reliability of the devices and the safety aspects. Pneumatic actuators are also highly durable, requiring less maintenance and long operating cycles.

7.5.1.4   Modern Actuators

Many new actuation methods and actuators have emerged in recent times. These include pneumatic tendons (Fig. 7.12) and other biologically inspired actuators, such as fish fins or octopus tentacles. Soft robotics is an emerging field that explores some of these developments. However, the compliance requirements and morphology of soft robots prevent the use of many conventional sensors seen in hard robots. As a result, there has been active research into stretchable electronic sensors. Elastomer sensors allow for minimal impact on the actuation of the robot.

Fig. 7.12 Pneumatic rubber muscles used in animating this giant robotic structure during a performance by the artist, Stelarc (Reclining StickMan, 2020 Adelaide Biennial of Australian Art: Monster Theatres, Photographer—Saul Steed, Stelarc)

7.6 Computer Vision in Robotics

Computer vision techniques have been the subject of heightened interest and rigorous research for decades now as a way of sensing the world in all its complexity. Computer vision attempts to achieve the function of understanding the scene and the objects of the environment. Furthermore, the increasing computational power and progress in computer vision methods have made making robots '*see*' a popular trend. As computer vision combines both sensors and algorithms, it deserves its own unique section within this chapter.

Computer vision in robotics refers to the capability of a robot to visually perceive and interact with the environment. Typical tasks are to recognise objects, detect ground planes, traverse to a given target location without colliding with obstacles, interact with dynamic objects, and respond to human intents.

Vision has been used in various robotic applications for more than three decades. Examples include applications in industrial settings, service, medical, and underwater robotics, to name a few. The following section will introduce some classic computer vision algorithms widely used in robotics, such as plane detection, optical flow, and visual odometry.

7.6.1   Plane Detection

For an autonomous mobile robot system, detecting the dominant plane is a fundamental task for obstacle avoidance and trajectory finding. The dominant plane can be
considered a planar area occupying the largest region on the ground towards which
the robot is moving. It provides useful information about the environment, particularly whether objects above the detected dominant plane and along the direction of
the robot's movement can be viewed as obstacles. A ground mobile robot or micro-
aerial vehicle operating in an unknown environment must identify its surroundings
before the system can conduct its mission. These vehicles should recognise obstacles
within their operating area and avoid detected obstacles or travel over them where
possible. There are various plane detection techniques such as RANSAC and the
region growth method.

7.6.1.1   RANSAC

The random sample consensus (RANSAC) (Fischler & Bolles, 1981) method is
an iterative method to estimate parameters of a mathematical model from a set of
observed data that contains outliers. It is a very useful tool to find planes, with its
principle to search for the best plane among three-dimensional (3D) point clouds. At
the same time, it is computationally efficient even when the number of points is vast.
Plane detection using RANSAC starts by randomly selecting three points from the
point cloud and calculating the parameters of the corresponding plane. The next step
detects all the points of the original cloud belonging to the calculated plane based on
the given threshold. Repeating this procedure for N rounds, each time, it compares
the obtained result with the last saved one, and if the new one is better, it replaces
the saved one (see Algorithm 1).

The four types of data needed as input for this algorithm are:

- a 3D point cloud which is a matrix of the three coordinate columns X, Y, and Z;
- a tolerance threshold of distance t between the chosen plane and other points;
- a probability (α) which lies typically between 0.9 and 0.99 and is the minimum
  probability of finding at least one good set of observations in N rounds; and
- the maximum probable number of points belonging to the same plane.

```
// Algorithm 1: RANSAC for plane fitting
Input: 3D points data;

For all N_rounds DO

    Select three random points (P_1, P_2, P_3) from the input data;

    Fit a plane through the points;

    Set: N_inliers = 0;

    For all pints do

        If point distance to plane is within a threshold then ;

        Increment N_inliers;

        If N_inliers is larger than the best plane then;

        Update plane estimation using all points;

        Update best plane;

    End

End

Output: Dominant plane;
```

As one of the most well-known methods for plane detection, RANSAC has been shown to be capable of detecting planes in both 2D and 3D. For example, in Fig. 7.13,

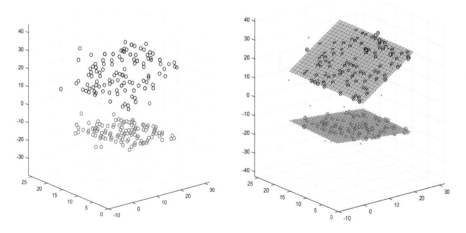

Fig. 7.13 Two groups of 3D points representing two planes detected using the RANSAC method

two groups of noisy 3D points (blue and red) with two planes detected successfully using the RANSAC method.

7.6.1.2   Region Growth Method

The region growth method for plane detection was first introduced by Hähnel et al., (2003) with the goal of creating a low complexity model that can be implemented in real time. It works from a seed chosen randomly from the point cloud, which consists of sufficient information to fit a plane and adds more points based on specific selection conditions, such as if three points are needed or whether a point with a corresponding normal can be used. Then, when the neighbouring points are consistent with the plane, they are considered part of it. This procedure is repeated until no more points can be found, and then the algorithm stops and adds the plane if it contains enough points. Finally, the points are removed from the point set, and a new seed is selected. A brief outline of this algorithm is presented in Algorithm 2.

```
// Algorithm 2: Region growing method for plane fitting
   Input: 3D points data;

For all point DO

      S= two random neighbouring point P_1, P_2;

      Select nearest point P_n within the certain distance;

      If the new added point P_n does not change the plane estimation, then add this point
      P_n into the selected date set S;

      If the size of S > certain threshold;

      The add plane estimate to set of planes;

End

Output: Set of planes;
```

7.6.2   Optical Flow

Optical flow is the pattern of apparent motion of objects, surfaces, and edges in a visual scene caused by the relative motion between an observer and the scene. It is believed that insects and birds frequently use optical flow for short-range navigation and obstacle avoidance. For example, biologists have reported that birds use optical flow to avoid obstacles and manoeuvre landings. In addition, many mammals possibly use optical flow to detect the motions of objects. All these discoveries regarding optical flow provide new ideas for roboticists to develop visual-based robots with the capability to navigate safely and quickly in unknown environments.

Optical flow can be treated as the apparent motions of objects, brightness patterns or feature points observed by eyes or cameras. Based on this definition, it can be computed from the difference between two sequences, which is usually expressed as:

$[\dot{u}, \dot{v}]^T = f(u, v)$, where its unit is pix/sec or pix/frame.

Optical flow can also be defined as the projection of the relative 3D motion between an observer and scene into the image plane. As an image consisting of many pixels with unique coordinates, it can be described as a two-dimensional (2D) vector in image sequences. Therefore, the motion field model can be described as:

$$OF = \frac{V}{d}$$

where OF is the optical flow field, V is the observer velocity vector, and d is the distance between the observer and the image plane with the unit normally rad/s or °/s. The two definitions above mentioned are the same for an ideal situation after a coordinate transformation.

For a short duration, the intensity structures of local time-varying image regions are approximately constant. Based on this assumption, if $I(x, t)$ is the image intensity function, we have:

$$I(x, t) = I(x + \delta_x, y + \delta_y),$$

where δ_x is the displacement of the local image region at (x, t) at time $t + \delta_t$. This equation expanded in a Taylor series yields:

$$I(x, t) = I(x, t) + \nabla_I \cdot \delta_x + \delta_t I_t + O^2$$

where $\nabla_I = (I_x, I_y)$ and I_t are the first-order partial derivatives of $I(x, t)$ and O^2 the second-and higher-order terms, which are negligible. The previous equation can be rewritten as:

$$\nabla_I \cdot V + I_t = 0$$

dividing by δ_t, where $\nabla_I = (I_x, I_y)$ is the spatial intensity gradient, and $V = (u, v)$ is the image velocity. This is known as the optical flow constraint equation, which defines a single local constraint on image motion (Fig. 7.14).

Many methods have been proposed for detecting the optical flow. Some techniques are briefly discussed next.

Fig. 7.14 Detected optical flow indicated by red arrows, longer the arrow length faster movement of the pixel patch (translation on the left, rotation on the right)

7.6.2.1 Lucas–Kanade Method and Horn–Schunck Method

The Lucas–Kanade method (Lucas & Kanade, 1981) and Horn–Schunck method (Horn & Schunk, 1981) are widely used classical differential methods for optical flow estimation. Lucas–Kanade method assumes that the flow is constant in a local neighbourhood of the pixel under consideration and solves the basic optical flow equations for all the pixels in that neighbourhood by the least-squares criterion. By combining information from several nearby pixels, the Lucas–Kanade method can often resolve the inherent ambiguity of the optical flow equation. It is also less sensitive to image noise compared with other methods.

Horn–Schunck method is another classical optical flow estimation algorithm. It assumes smoothness in the flow over the whole (global) image. Thus, it tries to minimise distortions in flow and prefers solutions that show more smoothness. As a result, it is more sensitive to noise than the Lucas and Kanade method. Many current optical flow algorithms are built upon these frameworks.

7.6.2.2 Energy-Based Methods

Energy-based optical flow calculation methods are also called frequency-based methods because they use the energy output from velocity-tuned filters. Under certain conditions, these methods can be mathematically equivalent to differential methods mentioned previously. However, it is more difficult for differential and correlation methods to deal with sparse patterns of moving dots than energy-based methods.

7.6.2.3 Phase-Based Methods

A phase-based technique is a classical method calculating the optical flow using the phase behaviours of band-pass filter outputs. It was first introduced by Fleet and Jepson (1990) and has been shown to be more accurate than other local methods mainly because phase information is robust to changes, in contrast, scale orientation and speed (Fleet & Jepson, 1990). However, the main drawback of phase-based techniques is the high computational load associated with their filtering operations.

Correlation methods

Correlation-based methods find matching image patches by maximising some similarity measure between them under the assumption that the image patches have not been overly distorted over a local region. Such methods may work in cases of high noise and low temporal support where numerical differentiation methods are not as practical. These methods are typically used for finding stereo matches for the task of recovering depth.

7.6.3 Visual Odometry

Visual odometry (VO) is a method for estimating the position and orientation of mobile robots, such as a ground robot or flying platform, using the input from a single or multiple cameras attached to it (Scaramuzza & Fraundorfer, 2011). It estimates a position by integrating the displacements obtained from consecutive images observed from onboard vision systems. It is vital in environments in which a GPS is not available for absolute positioning (Weiss et al., 2011).

Many conventional odometry solutions produce unpredictable errors in the measurements delivered by gyroscopes, accelerometers, and wheel encoders. It has been found that, for Mars exploration Rovers experiencing small translations over the sandy ground, large rocks or steep slopes, the visual odometry needs to be corrected for errors arising from motions and wheel slip (Maimone et al., 2007). A vehicle's position can be estimated by either stereo or monocular cameras using feature matching or tracking technologies. In Garratt and Chahl (2008), the translation and rotation are estimated using the image interpolation algorithm with a downward-facing camera. Methods for computing ego-motion directly from image intensities have also been suggested (Hanna, 1991; Heeger & Jepson, 1992). The issue with using just one camera is that only the direction of motion, not the absolute velocity scale, can be determined, known as the scaling factor problem. However, using an omnidirectional camera can solve this problem; for example, safe corridor navigation for a micro air vehicle) (MAV) using an optical flow method is achieved in Conroy et al., (2009), but this operation requires a great deal of computational time.

7.7 Review Questions

- What is the difference between an AC motor and a DC motor?
- What is the difference between a camera and an RGB-D sensor?
- A typical rotary encoder used in a wheeled mobile robot to measure the distance it travels has 40 slots. The robot's wheel to which this sensor is mounted has a diameter of 7 cm. If the sensor gives out a steady 7 Hz square pulse, what is the robot's speed in cm/s?

7.8 Further Reading

Although a little dated, the *Sensors for Mobile Robots* by Everett and *Robot Sensors and transducers* by Ruocco provide comprehensive coverage of classical sensors used in robotics. *Computer Vision: Algorithms and Applications* by Szeliski is an excellent introductory book on computer vision in general. For more robotics-related concepts in computer vision as well for those interested in reading more advanced topics in robotics, Corke's *Robotics, Vision and Control* are highly recommended. The book includes many code samples and associated toolboxes in Matlab®. *Programming Computer Vision with Python: Tools and algorithms for analysing images* by Solem provide many Python-based examples of vision algorithm implementations. *Algorithms* by Sedgewick and Wayne is one of the best books on the topic.

References

Aboelmagd, N., Karmat, T. B., & Georgy, J. (2013). *Fundamentals of inertial navigation, satellite-based positioning and their integration.* Springer.

Chum, O., & Matas, J. (2005). Matching with prosac-progressive sample consensus. In *IEEE Computer Society Conference on Computer Vision and Pattern Recognition, 2005. CVPR 2005* (Vol. 1, pp. 220–226). IEEE.

Conroy, J., Gremillion, G., Ranganathan, B., & Humbert, J. (2009). Implementation of wide-field integration of optic flow for autonomous quadrotor navigation. *Autonomous Robots, 27*(3), 189–198.

Fischler, M. A., & Bolles, R. C. (1981). Random sample consensus: A paradigm for model fitting with applications to image analysis and automated cartography. *Communications of the ACM, 24*(6), 381–395.

Fleet, D. J., & Jepson, A. D. (1990). Computation of component image velocity from local phase information. *International Journal of Computer Vision, 5*(1), 77–104.

Garratt, M. A., & Chahl, J. S. (2008). Vision-based terrain following for an unmanned rotorcraft. *Journal of Field Robotics, 25*(4), 284.

Hähnel, D., Burgard, W., & Thrun, S. (2003). Learning compact 3D models of indoor and outdoor environments with a mobile robot. *Robotics and Autonomous Systems, 44*(1), 15–27.

Hanna, K. (1991). Direct multi-resolution estimation of ego-motion and structure from motion. In *Proceedings of the IEEE Workshop on Visual Motion* (pp. 156–162). IEEE.

Heeger, D. J., & Jepson, A. D. (1992). Subspace methods for recovering rigid motion I: Algorithm and implementation. *International Journal of Computer Vision, 7*(2), 95–117.

Horn, B. K. P., & Schunk, B. G. (1981). Determining optical flow. *Artificial Intelligence, 17*, 185–203.

Jiménez, A., & Seco, F. (2005). *Ultrasonic localisation methods for accurate positioning.* Instituto de Automatica Industrial.

Kreczmer, B. (2010). *Objects localisation and differentiation using ultrasonic sensors.* INTECH Open Access Publisher.

Lucas, B., & Kanade, T. (1981). An iterative image registration technique with an application to stereo vision. In *Proceedings of DARPA IU Workshop* (pp. 121–130).

Maimone, M., Cheng, Y., & Matthies, L. (2007). Two years of visual odometry on the mars exploration rovers. *Journal of Field Robotics, 24*(3), 169–186.

Matas, J., & Chum, O. (2005). Randomised RANSAC with sequential probability ratio test. In *Tenth IEEE International Conference on Computer Vision, 2005. ICCV 2005* (Vol. 2, pp. 1727–1732). IEEE.

Scaramuzza, D., & Fraundorfer, F. (2011). Visual odometry [tutorial]. *IEEE Robotics & Automation Magazine, 18*(4), 80–92.

Schnabel, R., Wessel, R., Wahl, R., & Klein, R. (2008). Shape recognition in 3D point-clouds. In *The 16th International Conference in Central Europe on Computer Graphics, Visualization and Computer Vision* (Vol. 8). Citeseer.

Sutton, M., Wolters, W., Peters, W., Ranson, W., & McNeill, S. (1983). Determination of displacements using an improved digital correlation method. *Image and Vision Computing, 1*(3), 133–139.

Takahashi, T. (2007). *2D localisation of outdoor mobile robots using 3D laser range data* (Doctoral dissertation). Carnegie Mellon University.

Tarsha-Kurdi, F., Landes, T., & Grussenmeyer, P. (2007). Hough-transform and extended RANSAC algorithms for automatic detection of 3D building roof planes from lidar data. In *ISPRS Workshop on Laser Scanning 2007 and SilviLaser 2007* (Vol. 36, pp. 407–412).

Weiss, S., Scaramuzza, D., & Siegwart, R. (2011). Monocular-SLAM-based navigation for autonomous micro helicopters in GPS-denied environments. *Journal of Field Robotics, 28*(6), 854–874.

Jiefei Wang research focuses on sensing, real-time image processing, guidance, and control for autonomous systems. He received the master's degree in electrical engineering from Australian National University in 2011, and the Ph.D. degree in electrical engineering from the University of New South Wales in 2016. His research interests include sensing and image processing, scene understanding for obstacle avoidance, control of autonomous systems, and aerial robotics.

Damith Herath is an Associate Professor in Robotics and Art at the University of Canberra. He is a multi-award winning entrepreneur and a roboticist with extensive experience leading multidisciplinary research teams on complex robotic integration, industrial and research projects for over two decades. He founded Australia's first collaborative robotics startup in 2011 and was named one of the most innovative young tech companies in Australia in 2014. Teams he led in 2015 and 2016 consecutively became finalists and, in 2016, a top-ten category winner in the coveted Amazon Robotics Challenge—an industry-focused competition among the robotics research elite. In addition, he has chaired several international workshops on Robots and Art and is the lead editor of the book 'Robots and Art: Exploring an Unlikely Symbiosis'—the first significant work to feature leading roboticists and artists together in the field of robotic art.

Chapter 8
How to Move? Control, Navigation and Path Planning for Mobile Robots

Jiefei Wang and Damith Herath

8.1 Learning Objectives

You will learn about:

- *Controllers and control techniques used in robotics, including the PID controller*
- *Mobile robot locomotion types*
- *Robot path planning and obstacle avoidance.*

8.2 Introduction

When we think of robots, we think of them as manipulators, such as in manufacturing facilities where they are fixed to a location or robots that are moving about (Fig. 8.1). Robots that move around in the environment are called *mobile robots*. This chapter looks at mobile robots, how to control them, different locomotion types and algorithms used for planning paths, and obstacle avoidance while navigating.

J. Wang (✉)
The School of Engineering and Information Technology, University of New South Wales, Canberra, Australia
e-mail: Jiefei.wang@adfa.edu.au

D. Herath
Collaborative Robotics Lab, University of Canberra, Canberra, Australia
e-mail: Damith.Herath@Canberra.edu.au

© The Author(s) 2022
D. Herath and D. St-Onge (eds.), *Foundations of Robotics*,
https://doi.org/10.1007/978-981-19-1983-1_8

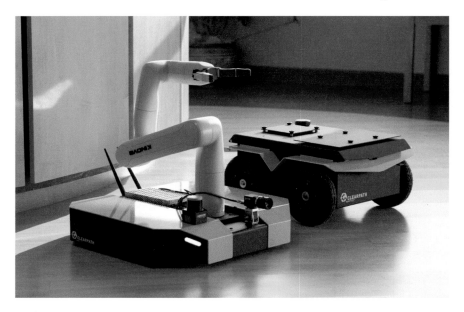

Fig. 8.1 A Kinova Gen3 lite robot arm mounted on a Clearpath Dingo Indoor mobile robotic platform (left) alongside a Jackal Unmanned Ground Vehicle used for outdoor navigation. (right) (Credits Clearpath/Kinova)

An Industry Perspective

Dana Leslie
Former Clearpath Robotics' Employee

Like many young engineers, I have my parents to thank for enabling me to explore the world through robotics. The enjoyment of playing with lego, electronics kits, and computer programming at a young age, was undoubtedly the catalyst that resulted in my career trajectory.

After studying electrical engineering at the University of Victoria, I was fortunate to get a start in the industry by landing my first job at Cellula Robotics, a subsea robotics company. It was here that our team designed, manufactured, and deployed robots to the darkest depths of the ocean, studying and learning about the undersea world!

From water onto land, the robots I've helped design continued to evolve; developing wheeled terrestrial systems at Clearpath Robotics in Ontario, and most recently legged humanoids at Agility Robotics in Oregon.

During the design of a mobile robot, diodes were incorporated into the power system to enable battery hot-swapping. Consequently, the energy generated by back-EMF from the motors (while braking or being pushed) could not be absorbed by the battery. The result was an uncontrolled increase in voltage, causing various subsystems to glitch, with the robot lifelessly rolling to a halt…

This type of challenge is trivial to conceptualise, but much harder to quantify. It's only apparent in a fully integrated system, is correlated to things outside of your control, and is intensified when carrying heavy payloads or traveling down ramps. (Increased mechanical to electrical energy conversion.)

In the end, through comprehensive and iterative testing, the solution was a combination of reducing deceleration rates, varying system capacitance, and utilising transient-voltage-suppression diodes.

It's nice to be able to power your robot by giving it a push, but it's critical that your robot behaves when it's in a hurry to stop.

Innovation in embedded sensing, processing, power electronics, and battery chemistries have collectively advanced the robotics industry throughout my career.

Precise and energy-dense servo-actuators have recently enabled cutting-edge humanoid robotic development that is poised to redefine the workforce; automating the dullest and dangerous of human tasks.

These same actuators have advanced robotic manipulation, enabling the technology to emerge from the factory line and onto the front lines. Robotic arms are no longer just being used to assemble cars, they're being used to flip hamburgers and pack your groceries!

8.3 Mobile Robots

Mobile robots have received much attention in the last few decades due to their ability to explore complex environments such as space, rescue operations, and accomplish tasks autonomously without human effort. Mobile robots can be broadly categorised as wheeled, legged, and flying robots.

8.3.1 Wheeled Robots

Wheeled robots traverse around the ground using motorised wheels to propel themselves and a comparatively easier to design, build, and operate for movement in flat or rocky terrain than robots that use legs or wings. They are also better controlled as they have fewer degrees of freedom than flying robots. One of the challenges of wheeled robots is that they cannot operate well over certain ground surfaces, such as sharp declines, rugged terrain, or areas with low friction. Nevertheless, wheeled robots are the most popular in the consumer market due to the low cost and simplicity of differential steering mechanisms they employ. Although wheeled robots can have any number of wheels, the mechanisms need to be modified to keep dynamic balance based on the number of wheels. Three or four wheels are the most popular and sufficient for static and dynamic balance among all wheeled robots, which are widely used in research projects.

8.3.1.1 Kinematic Modelling

This book primarily discusses two types of robots and their motions, mobile robots and arm type robots. In either type, we need to understand how the movements generated by the actuators translate into complex body movements. To design a robot to act in the environment, we need to understand these geometric relationships of motion.

Kinematics is the study of motions of points, bodies, and systems of bodies (such as robots) without considering the forces acting on these systems. In this chapter, we will discuss some common wheel configurations and their respective kinematic models used in mobile robots that use motors to drive them around. Then, Chap. 10 will delve into modelling kinematics of arm-type robots.

8.3.1.2 Holonomic Drive

Holonomic refers to the relationship between controllable and total degrees of freedom of a robot. If the controllable degree of freedom is equal to the total degrees of freedom, then the robot is said to be Holonomic. A robot built on castor wheels or omniwheels is a good example of a holonomic drive. It can freely move in any direction, and the controllable degrees of freedom is equal to total degrees of freedom.

If the controllable degree of freedom is less than the total degrees of freedom, it is known as *non-holonomic* drive. For example, a car has three degrees of freedom: its position in two axes and orientation. However, there are only two controllable degrees of freedom: acceleration (or braking) and the turning angle of the steering wheel. This makes it difficult for the driver to turn the car in any direction (unless it skids or slides).

For a typical differential drive robot (see Fig. 8.4), the non-holonomic constraint could be written as:

$$\dot{x}\sin\phi - \dot{y}\cos\phi = 0$$

8.3.1.3 Three-Wheeled Robots

One of the most common actuator configurations to drive a mobile robot is the three-wheeled configuration (also known as the tricycle model).

There are two types of three-wheeled robots:

- Differentially steered—two separately powered wheels with an extra free rotating wheel. The robot direction can be changed by varying the relative rate of rotation of the two separately driven wheels. If both the wheels are driven in the same direction and speed, the robot will go straight. Otherwise, depending on the speed of rotation and its direction (Fig. 8.2).
- Two wheels powered by a single actuator and a powered steering wheel.

The centre of gravity in this type of robot has to lay inside the triangle formed by the wheels. If too much weight is allocated to the side of the free rotating wheel, it will cause an imbalance that could make the robot tip over.

Let us now explore how a differentially steered three-wheeled robot could be modelled kinematically.

The model presented in Fig. 8.3 introduces a virtual wheel for the front set of differential drive wheels. The two wheels along the centreline of the robot essentially represent the whole system. With the said constraints, the robot can only exercise two degrees of freedom. Thus, the derivation of the kinematic model refers to the robot's simplified model. The *instantaneous centre of rotation* (also known as the instantaneous velocity centre) in this model refers to an imaginary point attached to the robot where at a given point in time has zero velocity while the rest of the robot body is in planar motion. You could imagine the robot to be rotating around this point at the time instance being considered.

It can be shown that the continuous time form of the vehicle model (with respect to the centre of the front wheel) can be derived as follows:

$$\dot{x}(t) = V(t)\cos(\phi(t) + \gamma(t))$$
$$\dot{y}(t) = V(t)\sin(\phi(t) + \gamma(t))$$
$$\dot{\varphi}(t) = \frac{V(t)\sin(\gamma(t))}{B}$$

where $x(t)$ and $y(t)$ denote the position of the vehicle, the angle $\phi(t)$ is the orientation of the robot with respect to the x-axis, and $V(t)$ represents the linear velocity of the

Fig. 8.2 Differentially steered three-wheeled robot. The front two wheels (top) are powered by two DC motors. A back castor wheel is free to rotate around and is not powered

front wheel. The angle γ is defined as the steer angle of the vehicle. B is the base length between the two sets of wheels.

A simpler kinematic model can be derived from the model discussed earlier in many simple robot configurations where the system makes the velocity of the robot $V(t)$ and the angular velocity of the robot $\dot{\phi}(t)$ directly available (e.g. via wheel encoders). Then the process model for the corresponding system can be represented as follows (Fig. 8.4):

Following simpler equations can be derived then:

$$\dot{x}(t) = V(t)\cos(\phi(t))$$
$$\dot{y}(t) = V(t)\sin(\phi(t))$$
$$\dot{\varphi}(t) = \omega(t)$$

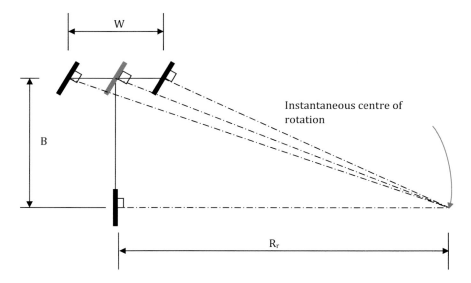

Fig. 8.3   Vehicle geometry of a typical three-wheeled robot

Fig. 8.4   Simplified robot
model

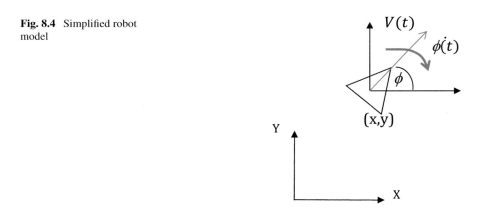

8.3.1.4   Two-Wheeled Robots

Two-wheeled robots are harder to balance than other types because they must keep moving to maintain upright. The centre of gravity of the robot body is kept below the axle. Usually, this is accomplished by mounting the batteries below the body. They can have their wheels parallel to each other, and these vehicles are called dicycles, or one wheel in front of the other, tandemly placed wheels (bicycle). Two-wheeled robots must keep moving to remain upright, and they can do this by driving in the direction the robot is falling. To balance, the base of the robot must stay under its centre of gravity. For a robot that has left and right wheels, it needs at least two

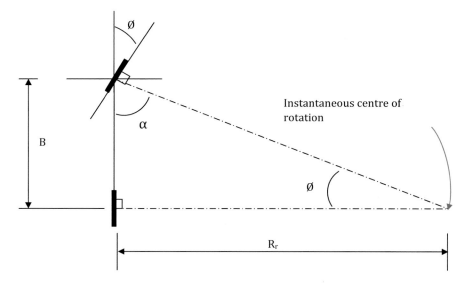

Fig. 8.5 Wheel configuraiton of a two-wheeled bicycle robot

sensors. A tilt sensor is used to determine tilt angle and wheel encoders that keep track of the position of the robot's platform (Fig. 8.5).

where $R_{rr} = \frac{B}{\tan \emptyset}$, $\alpha + \emptyset + 90° = 180°$.

8.3.1.5   Four-Wheeled Robots

There are several configurations possible with four wheels.

- *Two powered and two free rotating wheels*

Same as the differentially steered ones mentioned previously but with two free rotating wheels for extra balance.

Four-wheeled robots are more stable than three-wheeled ones as the centre of gravity has to remain inside the rectangle formed by the four wheels instead of a triangle. Still, it is advisable to keep the centre of gravity to the middle of the rectangle as this is the most stable configuration, especially when taking sharp turns or moving over a non-even surface.

- *Two-by-two powered wheels for tank-like movement*

This type of robot uses two pairs of powered wheels, and each pair turns in the same direction. The tricky part of this kind of propulsion is getting all the wheels to turn with the same speed. If the wheels in a pair are not running at the same speed, the

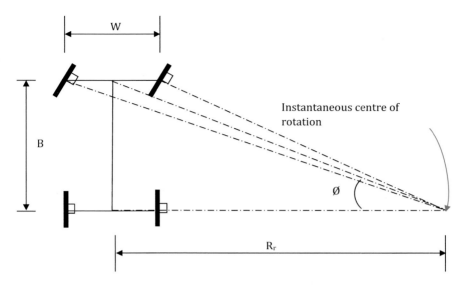

Fig. 8.6 Ackerman drive

slower one will slip. If the pairs do not run at the same speed, the robot is not able to drive straight. A good design has to incorporate some form of car-like steering.

- *Car-like steering (Ackerman drive)*

This method allows the robot to turn the same way a car does (Fig. 8.6). However, this system does have an advantage over previous methods where it only needs one motor to drive the rear wheels and a servo for steering. The previous methods would require either two motors or a highly complex gearbox since they require two output axles with independent speed and direction of rotation.

where $R_{rr} = \frac{B}{\tan \emptyset}$.

8.3.1.6 Omnidirectional Wheels

Omnidirectional (Omni) wheeled robots fall under a class of unconventional mobile robots (Fig. 8.7).

An omniwheel could be thought of as having many smaller wheels making up a large one, and the smaller ones are mounted at an angle to the axis of the core wheel. This allows the wheels to move in two directions and move holonomically, which means it can instantaneously move in any direction, unlike a car, which moves non-holonomicallly and has to be in motion to change heading. In addition, omniwheeled robots can move in at any angle in any direction without rotating beforehand. Some omniwheel robots use a triangular platform, with the three wheels spaced at 60-degree angles. The advantage of using omniwheels is that they make it easier for

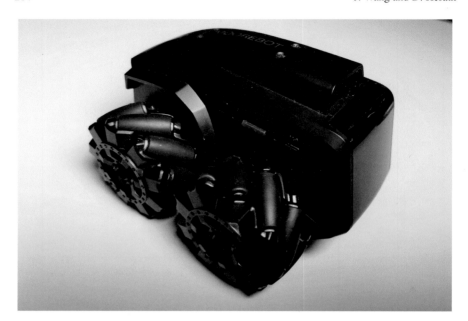

Fig. 8.7 A set of Mecanum wheels (a type of omniwheel) on a home robot

robots to be designed with wheels mounted on an unaligned axis. The disadvantage of using omniwheels is that they have poor efficiency due to not all the wheels rotating in the direction of movement, which also causes loss from friction, and are more computationally complex because of the angle calculations of movement.

8.3.2 Walking Robots

Legged robots are inspired by human beings, legged animals or insects which use leg mechanisms to provide locomotion. Compared with wheeled robots, they are more versatile. They can traverse extreme environments such as unstructured, uneven, unstable, rugged terrain and complex confined spaces such as underground environments and industrial structures.

Legged robots can be categorised by the number of limbs they use. Robots with more legs tend to be more stable, while fewer legs lend themselves to greater manoeuvrability. For a legged robot to keep its balance, it requires maintaining its centre of gravity within its polygon of stability. The polygon of stability is the horizontal surface defined by the leg-ground contact points made by the robot. These multidegrees of freedom legs are usually modelled as kinematics chains which is covered in Chap. 10.

8.3.2.1   Robot Gait

The periodic contact of the robot's legs with the ground is called the gait of the walker. The specific gait depends on the leg configuration of the robot and parameters such as the speed, terrain the robot is moving, intended task and power limitations of the robot. Milton Hildebrand was one of the earliest zoologists to study animal gaits. Various researchers have since adopted his method for gait-pattern specification in robotics, providing a formal method for studying and improving robot gait.

8.3.2.2   Two-Legged Robots

Two-legged robots are also called bipedal robots. The fundamental challenges for two-legged robots are stability and motion control, which refers to balance and movement control. In advanced systems, accelerometers or gyroscopes provide dynamic feedback to control the balance. Such sensors are also used for motion control, walking, jumping, and even running, combined with technologies such as machine learning. On the other hand, the *passive walker* is a bipedal mechanism that "walks" without actuation, simply using gravity as its energy source (Fig. 8.8).

Fig. 8.8   A bipedal robot

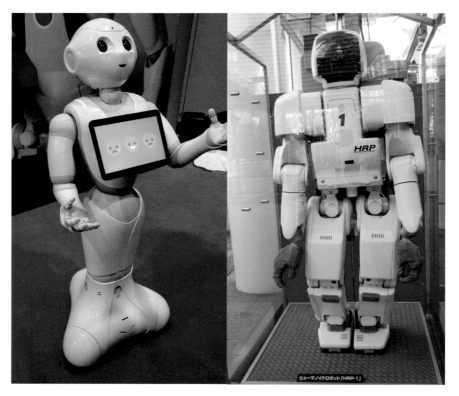

Fig. 8.9 Pepper robot (left)—a wheeled semi-humanoid robot used in retail marketing. The HRP-1 (right)—an early Humanoid Robot Prototype developed by the National Institute of Advanced Industrial Science and Technology (AIST), Japan, on public display at its premises

8.3.2.3   Humanoid Robots

If you close your eyes and think about a robot, what would you picture in your mind? Most likely a fictional creature like Arnold Schwarzenegger in the Terminator series movies or C-3PO from Star Wars. It is likely a *humanoid*—a humanlike robot with a head and body with arms and legs, probably painted metallic silver. Humanoid robots are expected to imitate human motion and interaction (Fig. 8.9) and have their roots in longing and mythmaking, as discussed in our first chapter. With years of research, they are becoming commercially available in several application domains, including in competitive game-playing (such as in the RoboCup humanoid league[1]) and social and interactive robots such as the Pepper (Fig. 8.9) by Softbank Robotics. Strictly speaking, Pepper is a semi-humanoid robot with a wheeled robot base and not a bipedal robot. As mentioned earlier, a wheeled robot is much simpler, stable

---

[1] https://humanoid.robocup.org/.

and economical to produce. How these robots are deployed are constantly expanding, and with the development of new technology, the market will follow suit.

8.3.2.4   Four-Legged Robots

Four-legged robots are also called quadruped robots. They have better stability compared to two-legged robots during movement. Also, the lower centre of gravity and four legs keep them well balanced when they are not moving. They can move either by moving one leg at a time or by moving the alternate pair of legs (Fig. 8.10).

Types of Gait for Four-Legged Robots

Four-legged robots can walk with statically and dynamically stable gaits. In the *statically stable gait*, each leg of the robot is lifted up and down sequentially, and there are three stance legs at least at any moment. This type of gait is called creeping gait (Zhao et al., 2012). *Dynamically stable* gaits are often used in four-legged robots to walk and run due to their efficiencies, such as trotting, pace, bounce, and gallop gait (Fukuoka & Kimura, 2009). In trotting gait, two of the legs are in the same diagonal lift, and the two legs are in contact with the ground until the other two legs lift off, and then repeat the motion two by two in order.

Fig. 8.10 Sony Aibo robot dog—One of the early versions of Sony's four-legged robot dog series

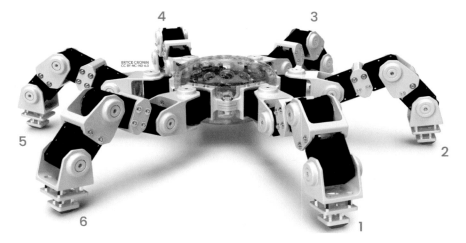

Fig. 8.11 A robot hexapod (Credit Bryce Cronin)

8.3.2.5   Six-Legged Robots

Six-legged robots are also called hexapods. They are designed to mimic the mechanics of insects. Their legs move in a "wave" form from the back to the front. As a result, six-legged robots offer greater stability while moving and standing, they can operate just on three legs, and the remaining legs provide flexibility and increase their capabilities. In Chaps. 12 and 17, you will explore the design and implementation of a hexapod robot (Fig. 8.11).

Types of Gait for Six-Legged Robots

One by one is the simplest gait, which moves each leg forward one after the other in a clockwise or anticlockwise direction while the remaining five legs are in the *stance phase*—not moving. For *a quadruped gait* (Fig. 8.11), the robot moves the front two legs (1 and 2) forward, and the rest (3, 4, 5, 6) support the body, then the robot moves the middle two legs (3, 6) to push the body forward while the rest of the legs (1, 2, 4, 5) support, then swing the last two legs (4 and 5) forward while the other legs support (1, 2, 3, 6) the robot. The pattern is then repeated. The *tripod gait* uses two legs on one side and another on the other side (e.g. 1, 5, and 3), as in a tripod, to hold the robot steady while moving the three remaining legs forward (2, 4, and 6) together.

8.3.2.6   Eight-Legged Robot

Spiders and other arachnids inspire eight-legged robots. Compared with other legged robots, eight-legged robots offer the greatest stability with potential use in more challenging environments such as in hazardous areas to perform reconnaissance, identify structural damages, and perform maintenance tasks.

8.3.3 Flying Robots

Much effort has been devoted to improving the flight endurance and payload of Unmanned Aerial Vehicles (UAVs), commonly known as drones, which has resulted in various configurations in different sizes, capabilities, and endurance. Unlike legged and wheeled robots, flying robots are free to utilise the full six degrees of freedom, allowing for different types of flight for a drone. These are known as Yaw, Pitch, and Roll (Fig. 8.12).

Yaw (ψ) – This is the rotation of the drone's head to either right or left. It is the basic movement to spin the drone. In a remotely piloted drone, this is usually achieved using the left throttle stick by moving to either the left or right.

Pitch (θ) – This is the drone's movement, either forward or backward. The forward pitch is generally achieved in a remotely piloted drone by pushing the throttle stick forward, making the drone tilt and move forward, away from you. Backward pitch is achieved by moving the throttle stick backwards.

Roll (Ø) – Roll makes the drone fly sideways to either left or right. The right throttle stick controls the roll in a remotely piloted drone.

8.3.3.1   Multicopters

A multicopter is a type of flying vehicle with propellers driven by motors (Fig. 8.13). The main rotor blade(s) produces a forceful thrust used for both lifting and propelling the vehicle. Multirotor uncrewed aerial vehicles are capable of vertical take-off and landing (VTOL) and may hover at a place, unlike fixed-wing aircraft. Their hovering

Fig. 8.12 Roll, pitch, and yaw

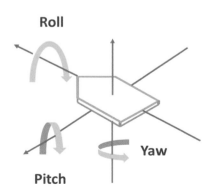

Fig. 8.13 Different types of multicopper (clockwise from top left—A quadrotor—DJI MAVIC PRO, A hexacopter—Custom built model, DJI Phantom Model and An octocopter—Custom built model)

capability and ability to maintain speed make them ideal for civilian fields, monitoring, surveillance, and aerial photography work. One of the challenges with multicopters is that they consume more power, leading to limited endurance. Also, multicopters, unlike fixed-winged counterparts, are inherently aerodynamically unstable and requires an on-board flight controller (an autopilot) to maintain stability.

Multicopters can be divided into specific categories based on the number and positioning of motors, and each category has its own mission (Fig. 8.14). And based on the mission requirements, they are classified in various configurations such as *Monocopter* (1 rotor), Tricopter (3 rotors), quadcopter (4 rotors), *hexacopter* (6 rotors) (X/ + configurations), *Octacopter* (8 rotors) (X/ + configurations), X8-rotor, and Y6-rotor. A quadrotor is a multirotor helicopter lifted and propelled by four rotors. It is a useful tool for university researchers to test and evaluate new ideas in several fields, including flight control theory, navigation, real-time systems.

8.3.3.2 A Quadrotor Example

A quadrotor (drone) is able to perform three manoeuvres in the vertical plane: hover, climb, or descend.

Hover—To hover, the net thrust of the four rotors push the drone up and must be exactly equal to the gravitational force pulling it down.

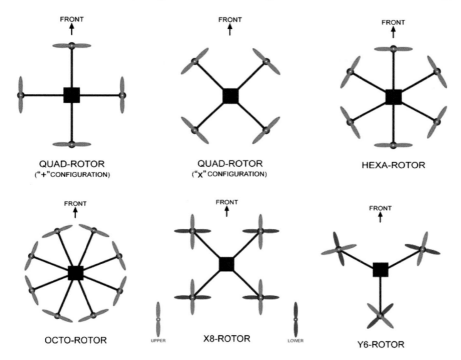

Fig. 8.14 Various configurations possible with the hoverfly multirotor control board (Ed Darack, 2014)

Climb (Ascend)—Increasing the thrust (speed) of the four rotors so that the upward force is greater than the weight and pull of gravity.

Descend—Dropping back down requires doing the exact opposite of the climb, decreasing the rotor thrust (speed) so the net force is downward.

To fly forward, an increase in the quadcopter motor rpm (rotation rate) of rotors 3 and 4 (rear motors) and a decrease in the rate of rotors 1 and 2 (front motors) is required. The total thrust force will remain equal to the weight so that the drone will stay at the same vertical level. To rotate the drone without creating imbalances, a decrease in the spin of motors 1 and 3 with an increase in the spin of rotors 2 and 4 is required (Fig. 8.15).

Mathematical Model of a Quadcopter

The structure of the quadcopter is presented in the below figure, including the corresponding angular velocities, torques and forces created by the rotors (Fig. 8.16).

The absolute linear position ξ of the quadcopter is defined in the inertial frame. Angular position is defined with three Euler angles η. Vector q contains the linear and angular position vectors.

Fig. 8.15 A quadcopter
rotor configuration

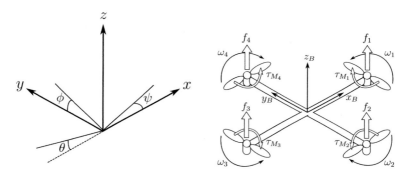

Fig. 8.16 Inertial and body frames of a quadcopter

$$\xi = \begin{bmatrix} x \\ y \\ z \end{bmatrix}, \quad \eta = \begin{bmatrix} \emptyset \\ \theta \\ \psi \end{bmatrix}, q = \begin{bmatrix} \xi \\ \eta \end{bmatrix}$$

The origin of the body frame is in the centre of mass of the quadcopter. In the body frame, the linear velocities are determined by V_B and the angular velocities by υ

$$V_B = \begin{bmatrix} \upsilon_{x,B} \\ \upsilon_{y,B} \\ \upsilon_{z,B} \end{bmatrix}, \quad \upsilon = \begin{bmatrix} p \\ q \\ r \end{bmatrix}$$

The rotation matrix from the body frame to the inertial frame is

$$R = \begin{bmatrix} C_\psi C_\theta & C_\psi S_\theta S_\emptyset - S_\psi C_\theta & C_\psi S_\theta C_\emptyset + S_\psi S_\theta \\ S_\psi C_\theta & S_\psi S_\theta S_\emptyset + C_\psi C_\theta & S_\psi S_\theta C_\emptyset - C_\psi S_\theta \\ -S_\theta & C_\theta S_\emptyset & C_\theta C_\psi \end{bmatrix}$$

where $S_x = \sin(x)$ and $C_x = \cos(S)$. The rotation matric R is orthogonal thus $R^{-1} = R^T$ which is the rotation matrix from the inertial frame to the body frame. The transformation matric for angular velocities from the inertial frame to the body frame is W_η, and from the body frame to the inertial frame is W_η^{-1}:

$$\dot{\eta} = W_\eta^{-1} v \quad \text{then} \quad v = W_\eta \dot{\eta},$$

The quadcopter is assumed to have a symmetric structure with the four arms aligned with the body x- and y-axes. Thus, the inertia matrix is diagonal matrix I in which $I_{xx} = I_{yy}$

$$I = \begin{bmatrix} I_{xx} & 0 & 0 \\ 0 & I_{yy} & 0 \\ 0 & 0 & I_{zz} \end{bmatrix}$$

The inverse of the following equation could be used to solve for the required rotor speeds to achieve the desired thrust (T_Σ) and moments $\tau = (\tau_1, \tau_2, \tau_3)$ of the quadcopter (Mahony et al., 2012);

$$\begin{pmatrix} T_\Sigma \\ \tau_1 \\ \tau_2 \\ \tau_3 \end{pmatrix} = \begin{bmatrix} C_T & C_T & C_T & C_T \\ 0 & dc_T & 0 & -dc_T \\ -dc_T & 0 & dc_T & 0 \\ -C_q & C_q & -C_q & C_q \end{bmatrix} \begin{pmatrix} \omega_1^2 \\ \omega_2^2 \\ \omega_3^2 \\ \omega_4^2 \end{pmatrix}$$

where C_T (>0) and C_q are two coefficients that can be experimentally determined for the considered quadcopter using thrust tests.

8.3.3.3   Fixed Wings

Fixed-wing UAVs require a runway for take-off and landing and also, unlike multi-copters, cannot hover and maintain flight at low speeds. However, they have longer endurance and can fly at high cruising speeds because of the successful generalisation of larger fixed-wing planes with slight modifications and improvements.

Fixed wings are the main lift generating elements in response to forward accelerating speed. The velocity and steeper angle of air flowing over the fixed wings controls the lift produced. Fixed-wing drones require a higher initial speed and a thrust to load ratio of less than 1 to initiate a flight. If fixed-wing and Multirotor are compared for the same amount of payload, fixed-wing drones are more comfortable with less power requirement and thrust loading of less than 1. Rudder, ailerons, and elevators control aircraft orientation in yaw, roll, and pitch angles.

8.3.3.4   Other Flying Robots

There are also some non-conventional configurations of UAVs used for scientific research. They include hybrid, convertible and flapping wing drones that can take off vertically or act as an insect for spying missions. Flapping wing drones inspired by insects such as small dragonflies[2] and birds[3] have regularly appeared in the research literature and at times as commercial prototypes. Due to the lightweight and flexible wings, the flapping drones can contribute well to stable flight in a windy environment. A large amount of research work on flapping wing drones has been carried out by researchers and biologists because of their exclusive manoeuvrability benefits. Blimps and airships are other categories of flying robots that utilise a lifting gas that is less dense than the environment it is operating.

8.4   Controlling Robots

Using the Sense, Think, Act framework, the robot's *controller* can be thought of as the component within the Think element responsible for the robot's movements. It is usually a microcontroller or an onboard computer or a mix of these used to store information about the robot and its surrounding environment and execute designated programmes that operate the robot. The *control system* includes data processing, control algorithms, logic analysis, and other processing activities which enable the robot to perform as designed. Based on the different requirements, more sophisticated robots have more sophisticated control systems.

The control system involves all three aspects of the sense, think, and act loop during execution. First, the perception system provides information about the environment, the robot itself, and the relationship between the robot and the environment. Based on the information from the sensors and the robot's objectives, the cognition and control system must then decide on how to act and what to do to achieve its objectives. The appropriate commands are then sent to the actuators, which move the mechanical structure. The control system coordinates all the input data and plans the robot's motion towards the desired goal.

Various control techniques have been proposed and are being researched. The control strategies of mobile robots can be divided into open-loop and closed-loop feedback strategies. When it comes to *open-loop* control, human operators are involved in sending instructions. The robot relays information to the operator only to perform as instructed. An example of such a system is piloting a drone using a drone controller. The robot's success in achieving its mission is essentially dependent on your piloting skills—the controller simply relays your "intent" to the drone. Most of the time, control commands such as velocities or torques are calculated beforehand, based on the knowledge of the initial and end position ("Goal pose") of the

---

[2] https://spectrum.ieee.org/somehow-an-incredible-robotic-dragonfly-is-now-on-indiegogo.

[3] https://spectrum.ieee.org/festo-bioinspired-robots-bionicswift.

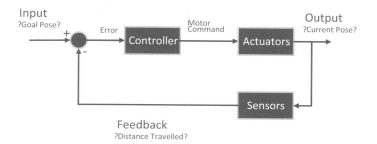

Fig. 8.17 A typical closed-loop feedback controller

robot. However, this strategy cannot compensate for disturbances and model errors ("Error").

On the other hand, closed-loop control strategies could provide the required compensation since the inputs are functions of the actual state of the system and not only of the initial and endpoints. Therefore, disturbances and errors causing deviations from the predicted state are compensated by real-time sensor data ("Feedback") (Fig. 8.17). Formally, we could define a feedback controller as enabling a robot to reach and maintain the desired state (called a *set point*) by repeatedly comparing its current state with the desired goal state. Here, *feedback* refers to the information that is literally "fed back" into the system's controller. When a system is operating at the desired state, it is said to be operating at the *steady state*.

8.4.1 PID Controllers

A *PID controller* is a control loop feedback mechanism that calculates the difference between a desired value (*setpoint*) and the actual output from a process and use that result to apply a correction to the process. The term PID stands for **P**roportional–**I**ntegral–**D**erivative feedback control, and it is one of the most commonly used controllers in the industry. It is the best starting point when designing an autonomous control system and is very popular in commercial autopilot systems and open-source developments.

The main goal of this process is to maintain a specified setpoint value. For example, you may want a DC motor to maintain a setpoint value $r(t)$ of 600 encoder pulses per second. The actual motor speed $y(t)$, called the process variable, is subtracted from the setpoint value 600 to find the error value $e(t)$. The PID controller then computes the new control value $u(t)$ to apply to the motor based on the computed error value. In the case of a DC motor, the control value would be a pulse-width-modulated (PWM) signal. The (t) represents a time parameter being passed into the process (Fig. 8.18).

Let us now look at how each of the three elements, P, I, D, contributes to the overall controller.

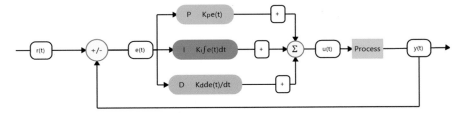

Fig. 8.18 A PID controller—$r(t)$ is the reference setpoint, $e(t)$ is the difference between the process output and the desired setpoint, $u(t)$ is the process input control value, $y(t)$ is the process output

8.4.1.1 Proportional Control (P)

This element takes some proportion of the current error value. The proportion is specified by a constant called the gain value, and a proportional response is represented by the letters Kp. As an example, Kp may be set to 0.25, which will compute a value of 25% of the error value. This is used to compute the corrective response to the process. Since it requires an error to generate the proportional response, there is no proportional part of the corrective response if there is no error. For example, when controlling a drone autonomously, increasing the P gain Kp typically leads to shorter rise time (i.e. the drone reaches the required altitude quickly) and larger overshoots. Although it can decrease the system's settling time, it can also lead the drone to display highly oscillatory or unstable behaviour (Fig. 8.19).

8.4.1.2 Derivative Control (D)

The derivative term is used to estimate the future trends of the error based on its current rate of change. It is used to add a dampening effect to the system such that the quicker the change rate, the greater the controlling or dampening effect. In that sense, increasing the D gain Kd typically leads to smaller overshoot and a better-damped behaviour. However, increasing Kd could lead to larger steady-state errors (Fig. 8.20).

8.4.1.3 Integral Control (I)

Element I takes all past error values and integrates them over time. The term integrates simply means to accumulate or add up. This results in the integral term growing until the error goes to zero. When the error is eliminated, the integral term will stop growing. If an error still exists after the application of proportional control, the integral term tries to eliminate the error by adding in its accumulated error value. This will result in the proportional effect diminishing as the error decreases, and the growing integral effect compensates for this. Increasing the I gain Ki leads to a

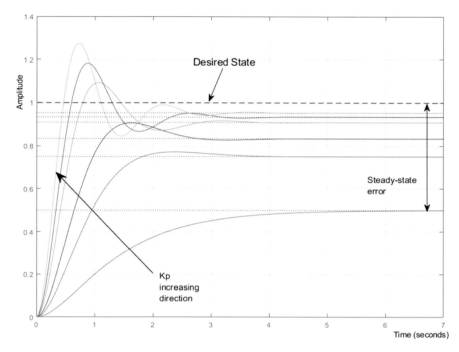

Fig. 8.19 An example showing the effects of increasing Kp—shorter rise time but oscillatory behaviour increasing. (No Integral and Derivative control)

reduction in the steady-state error (often elimination) but also could lead to larger oscillations (Fig. 8.21).

Another issue to be mindful of when using the integral term in a controller refers to *Integral windup*. This is common in most physical systems (nonlinear systems), where a significant change in the setpoint (either positive or negative) results in the integral term accumulating significant errors that cannot be offset by errors in the opposite direction leading to a loss of control. Researchers have developed several anti-windup techniques over the years to counter the phenomenon. One common technique is setting boundaries for the integral term depending on the known system limitations, such as actuator operational range.

8.4.1.4   Tuning a PID Controller

As understood from this brief overview of the role of each element of the PID controller, it is not possible to independently tune the three different gains. Each of them aims to offer the desired response characteristic (e.g. faster response, damped and smooth oscillations, near-zero steady-state error) but has a negative effect that must be compensated by re-tuning another gain. Therefore, PID tuning is a highly

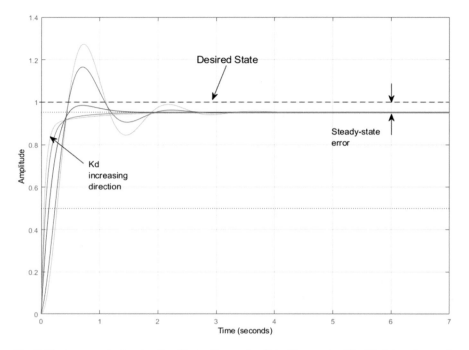

Fig. 8.20 An example showing the effects of increasing Kd with a constant Kp (No Integral control)

coupled and iterative procedure. The PID controller consists of the additive action of the Proportional, the Integral, and the Derivative component. Not all of them have to be present; therefore, we often employ P controllers, PI controllers or PD controllers when a simpler controller yields the desired result.

8.4.2 Fuzzy Logic Controllers

The fuzzy logic theory was developed in the mid-1960s as a way to deal with the imprecision and uncertainty inherent to perception systems. Since then, it has been used in many engineering applications. Designers consider it one of the simpler solutions available for many nonlinear control problems, including most robotics navigation and control problems. Fuzzy logic is more advantageous than traditional solutions because it allows computers to act more like humans, responding effectively to complex inputs to deal with linguistic notions such as "too hot", "too cold" or "just right". Furthermore, fuzzy logic is well suited to low-cost implementations based on cheap sensors, low-resolution analog-to-digital converters, and 4-bit or 8-bit microcontroller chips. Such systems can be easily upgraded by adding new rules to improve performance or by adding new features. In many cases, fuzzy control can

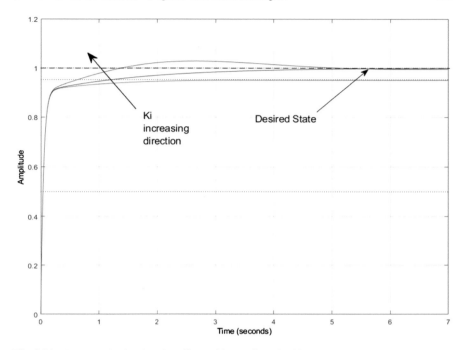

Fig. 8.21   An example showing the effects of increasing Ki with a constant Kp and Kd

improve existing traditional control systems by adding an extra layer of intelligence to the current control method.

8.4.2.1   A Simple Example

Consider a ground robot moving towards a target.

The fuzzy logic controller (FLC) used has two inputs: error in distance (Δe_d) and error in the angle of orientation (Δe_a) of the robot. The controller's output (that is, the control signals) would be pulse-width-modulated signals to control the angular velocity of the two servo wheels. Therefore, the fuzzy logic controller is a two-input, two-output system. The block diagram of the robotic system is shown in Fig. 8.22.

8.5   Path Planning

Path planning is the means of finding a suitable (optimal) path for a moving platform to travel from its starting point to the goal point in a given environment. Early work on path planning focused on planning paths for robotic manipulators, where a perfect

Fig. 8.22 Fuzzy logic control system

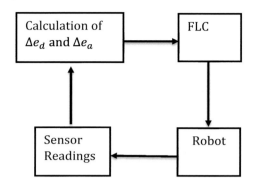

world model and precise knowledge of the joint angles were assumed. However, these assumptions cannot be made for mobile robots operating in partly known or unknown environments and with localisation uncertainties.

Classical algorithms, such as Dijkstra's algorithm (Dijkstra, 1959), A and A* algorithms (Hart et al., 1968), apply a global graph search to find the least-cost path from the starting point to the target point. There are also other methods for sampling the local environment to determine the least-cost path (Kuffner & LaValle, 2000). The main purpose of obtaining the best path is to find the shortest path with minimal energy usage and maximum coverage of an area or optimised predicted perception quality. In some situations, it is beneficial to choose from a given set of trajectories that can be followed by the robot's controller rather than planning a specific and maybe impossible path (Dey et al., 2011). Therefore, different path planning algorithms are used for different situations, with most algorithms relying on heuristic and probabilistic techniques.

8.5.1 Heuristic Path Planning Algorithms

Heuristic methods use an estimated cost function for target-oriented path searching which considerably reduces the computational time. These algorithms calculate the path based on the fewest number of grid cells in the queue by assigning a cost to each node with respect to the difference of its distance from that of the minimal distance between the starting and goal nodes.

8.5.1.1 A* Algorithm

The most well-known path planning algorithm is the A* algorithm (Hart et al., 1968) which uses a best-first search method to find the least-cost path from the starting to the goal node. Unlike other path planning techniques, we can consider that the

A* algorithm has a "brain" that can do the calculations. It is widely used for games and web-based maps to find the shortest path in a very efficient way. The vehicle traverses towards the goal node until it either reaches it or determines that there is no available path with a heuristic function used to evaluate the goodness of each node.

Considering a graph map with multiple nodes, what the A* algorithm does is that at each step, it picks the node according to the value "f", which is equal to the sum of "g" and "h". At each step, it picks the node having the lowest "f" value and proceeds to the next until it finds the goal point.

$$f(\text{node}) = g(\text{node}) + h(\text{node})$$

where:

$g(\text{node})$ is the travelling cost from the initial point to the current point; $h(\text{node})$ is the heuristic function that includes the cost from the starting node to the current location, $c(n, n')$ and estimated cost from the current location to goal $h(n')$.

A* (star) Pathfinding Pseudocode

```
// Initialise both open and closed list
let the openList and closedList equal empty list of nodes

// Add the start node
put the startNode on the openList (leave it's f at zero)

// loop until find the end
while the openList is not empty

// Get the current node
let the currentNode equal the node with the least f value
remove the currentNode from the openList
add the currentNode to the closedList

// Found the goal
if currentNode is the goal
Goal found! Backtrack to get path

// Generate children
let the children of the currentNode equal the adjacent nodes
for each child in the children
```

```
// Child is on the closedList
if child is in the closedList
continue to beginning of for loop

// Create the f, g, and h values
child.g = currentNode.g + distance between child and current
child.h = distance from child to end
child.f = child.g + child.h

// Child is already in openList
if child.position is in the openList's nodes positions
if the child.g is higher than the openList node's g
continue to beginning of for loop

// Add the child to the openList
add the child to the openList
```

For example:

We would like to find the shortest path between A to K in the following map. The number written with red is the distance between the nodes, and the number in the blue circle written in black is the heuristics value. A* uses $f(n) = g(n) + h(n)$ to find the shortest path.

Let's start with start point A. A has three nodes: B, E, and F, then we can start calculate $f(B), f(E)$, and $f(F)$:

$$f(B) = 3 + 8 = 11$$
$$f(E) = 1 + 1 = 2$$
$$f(F) = 5 + 4 = 9$$

$f(E) < f(F) < f(B)$, so we will choose E as the new start node.

For node E, it two nodes F and $H, f(F) = 7 (1 + 6) + 4 = 11, f(H) = 3 (1 + 2) + 4 = 7, f(H) < f(F)$, so we will choose H as the new start node.

For node H, it has two nodes J and $I, f(J) = 5 (1 + 2 + 2) + 3 = 8, f(I) = 4 (1 + 2 + 1) + 2 = 6, f(I) < f(J)$, so we will choose I as the new start node.

For node I, it has two nodes D and $K, f(D) = 10 (1 + 2 + 1 + 6) + 5 = 15, f(K) = 6(1 + 2 + 1 + 2) + 0 = 6, f(K) < f(D)$, so we will choose K as the next node, as K is the goal point, the algorithm stop here.

The shortest path from A to K is $A—E—H—I—K$ (Fig. 8.23).

The A* algorithm is similar to *Dijkstra's algorithm* (Dijkstra, 1959), except that it guides its search towards the most promising states, which can save a significant amount of computational effort. The limitation of the above approaches is that they need a complete map of the area under exploration. However, when operating in real-world scenarios, as new information might be added to the map, replanning is

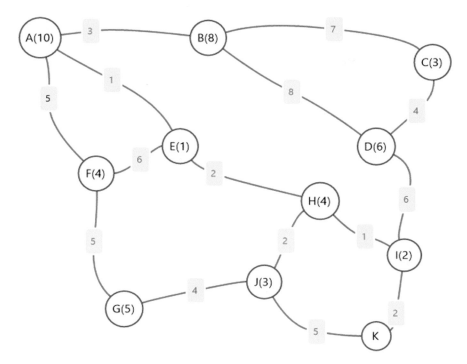

Fig. 8.23 A* algorithm example

essential. While A ∗ could be used to plan from scratch for every update, this is computationally expensive.

Instead, the D* Lite (Koenig & Likhachev, 2005) and Focussed Dynamic A* (D*) (Stentz et al., 1995) algorithms search for a path from the goal towards the start and update nodes only when changes occur. An updated path is calculated based on the previous path, which is much more effective than the A ∗ algorithm and Dijkstra's algorithm. *D* Lite algorithm* is one of the most popular goal-directed navigation algorithms that is widely used for mobile robot navigation in unknown environments. It is a reverse searching method and can replan from the current position when new obstacles are blocking the path.

Finally, *Field D** is an interpolation based path planning and replanning algorithm (Ferguson & Stentz, 2006). In contrast to other methods in which nodes are defined as the centres of grids, it defines nodes on the corners of grids. Then linear interpolation is used to create waypoints along the edges of grids which allows the planning of direct, low-cost, smooth paths in non-uniform environments. D* and its variants are widely used for autonomous robots, including Mars rovers and autonomous cars (Stentz & Hebert, 1995; Urmson et al., 2008).

8.5.2 Probabilistic Path Planning Algorithms

Probabilistic approaches sample the configuration space randomly, which helps to decrease the path planning time and memory usage. However, their main disadvantage is that they cannot always be guaranteed to find the optimal path.

Much work has been conducted based on probabilistic path planning methods. One of the most popular approaches is the *probabilistic roadmap* (PRM) algorithm (Kavraki & Latombe, 1998; Kavraki et al., 1996) which generally consists of two phases: firstly, it randomly samples points in the configuration space to build a roadmap graph and then connects the sampled configurations to their neighbours; and secondly, in the query phase, the starting and goal nodes are connected to their neighbours in the graph and the path calculated using a heuristic method. Although any existing path can be found if there is a sufficiently increasing number of samples, as situations such as narrow corridors in large environments can rapidly increase the path planning time, deliberate sampling strategies are necessary. While multiple queries can be executed on the same graph-based on PRMs, some pre-processing is needed during which, in some cases, obstacles are defined.

8.6 Obstacle Avoidance

In mobile robotics, the goal of obstacle avoidance is generally to navigate from one location to the goal location while avoiding collisions with obstacles during the robot motion in a known or unknown environment. Therefore, obstacle avoidance is almost always is combined with path planning. The process requires an understanding of the environment, such as a full map or partial map, a target location and robot's location (localisation) (discussed in the next chapter), and sensors such as cameras or laser sensors to provide obstacle information.

Obstacle avoidance is always comprised of obstacle detection and collision avoidance. There are varieties of algorithms that use different kinds of sensors and techniques to achieve the goal of obstacle detection. The processed data received from sensors are then sent to the controller to operate the robot to avoid obstacles. There are some widely used obstacle avoidance algorithms such as bug algorithms, VFH, and other proximity-based techniques (e.g. sonar, bumper sensors).

8.6.1 Bug Algorithm

The bug algorithms are the simplest obstacle avoidance method among all obstacle avoidance methods. In the bug algorithm, the main idea is to track the contour of the obstacles found in the robot's path and make the robot circumnavigate it (Lumelsky,

2005; Lumelsky & Stepanov, 1987). There are several modified versions of the bug algorithm, such as Bug 1, Bug 2, DistBug, and Tangential Bug algorithm.

Bug 1 algorithm is the simplest of all Bug algorithm variations. It reaches the goal almost all the time with high reliability. But the matter of concern with this method is efficiency. The robot moves on the shortest path joining the robot's position X and goal location until it encounters a hurdle in the path. When an obstacle confronts it, it starts revolving around its surface and calculates the distance from the destination point. After one complete revolution, it figures out the point of departure closest to the goal. Then, it maintains or changes the direction of motion depending on the distance of leaving point from the hit point. This method can be illustrated in the following steps:

- Head towards the goal
- If an obstacle is encountered, circumnavigate it and remember how close you get to the goal
- Return to that closest point and continue

Robot revolves around every obstacle on the way towards the goal, increasing the computational efforts. But ease of implementation makes it worth it when only completion of the task is required irrespective of time.

Generally speaking, the bug algorithms work well with single obstacle avoidance. However, these bug algorithms are not very reliable in a more complex and cluttered environment, and in some tricky conditions, one version works better than the other version.

8.6.2 The Vector Field Histogram (VFH)

Vector field histogram is a real-time obstacle avoidance method for mobile robots developed by Borenstein and Koren (1991). This method contains three major components that help to achieve obstacle avoidance. Firstly, the robot generates a two-dimensional sensory histogram around its body or within a limited angle and starts updating the histogram data at every stage. Secondly, the two-dimensional histogram data are converted into a one-dimensional polar histogram. Finally, it selects the lower polar dense area and moves the vehicle, calculating the direction.

This approach overcomes the issue of sensor noise. A histogram is a graph between probabilities of the presence of obstacles to the angle associated with the sensor reading. The probabilities are obtained by creating a local occupancy grid map (see Chap. 9) of the environment of the robot's surroundings. The histogram is used to discover all the passages large enough to allow the robot to pass through. The selection of path is based on a cost function which is a function of the alignment of the robot's path with the goal and on the difference between the current wheel orientation and the new direction. A minimum cost function is desirable. One of the advantages of using VFH is that it conquers the problem of sensor noise by making a polar histogram that represents the probability of obstacle of a particular angular direction. Some demerits need to be taken into consideration when using this technique, such as VFH does not guarantee the completeness, which can lead to an unfinished task. It can be problematic to pass through a narrow passage using this method. Moreover, it does not consider the robot's dynamics and its environment, making it not ideal for use in a complex dynamic environment.

8.7 Chapter Summary

Robots that move around in the environment instead of being fixed to a single location are called mobile robots. These can be categorised according to the type of locomotion they utilise, such as wheeled, legged, or flying.

A robot controller essentially provides the controlling commands to its actuators to drive the robot towards the desired goal. A common control loop is the PID (proportional–integral–derivative) controller, which uses sensor feedback to update the control signal in a repeated manner. Essentially the controller applies a correction to a control function where the correction could be proportional to the error (P) or reflective of the cumulative error (I) or the change in the error rate (D). A PID controller requires tuning of its parameters, which usually requires an iterative trial and error approach or sophisticated tuning algorithms to realise optimal performance.

For a robot to move from a given point to the desired goal point, it needs to plan a path between the two points using some optimal criteria, for example, shortest distance, the lowest energy consumption, or the largest area coverage. Many techniques have evolved over the years, including heuristic and probabilistic techniques, each having its own merits and concerns. Additionally, a complimentary problem in path planning is the obstacle avoidance problem. Again, researchers have come up with various strategies and techniques to solve the problem.

As a roboticist developing a mobile robot, your task is to select, develop, and implement techniques, algorithms, and platforms based on the ideas discussed in this chapter to suit the requirements of the job at hand.

8.8 Review Questions

- If using a PID controller for a drone, increasing the P gain Kp typically leads to shorter or longer rise times?
- If using a PID controller for a drone, increasing the I gain KI, would it result in smaller or larger oscillations?
- Comparing two-wheeled, three-wheeled, four-wheeled robots, which one is the most unstable type?
- What does pitch, yaw and roll mean in a drone?
- What is the difference between classic and heuristic path planning algorithms?

8.9 Further Reading

The chapter covered introductory material on several related topics. Once the basic concepts are well understood, you can explore these topics in more depth and expand onto advanced topics. Following titles, *Introduction to Robotics: Mechanics and Control* (3rd Edition) by John Craig, *Modern Robotics Mechanics, Planning, and Control* by Kevin M. Lynch and *Robotics Modelling, Planning and Control* by Bruno Siciliano provide some excellent reading. Another highly recommended book on mobile robots is the book by Roland Siegwart, *Introduction to autonomous mobile robots.*

References

Borenstein, J., & Koren, Y. (1991). The vector field histogram-fast obstacle avoidance for mobile robots. *IEEE Transactions on Robotics and Automation, 7*(3), 278–288.

da Silva, L. R., Flesch, R. C. C., & Normey-Rico, J. E. (2018). Analysis of anti-windup techniques in PID control of processes with measurement noise. *IFAC-PapersOnLine 51*(4), 948–953.

Dey, D., Liu, T. Y., Sofman, B., & Bagnell, D. (2011). Efficient optimisation of control libraries. Technical report, DTIC Document.

Dijkstra, E. W. (1959). A note on two problems in connexion with graphs. *Numerische Mathematik, 1*(1), 269–271.

Ed, Darack, https://www.airspacemag.com/flight-today/build-your-own-drone-180951417, 2014.

Ferguson, D., & Stentz, A. (2006). Using interpolation to improve path planning: The field D* algorithm. *Journal of Field Robotics, 23*(2), 79–101.

Fukuoka, Y., & Kimura, H. (2009). Dynamic locomotion of a biomorphic quadruped "Tekken" robot using various gaits: Walk, trot, free-gait and bound. *Applied Bionics & Biomechanics, 6*(1), 63–71.

Hart, P. E., Nilsson, N. J., & Raphael, B. (1968). A formal basis for the heuristic determination of minimum cost paths. *IEEE Transactions on Systems Science and Cybernetics, 4*(2), 100–107.

https://en.wikibooks.org/wiki/Robotics/Types_of_Robots/Wheeled, 2021.

Kavraki, L. E., Svestka, P., Latombe, J.-C., & Overmars, M. H. (1996). Probabilistic roadmaps for path planning in high-dimensional configuration spaces. *IEEE Transactions on Robotics and Automation*, 12(4), 566–580.

Kavraki, L. E., & Latombe, J. -C. (1998). Probabilistic roadmaps for robot path planning.

Koenig, S., & Likhachev, M. (2005). Fast replanning for navigation in unknown terrain. *IEEE Transactions on Robotics, 21*(3), 354–363.

Kuffner, J. J., & LaValle, S. M. (2000). Rrt-connect: An efficient approach to single-query path planning. In *Proceedings. ICRA'00 IEEE international conference on robotics and automation, 2000,* (vol 2, pp. 995–1001). IEEE.

Lumelsky, V. J. (2005). *Sensing, Intelligence, Motion: How Robots and Humans Move in an Unstructured World*. John Wiley & Sons.

Lumelsky, V. J., & Stepanov, A. A. (1987). Path-planning strategies for a point mobile automaton moving amidst unknown obstacles of arbitrary shape. *Algorithmica, 2*, 403–430.

Mahony, R., Kumar, V., & Corke, P. (2012). Multirotor aerial vehicles: Modeling, estimation, and control of quadrotor. *IEEE Robotics & Automation Magazine, 19*(3), 20–32. https://doi.org/10.1109/MRA.2012.2206474

Stentz, A., et al. (1995). The focussed D* algorithm for real-time replanning. *In IJCAI, 95*, 1652–1659.

Stentz, A., & Hebert, M. (1995). A complete navigation system for goal acquisition in unknown environments. *Autonomous Robots, 2*(2), 127–145.

Urmson, C., Anhalt, J., Bagnell, D., Baker, C., Bittner, R., Clark, M., Dolan, J., Duggins, D., Galatali, T., Geyer, C., et al. (2008). Autonomous driving in urban environments: Boss and the urban challenge. *Journal of Field Robotics, 25*(8), 425–466.

Zhao, D., Jing, X., Dan, W., et al. (2012). Gait Definition and successive gait-transition method based on energy consumption for a quadruped. *Chinese Journal of Mechanical Engineering, 25*(1), 29–37.

Jiefei Wang 's research focuses on sensing, guidance, and control for autonomous systems. He received the master's degree in electrical engineering from Australian National University in 2011, and the Ph.D. degree in electrical engineering from the University of New South Wales in 2016. His research interests include sensing and image processing, scene understanding for obstacle avoidance, control of autonomous systems, and aerial robotics.

Damith Herath is an Associate Professor in Robotics and Art at the University of Canberra. Damith is a multi-award winning entrepreneur and a roboticist with extensive experience leading multidisciplinary research teams on complex robotic integration, industrial and research projects for over two decades. He founded Australia's first collaborative robotics startup in 2011 and was named one of the most innovative young tech companies in Australia in 2014. Teams he led in 2015 and 2016 consecutively became finalists and, in 2016, a top-ten category winner in the coveted Amazon Robotics Challenge—an industry-focused competition amongst the robotics research elite. In addition, Damith has chaired several international workshops on Robots and Art and is the lead editor of the book *Robots and Art: Exploring an Unlikely Symbiosis*—the first significant work to feature leading roboticists and artists together in the field of Robotic Art.

Chapter 9
Lost in Space! Localisation and Mapping

Damith Herath

9.1 Learning Objectives

In this chapter, you will learn about:

- The robot localisation problem
- The robot mapping problem
- The Simultaneous Localisation and Mapping (SLAM) problem
- Common probabilistic state estimation techniques
- The Kalman filter and the role of the extended Kalman filter as a recursive state estimator in nonlinear systems.

9.2 Introduction

Imagine you are visiting a new city or country. Perhaps, if you are like me, one of the first things you might do is download or print a copy of the local area map. Or, perhaps make sure that the navigation app or the GPS on your phone is up to date with the latest map. But, while you are travelling across the new city, do you remember the time when you got lost? Even with the latest maps?

Similarly, have you ever wondered how a self-driving car knows where it is going?

A typical mobile robotic system architecture is shown in Fig. 9.1. It consists of several sensors, planning and control modules and actuators. While specific instantiations of these components will be application and platform-dependent, a typical mobile robotic system requires these building blocks to function. First of all, internal and external sensors provide information about the robot and the physical world it inhabits. Next, this information is interpreted by various algorithms to estimate the

D. Herath (✉)
Collaborative Robotics Lab, University of Canberra, Canberra, Australia
e-mail: Damith.Herath@Canberra.edu.au

© The Author(s) 2022
D. Herath and D. St-Onge (eds.), *Foundations of Robotics*,
https://doi.org/10.1007/978-981-19-1983-1_9

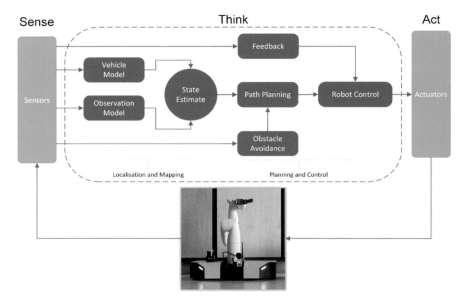

Fig. 9.1 A high-level overview of a mobile robot system from the sensors to algorithms to actuators

vehicle's state and its environment. The state estimate is then used to plan the robot's actions and generate commands for the actuators. We looked at sensors, control, path planning and obstacle avoidance in the previous chapters.

This chapter explores localisation and mapping, once considered the holy grail of robotics, which are two fundamental capabilities that any autonomous mobile robot requires to navigate in the wild, including self-driving cars (and Mars rovers, too!).

An Industry Perspective

Guillaume Charland-Arcand

ARA Robotics

I was exposed first during my CEGEP years in a small club where we were building sumo robots for robotics competitions. The competition was simple, 2 robots faced each other on a circular black ring with a small white bar that delimited the edge, the goal was for one robot to push the other one out of the ring. My robot was very simple, big motors, big wheels, a few sensors, an 8-bit microcontroller, my own circuit board, and a few lines of codes. I was not very successful in the competitions, but building a thing on your own, mixing mechanics, electronics, software, and seeing something move on its own, was pretty cool. But I felt I did not know enough, so at the university, I decided to join a scientific club focusing on multirotor UAVs, which was pretty new at the time. I fell in love with this branch of robotics instantly. It was mobile robots, like my good old sumo robot, but on steroids. Everything was harder; more vibration, complicated nonlinear dynamics, limited payload capacity and it's flying!

At the time of doing my master's, working with UAV was hard because of the lack of resources. I had convinced my supervisor to buy equipment, but based on the budget, we could only afford 1 UAV. This made things a lot more complicated for me, because, one mistake, one line of code in error, and the UAV crashes. Working on control law design made this even more problematic. Everybody that worked a bit in control theory has experienced this: it's always fine on paper and in simulations, but there is always the small caveat of finding the controller gains, which is done through experimentation typically. The challenge was to tune my controller and validate my controller software without breaking the only UAV I had. This is where I got introduced to safety-critical engineering and its practices, i.e., how to design software and hardware in a systematic fashion to guarantee that it won't fail. I did not go as far as following DO-178 standards, but it provided new insight on how to develop robotic products and applications.

When I started, SLAM was starting to be applied on UAVs. A few ROS packages existed, but it was mostly in 2D, using Hokuyo scanning laser rangefinder. There were also a few successful demonstrations of autonomous UAVs operating in GNSS denied environments, but it was mainly prototyped in experimental settings. Now, companies such as Exyn Technologies and Skydio, provide products for industrial applications that have a very high level of autonomy. These systems can generate extremely precise 3D maps, detect static and dynamic obstacles and plan paths through unknown environments at high velocities. I would say the last 5 years' technological improvements are major and are a great bedrock from the innovation to come because the task of autonomy is very complex and not completely solved yet.

9.3   Robot Localisation Problem

For any autonomous mobile robot to be successfully deployed, it requires to answer the question 'where am I?'. For example, in Fig. 9.2 a robot travelling in a local coordinate frame. In Fig. 9.2, a robot is shown at time $t = k$ at an unknown location. The robot localisation problem is to find the current coordinates and the heading (the direction which the robot is facing) of the robot with respect to a given local coordinate frame. The local coordinate frame is usually fixed at the location where

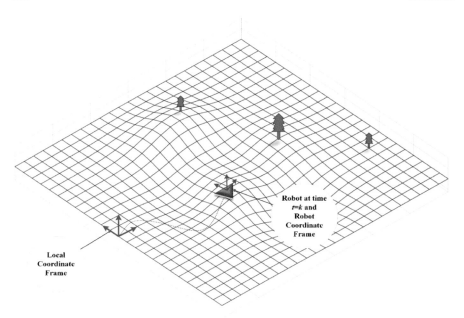

Fig. 9.2 A robot travelling in a local coordinate frame

the robot was at time $t = 0$, though this could be arbitrarily selected. The robot also carries its own coordinate frame. The heading is usually defined as the angular difference between the x-axis of the local coordinate frame and the x-axis of the robot coordinate frame. The localisation problem is said to be solved for a given time $t = k$ when the current (x, y, z) coordinates and the heading (Ψ) of the robot with respect to the given local coordinate frame are fully known.

In the example shown, the robot knows its starting position. The localisation problem could then be thought of as a *tracking problem*, where the requirement is to track the robot's movement from the beginning with respect to its initial position. On the other hand, if the initial position is not known, then the localisation problem is considered to be a *global positioning problem*. An example of this could be a robot being turned on at an arbitrary location without knowing its initial position.

Another related problem is the *'kidnapped'* robot problem, where a properly localised robot suddenly gets moved to a different location without being aware of the move. An example of this could be a vacuum cleaning robot starting from its charging station (known local coordinate frame) and suddenly being picked up by a user and placed in another room (who would do that?).

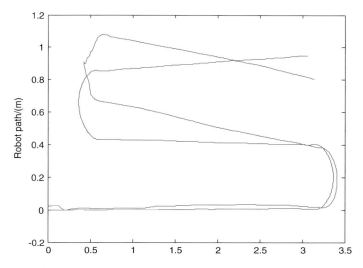

Fig. 9.3 True path of a robot (blue) compared to the odometry-based estimate of the path

9.3.1 Odometry-Based Localisation

One of the most common techniques used in robotics to extract the current location of a robot is to use wheel encoders. Odometry, the word derived from Greek roots (*odos*—street and *metron*—to measure) simply refers to any technique that uses motion information to derive relative position estimates of the robot. Similar to the odometer available on a vehicle that indicates the distance the vehicle has travelled (either the absolute distance since its manufacture or the relative distance from an arbitrary starting point), the wheel encoders could be used to calculate relative travel distances and heading of the robot based on the known wheel geometry and dynamics. This process is sometimes referred to as *dead reckoning*. However, due to sensor errors, slippage and other noise elements inherent in such systems, odometry accumulates errors over time. As shown in Fig. 9.3, a robot could lose track of its location relatively quickly if it only relied on odometry. Nevertheless, odometry is extensively used in robotics to acquire short-term localisation information.

9.3.2 IMU-Based Odometry

Inertial measurement units (IMU) are devices that integrate several sensors in a single package, including accelerometers and gyroscopes. IMUs could be used to measure the robot's linear acceleration in three dimensions (using a tri-axial accelerometer) and rotational rate (using gyroscopes). By appropriately integrating this information, it is possible to derive the robot's speed and travel distance so the robot's relative location and heading information can be worked out. However, IMU units suffer from 'drift' where they accumulate errors over time.

9.3.3 Visual Odometry

Visual odometry is an alternate technique that uses cameras and computer vision to derive odometric information. Various computer vision techniques have been used to estimate the motion of robots. Generally, these techniques attempt to understand the relative changes in images between subsequent frames due to movement. A common approach is to track a set of image features across frames (see Chap. 7). These techniques are sometimes coupled with an IMU to improve the estimates. Such systems are usually called visual inertial odometry.

9.3.4 Map-Based Localisation

The previous techniques described for solving the localisation problem only provide relative information. In other words, these techniques require the integration of a series of measurements to derive the robot's current location. An alternative technique is to use an external map to make direct observations to a series of external *landmarks* (also called *beacons* or *features*) using a sensor mounted on the robot to infer the robot's current location with respect to these landmarks based on *a-priori* map. That is if you know the locations of a set of previously identified landmarks in the environment with respect to a global/local coordinate frame (i.e. someone has already built a local map of the environment) and if you have a sensor capable of re-identifying and measuring relative location of these landmarks with respect to its position, then it is possible to localise your robot within the given coordinate frame using *triangulation*. Triangulation uses two known locations to 'triangulate' or work out the location of a third point using geometry. In order to triangulate, the distances and the angles to the known locations (landmarks) with respect to the unknown location must be measured (see Fig. 9.4).

Consider a robot observing n landmarks in the environment with known locations using a noisy sensor where the current location of the robot is unknown;

The current robot location (unknown) is: $\mathbf{x}_r = [x_r, y_r]$

Measured bearings to the known landmarks: $\boldsymbol{\theta} = \left[\theta_1, \theta_2, \ldots, \theta_n\right]^T$

Actual locations of the landmarks (known) $\mathbf{x}_i = [x_i, y_i]^T$

Actual bearings to the landmarks (unknown): $\overline{\theta}(\mathbf{x}_r) = \left[\overline{\theta}_1(\mathbf{x}_r), \overline{\theta}_2(\mathbf{x}_r), \ldots, \overline{\theta}_n(\mathbf{x}_r))\right]^T$

where $\tan\overline{\theta}_i(\mathbf{x}_r) = \frac{y_i - y_r}{x_i - x_r}$

Assuming the measurement error to be $(\delta\theta_i)$, we can write a relationship between the measured bearing and the actual bearing for each observed feature: $\theta_i = \overline{\theta}_i(\mathbf{x}_r) +$

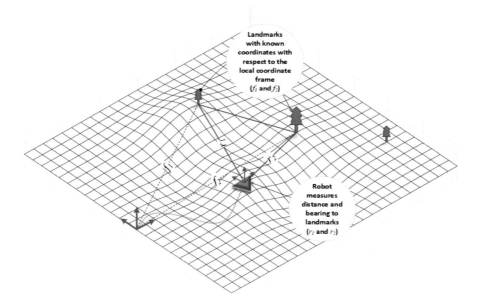

Fig. 9.4 Map-based localisation

$\delta\theta_i$ or

$$\boldsymbol{\theta} = \overline{\theta}(\mathbf{x}_r) + \delta\theta$$

where

$$\delta\theta = \left[\delta\theta_1, \delta\theta_2, \ldots, \delta\theta_n\right]^{\mathrm{T}}$$

and assuming the measurement noise to be zero-mean Gaussian and independent, the covariance matrix is given by

$$\Sigma = \mathrm{diag}\left(\sigma_1^2, \sigma_2^2, \ldots, \sigma_n^2\right)$$

We can then construct the maximum likelihood (ML) estimator of the robot location:

$$\hat{\boldsymbol{x}}_r = \mathrm{argmin}\frac{1}{2}\left[\overline{\theta}(\hat{\boldsymbol{x}}_r) - \boldsymbol{\theta}\right]^{\mathrm{T}}\Sigma^{-1}\left[\overline{\theta}(\hat{\boldsymbol{x}}_r) - \boldsymbol{\theta}\right]$$

This could now be solved recursively using a technique such as the Gauss–Newton algorithm as a nonlinear least-squares problem.

Most common localisation algorithms assume landmarks to be stationary during the entire robot operation. This is called the *static environment* assumption. An environment is considered *dynamic* if it contains map elements (except for the robot) that are moving, such as humans, vehicles and other robots. Obviously, such dynamic elements are not suitable as landmarks for localisation and are treated as noise. In robotics, such landmarks could be visually salient features naturally occurring in the given environment (e.g. corners and edges) or artificially placed (e.g. laser retroreflective beacons). A suitable sensor (e.g. a camera for the former and a laser range finder for the latter) along with relevant signal processing techniques should be used to detect and measure the distance and the angle (*bearing*) to these features with respect to the sensor coordinate frame.

The *Global Positioning System* (GPS) uses the distance information between the robot and the satellite (via time-of-flight and satellite-specific data) to localise, using a slight variation of the triangulation technique called *trilateration* which requires knowing only the distances to the landmarks (or the satellites). Ideally, having more than two landmarks will help to improve the accuracy of the location estimate. The same applies to the previously described triangulation scenario.

A related problem called the *data association* problem deals with the disambiguation of detected features and the correct association with the known map features. This problem does not arise with GPS, as each satellite on the constellation sends uniquely identifiable information to the receiver.

9.4   The Robot Mapping Problem

In the previous section, we discussed how objects within the robot's environment could be used to localise a robot. Of particular interest was the availability of maps. How are these maps generated? Figure 9.5 shows an ancient map of Taprobana, modern-day Sri Lanka, drawn by the ancient mathematician and cartographer Claudius Ptolemy. The map was drawn using geographical coordinates derived from tools and techniques available at the time. Today, such maps are drawn using modern surveying techniques using modern tools and GPS location data. The general idea, however, remains the same. Measurements are made to features of interest (e.g. contours, trees, structures) and are plotted against a fixed coordinate frame. As you would notice, Ptolemy's map has only a passing resemblance to today's maps of the country. During ancient times, in the absence of GPS to localise, sailors and cartographers relied on observing the sun and the stars using such instruments as the sextant resulting in significant errors in measurements and localisation. To create such maps accurately, one would need to make relative measurements to these features accurately and be able to localise the instrument that is making the measurements within the coordinate frame accurately.

When deploying robots, it is possible to access a priori maps on many occasions. For example, in indoor structures, it may be possible to refer to the architect drawn blueprints of the building or place artificial beacons (e.g. retroreflective markers that

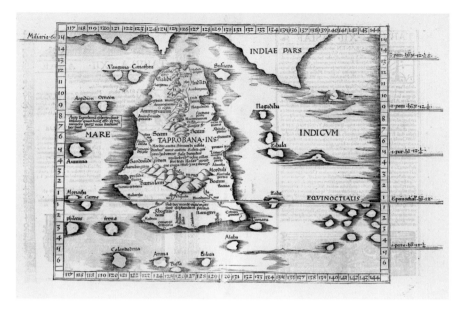

Fig. 9.5 Ptolemy's Taprobana as published in Cosmographia Claudii Ptolomaei Alexandrini, 1535 (modern-day Sri Lanka) (Image by Laurent Fries—(Bailey and Durrant-Whyte (2006)), Public Domain, https://commons.wikimedia.org/w/index.php?curid=16165526)

respond to laser light) at known and fixed locations. However, this assumes that the structure has not been changed from the original blueprints, which may not be accurate in the former and for large environments, placement and measurement of artificial beacons become a cumbersome proposition. In addition, there are semi-permanent elements within the building, such as furniture, that will change their locations over time, increasing the transient nature of maps. Also, it might be that the robot needs to be deployed in an environment where a pre-built map does not exist. In such situations, the robot is required to build a map. While the localisation problem is a relatively easier problem to tackle due to its low-dimensional nature, the map building problem is much harder, especially if the environment is large.

9.4.1 Occupancy Grid Maps

One of the simpler techniques used in robotics to create a map is the occupancy grid map. Occupancy grid maps are commonly used in 2D mapping scenarios to describe the floor plan of a robot's environment using a grid layout. Figure 9.6 shows a small occupancy grid map of an indoor environment. The grey shadings indicate where the sensor (a laser range finder in this case) has detected obstacles. Less dense areas of the 'map' indicate lower certainty of an obstacle at those locations. The blue line

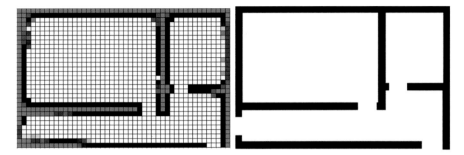

Fig. 9.6 An occupancy grid map (left) and the actual plan of the mapped area (right) (image credit: Rafael Gomes Braga)

indicates the robot's path during the mapping exercise. If you are wondering how the robot's path was generated, we used a separate localisation algorithm with known laser beacons placed in the environment. If we used the odometric data to generate the occupancy grid map, the results would have been as bad as Ptolemy's map.

9.4.2 Other Types of Maps

Maps such as the occupancy grid maps (Fig. 9.6) that represent the environment using a geometric representation are called *metric maps*. Other types of maps include *feature maps* that represent the environment using a set of salient features such as edges and corners; *semantic maps* that combine geometric information with high-level semantic information such as human identifiable objects in the environment (e.g. Books, tables) and their relationship to each other. Metric and feature maps are relatively less intuitive to humans. Semantic maps provide a more data-rich environment that humans can interpret. Such maps are helpful when humans and robots need to interact. For example, a semantic map implementation is better suited in a self-driving car situation, whereas a metric map might be more efficient for an underground mining application.

9.5 The Simultaneous Localisation and Mapping (SLAM) Problem

When the robot is provided with a priori map, it is possible to localise the robot in the environment—like our example at the beginning of the chapter of you visiting a new city with a map. Conversely, if the robot's pose is known, it is possible to construct a map of the environment—such as when surveyors create new maps using GPS location information. But, what happens if the robot does not have a map and does not know the location? This is the dreaded chicken and the egg problem in robotics. The question is, how do you construct a map while using the same map to localise simultaneously. The problem is commonly known as the Simultaneous Localisation and Mapping (SLAM) problem. In the late '90s, it was shown that it is indeed possible for a robot to start from an unknown location in an unknown environment to incrementally build a map of the environment while simultaneously computing the pose of the robot using the map being built.

9.5.1   An Estimation Theoretic Approach to the Localisation, Mapping and SLAM Problems

However, as we have seen in previous chapters, the sensors used in a robot could be noisy, and the environment could be unpredictable, resulting in inherent uncertainties in the measurements made about the environment. Therefore, it is required to consider these uncertainties in any canonical formulation of the problem.

In order to accommodate the underlying uncertainties of the system, it is possible to explore a class of algorithms that explicitly model system uncertainty using theories of probability. Thus, the localisation, mapping or the SLAM problem could be formulated as a multivariate *state estimation problem* with noisy measurements:

$$\hat{\mathbf{x}}_{\text{MAP}} \triangleq \underset{\mathbf{x}}{\operatorname{argmax}}\ p(\mathbf{x}|\mathbf{z})$$

where $\mathbf{x}$ represents the state variable and $\mathbf{z}$ is the measurement vector. We assume each multivariate state as a *normal distribution*.

If we can further assume a prior distribution p over $\mathbf{x}$ exists, then the above problem could be restated using *Bayes' theorem* (see Chap. 6):

$$\hat{\mathbf{x}}_{\text{MAP}} \triangleq \underset{\mathbf{x}}{\operatorname{argmax}}\ p(\mathbf{z}|\mathbf{x})p(\mathbf{x})$$

Here, $p(\mathbf{z}|\mathbf{x})$ is the measurement likelihood and $p(\mathbf{x})$ is the prior. This sets up the problem in a way that it is possible to solve it in a recursive manner, as was the case with the simple triangulation problem discussed earlier. We can now estimate the states of the robot and the map repeatedly from a given starting point. To do so, we should first define the relevant state vectors of the robot and the environment

The robot's state for a mobile robot operating in 2D could be expressed as

$$\mathbf{x}_r = \begin{bmatrix} x_r \\ y_r \\ \phi_r \end{bmatrix}$$

where x_r and y_r denote the current location of the origin of the robot's coordinate frame with respect to the local coordinate frame and ϕ_r is the robot's heading with respect to the x-axis of the robot's coordinate frame.

Assuming that our map is to be constructed using a metric feature map, the location of the ith map feature could be defined as the vector,

$$\mathbf{x}_{fi} = \begin{bmatrix} x_i \\ y_i \\ z_i \end{bmatrix}$$

where (x_i, y_i, z_i) are the 3D coordinates of the ith feature with respect to the local coordinate frame. If we assume that the entire map of the environment will constitute an n number of such features, then the map vector could be defined as,

$$\mathbf{x}_m = \begin{bmatrix} \boldsymbol{x}_{f1} \\ \vdots \\ \boldsymbol{x}_{fn} \end{bmatrix}$$

Then, depending on the specific problem, we can assemble the state vector $\mathbf{x}$ to be estimated for a given time $t = k$, as follows,

- For the localisation problem,

$$\mathbf{x}(k) = [\mathbf{x}_r(k)]$$

- For the mapping problem,

$$\mathbf{x}(k) = [\mathbf{x}_m(k)]$$

- For the Simultaneous Localisation and Mapping problem,

$$\mathbf{x}(k) = \begin{bmatrix} \mathbf{x}_r(k) \\ \mathbf{x}_m(k) \end{bmatrix}$$

Once the state vector is defined, the estimation process occurs in three steps at any given time step $t = k$ (see Fig. 9.8):

1.   The Prediction Stage

In the prediction stage, we will use a model of the robot's motion (control input) to predict how the states of the robot would evolve over the considered time step.

2.   The Observation Stage

In this stage, the robot uses its on-board sensors to make measurements to the map features in the environment. For example, a depth camera and computer vision algorithm could be used to detect salient features in the environment and measure their location with respect to the robot's coordinate frame. It is important to note that the algorithms used should be able to identify the features repeatedly and match them correctly over time (i.e. newly observed features should be correctly matched to features already on the map). This ability to match observations to corresponding map features is called data association (Fig. 9.7). Of course, if an observed feature is not already on the map, such a feature will need to be initialised first (i.e. added to the map vector).

These measurements are inherently noisy due to their physical nature and the limitations of the algorithms. The estimates of the states based on these sensor inputs are, therefore, limited by the accuracy of the sensors used for observations. However, during the final stage of the estimation process, such noise will be filtered out.

3.   The Update Stage

Fig. 9.7 A vision algorithm is used to detect salient features in the environment. The top and bottom images show two views of the same environment captured at two different time intervals of the robot's journey. The lines indicate matched features between the two images (data association). Can you spot any instances of failed data association?

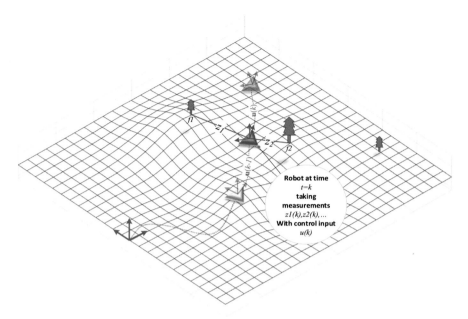

Fig. 9.8 Unknown states could be recursively estimated using sensor information within a Bayesian framework

During the update stage, the prediction information and observations are combined to produce an improved estimate of the states. The process could be represented as:

$$\hat{x} = (1 - w)\mathbf{x}_{\text{predicted}} + w\mathbf{x}_{\text{observed}}$$

where w is a weighing factor used to determine the relative importance of the observation and the prediction. If the sensors used in the robot provide highly accurate observation measurements, this parameter will be set closer to 1. If the sensors are noisier and you have to rely on the predictions, this value needs to be set closer to 0. In order to select the best value for the weighting factor, it is important to understand the nature of your observation measurements (called the sensor model) and the robot's motion model (also called the vehicle model or the control model). Suppose this is not appropriately 'tuned' to the system. In that case, the estimates could be either conservative, in which case the state estimates will carry large uncertainties or optimistic, therefore being overly confident about estimation where it is not warranted. Either scenario leads to state estimates that are not useful in the worst case, outright dangerous. An example of the latter could be observed when using GPS systems for localisation in densely constructed environments such as passing through a narrow pathway amidst tall buildings. Here, you would notice that your GPS location estimate suddenly starts to jump around and sometimes appears to be inside the buildings. However, it might still indicate high confidence in the estimate (usually denoted by a circle or an ellipse around the estimated location—the smaller the circle, the higher the confidence). This occurs because of the multi-path problem with GPS signals. Here, GPS signals are bounced off the tall structures before reaching the GPS receiver leading to large errors in the time-of-flight calculations. Since the built-in algorithms are unaware of what is happening outside, it continues to trust the observations leading to these erroneous state estimates. For a self-driving car or similar robot, this could be catastrophic, to say the least!

Various algorithms have been proposed to solve the multivariate state estimation problem. One of the most popular algorithms is the Kalman filter.

9.6 The Kalman Filter

The Kalman filter is a recursive linear statistical method for estimating the states of interest. The basic Kalman filter deals with linear systems, and nonlinear systems are treated by a linear approximation using the extended Kalman filter (EKF). Kalman filter has various applications in varying disciplines. For example, in robotic navigation and data fusion, Kalman filter is one of the methods frequently discussed in the literature with various adaptations and modifications.

9.6.1 Linear Discrete-Time Kalman Filter

For a linear system subject to Gaussian, uncorrelated, zero-mean measurement and process noises, the Kalman filter is the optimal minimum mean squared error estimator. To derive the filter for such a system, its model and the model of the observation must be defined. Then the problem can be stated as a recursive linear estimator with unknown gains. The gains can be determined using the minimum mean-squared error criterion (MMS).

The Kalman filter consists of the same three recursive stages discussed in the previous section.

1. Prediction stage
2. Observation stage
3. Update stage.

For a linear, discrete-time system, the state transition equation can be written as follows:

$$\mathbf{x}(k) = \mathbf{F}(k)\mathbf{x}(k-1) + \mathbf{B}(k)\mathbf{u}(k) + \mathbf{G}(k)\mathbf{v}(k)$$

where

- $\mathbf{x}(k)$ state at time k
- $\mathbf{u}(k)$ control input vector at time k
- $\mathbf{v}(k)$ additive motion noise
- $\mathbf{B}(k)$ control input transition matrix
- $\mathbf{G}(k)$ noise transition matrix
- $\mathbf{F}(k)$ state transition matrix

And the linear measurement equation can be written as follows:

$$\mathbf{z}(k) = \mathbf{H}(k)\mathbf{x}(k) + \mathbf{w}(k)$$

where

- $\mathbf{z}(k)$ observation made at time k
- $\mathbf{x}(k)$ state at time k
- $\mathbf{H}(k)$ measurement model
- $\mathbf{w}(k)$ additive observation noise

As mentioned earlier, system and measurement noise is assumed to be zero-mean and independent. Thus,

$$E[\mathbf{v}(k)] = E[\mathbf{w}(k)] = \mathbf{0}, \ \forall k$$

and,

$$E\left[\mathbf{v}(i)\boldsymbol{w}^T(j)\right] = 0, \ \forall i, j$$

Motion noise and the measurement noise will have the following corresponding covariance:

$$E\left[\mathbf{v}(i)\boldsymbol{v}^T(j)\right] = \delta_{ij}\mathbf{Q}(i), E\left[\mathbf{w}(i)\boldsymbol{w}^T(j)\right] = \delta_{ij}R(i)$$

The estimate of the state at a time k given all information up to time k is written as $\hat{\boldsymbol{x}}(k|k)$ and the estimate of the state at a time k given information up to time $k-1$ is written as $\hat{\boldsymbol{x}}(k|k-1)$ and is called the prediction. Thus, given the estimate at $(k-1)$th time step, the prediction equation for the state at kth time step can be written,

$$\hat{\boldsymbol{x}}(k/k-1) = \mathbf{F}(k)\hat{\boldsymbol{x}}(k-1|k-1) + \mathbf{B}(k)\mathbf{u}(k)$$

And the corresponding covariance prediction:

$$\mathbf{P}(k|k-1) = \mathbf{F}(k)\mathbf{P}(k-1|k-1)\mathbf{F}^T(k) + \mathbf{G}(k)\mathbf{Q}(k)\mathbf{G}^T(k)$$

Then the unbiased linear estimate is:

$$\hat{\mathbf{x}}(k|k) = \hat{\mathbf{x}}(k|k-1) + \mathbf{W}(k)[\mathbf{z}(k) - \mathbf{H}(k)\hat{\mathbf{x}}(k|k-1)]$$

Note that the conditional expected error between the estimate and the true state is zero.

$\mathbf{W}(k)$ is called the Kalman gain at time step k. This is calculated as:

$$\mathbf{W}(k) = \mathbf{P}(k|k-1)\mathbf{H}^T(k)\mathbf{S}^{-1}(k)$$

where $\mathbf{S}(k)$ is called the innovation variance at time step k and given by:

$$\mathbf{S}(k) = \mathbf{H}(k)\mathbf{P}(k|k-1)\mathbf{H}^T(k) + \mathbf{R}(k)$$

and the covariance estimate is:

$$\mathbf{P}(k|k) = (\mathbf{I} - \mathbf{W}(k)\mathbf{H}(k))\mathbf{P}(k|k-1)(\mathbf{I} - \mathbf{W}(k)\mathbf{H}(k))^T + \mathbf{W}(k)\mathbf{R}(k)\mathbf{W}^T(k)$$

Essentially, the Kalman filter takes a weighted average of the prediction $\hat{\mathbf{x}}(k|k-1)$, based on the previous estimate $\hat{\mathbf{x}}(k-1|k-1)$, and a new observation $\mathbf{z}(k)$ to estimate the state of interest $\hat{\mathbf{x}}(k|k)$.

Case study: We can illustrate this process with a simple 1-D toy example (Fig. 9.9). Let us assume that the robot can only move in one direction (x). At time $t = k-1$ the robot is at the location $x(k-1)$. Though the robot is not privy to this exact value, it has an estimate of its location, given by $\hat{\mathbf{x}}(k-1|k-1)$. As this is an estimate of the true value, it has uncertainty about its location, represented by the curve above it—the spread indicating the extent of the uncertainty. Now, the robot moves forward to the location $\mathbf{x}(k)$ at time $t = k$ using a control input $\mathbf{u}(k)$. Using these pieces of information, the robot can now predict its location at the new time step as $\hat{\mathbf{x}}(k|k-1)$. However, this leads to increased error in the estimated location as represented by

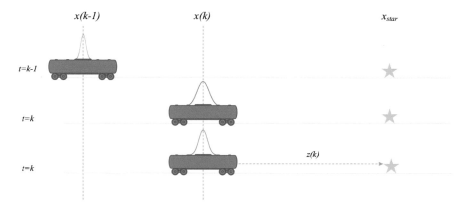

Fig. 9.9 A 1-D robot localisation example

the new curve. The robot observes a landmark (represented by the star) at this stage using its onboard sensor. Let us assume that the exact location of the star, x_{star} is known to the robot (i.e. a map is available). The robot measures the distance $\mathbf{z}(k)$ between the star and its current location using an internal sensor. As we now have an additional piece of information about the robot's location at time $t = k$, it is possible to integrate the new information to come up with a better estimate of the current location. If we trust the sensor that made the distance measure, we could weigh that information more. The previous derivation of the Kalman filter helps us make this decision and integrate the new sensor information weighted by the 'trust' we place on the sensor. Finally, an improved location estimate could be worked out as $\hat{\mathbf{x}}(k|k)$ resulting in a reduction in the uncertainty of the robot location. This new estimate then could be used to predict the robot's location at $t = k + 1$, and the process could be repeated.

9.6.2 The Extended Kalman Filter (EKF)

Although the Kalman filter is the optimal minimum mean squared error estimator for a linear system, any real robot is nonlinear. A solution is found in the extended Kalman filter (EKF), which uses a linearised approximation to nonlinear models. However, linear approximations for nonlinear functions must be treated with care, and if treated properly, the EKF will generate very good results in many applications. The extended Kalman filter algorithm is very similar to the linear Kalman filter algorithm with the substitutions:

$$F(k) \rightarrow \nabla \mathbf{f_X}(k) \text{ and } H(k) \rightarrow \nabla \mathbf{h_X}(k)$$

where $\nabla \mathbf{f_X}(k)$ and $\nabla \mathbf{h_X}(k)$ are nonlinear functions of both state and time step. These are called the Jacobians, or the partial derivates of the state transition and measurement functions, respectively (see Chap. 6). This implies that these functions are differentiable.

Then, the main equations can be summarised as follows:

1. Prediction equations:

$$\hat{\mathbf{x}}(k/k - 1) = \mathbf{f}((k - 1|k - 1), \mathbf{u}(k))$$

$$\mathbf{P}(k|k - 1) = \nabla \mathbf{f_X}(k)\mathbf{P}(k - 1|k - 1)\nabla^T \mathbf{f_X}(k) + \mathbf{Q}(k)$$

2. Update equations:

$$\hat{\mathbf{x}}((k|k) = \hat{\mathbf{x}}(k|k - 1) + \mathbf{W}(k)[\mathbf{z}(k) - \mathbf{h}(\hat{\mathbf{x}}(k|k - 1))]$$

$$P(k|k) = P(k|k-1) - \mathbf{W}(k)\mathbf{S}(k)\mathbf{W}^T(k)$$

where

$$S(k) = \nabla\mathbf{h}_X(k)\mathbf{P}(k|k-1)\nabla^T\mathbf{h}_X(k) + \mathbf{R}(k)$$

9.6.3   Data Association

One of the issues that arise during data fusion in a robotic navigation scenario is identifying the sensor observations with the observed. As mentioned earlier, this problem is commonly referred to as the data association problem. There are several methods available for discerning the observations. The most obvious way of doing this is to make the observations self-identifying. An example of this was presented earlier using computer vision, where a feature matching algorithm is used for the data association.

Statistical methods also exist to determine how likely a given observation is of the object thought to be observed. Derivation of equations for one such method referred to as the Mahalanobis distance is discussed below.

The difference between the observed and the predicted observation for a set of sensor data is called the innovation (v) and could be represented with the notations introduced earlier in the Kalman filter section as follows:

$$\mathbf{v}(k) = \mathbf{z}(k) - \hat{\mathbf{z}}(k|k-1)$$

where $\hat{\mathbf{z}}(k|k-1)$ is the predicted observation for time step k given the observation information up to time step $k-1$. It can be proven that the innovation is white with a mean of zero and variance $\mathbf{S}(k)$ given below:

$$S(k) = \mathbf{R}(k) + \mathbf{H}(k)\mathbf{P}(k|k-1)\mathbf{H}^T(k)$$

The above information can then be used in the problem of data association. The normalised innovation squared $\mathbf{q}(k)$ is defined as:

$$\mathbf{q}(k) = \mathbf{v}^T(k)\mathbf{S}^{-1}\mathbf{v}(k)$$

If the associated filter is assumed to be consistent, the above equation can be shown to be a χ^2 random distribution with m degrees of freedom, where $m = \dim(\mathbf{z}(k))$, which is the dimension of the observation sequence.

A confidence value can be chosen from the x^2 tables and compared against the value of $\mathbf{q}$ for each observation in the observation sequence. If the value of $\mathbf{q}$ for a given observation is less than the threshold, then that observation is likely to be associated with the correct object of observation. If multiple observations satisfy the above condition, it is safer to ignore such observations as improper data association could lead to unstable filter performance.

9.7 A Case Study: Robot Localisation Using the Extended Kalman Filter

Let us now consider a real-world application of the extended Kalman filter in solving the localisation problem.

9.7.1 Assumptions

The motion model used is a very important parameter in deciding the success of the filter to be used. Therefore, it needs proper consideration along with the choice of sensors. The algorithm used in this case study uses the rigid body and rolling motion constraints to simplify the analysis. The rigid body constraint assumes that the robot's frame is rigid, and the rolling motion constraint assumes that all points on the vehicle rotate about the instantaneous centre of rotation with the same angular velocity. This could be a reasonable model for a simple structure like the TurtleBot. In order to further simplify the analysis, it is assumed that there is no slip between tyres and the ground and that vehicle motion may be adequately be represented as a two-dimensional model whose motion is restricted to a plane.

9.7.2 Derivation of the EKF-Based Localisation Algorithm

Derivation follows the equations in Sect. 6. A state prediction for the $(k + 1)$ time step can be derived from the information available up to time step k:

$$\mathbf{X}(k + 1) = f(\mathbf{X}(k), \mathbf{u}(k))$$

Note the abbreviated representation used where $X(k + 1)$ represents the prediction of state at $(k + 1)$ time step given the information up to time step (k). $X(k)$ represents the best estimate available at time step (k). $u(k)$ represents the control inputs to the robot driver at time step (k):

$$u(k) = \begin{bmatrix} V(k) \\ \omega(k) \end{bmatrix}$$

Fig. 9.10 Sensor mounted
on the robot observes the *i*th
feature

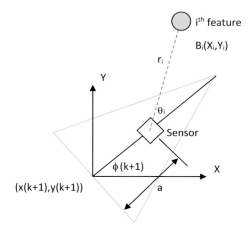

where $V(k)$ is the robot's forward velocity at time step (k) and $\omega(k)$ is the turning
rate (angular velocity) at time step (k). Thus, the complete process model can be
described as below:

$$\begin{bmatrix} x(k+1) \\ y(k+1) \\ \varphi(k+1) \end{bmatrix} = \begin{bmatrix} x(k) + \Delta T V(k) \cos(\varphi(k)) \\ y(k) + \Delta T V(k) \sin(\varphi(k)) \\ \varphi(k) + \Delta T \omega \end{bmatrix}$$

Assume that the robot makes an observation at time step $(k+1)$ with its onboard
sensor to a particular feature in the environment. The sensor is mounted on the robot
with an offset of a-units in the centreline, and the observation results in range (r_i) and
bearing (θ_i) information pertaining to the feature observed, as depicted in Fig. 9.10.

There are n-number of features scattered in the environment for which absolute
position coordinates are known a priori (i.e. the maps is given). A general observation
of the *i*th feature $B_i(X_i, Y_i)$ can be represented as follows:

$$r_i(k+1) = \sqrt{(X_i - x_r(k+1))^2 + (Y_i - y_r(k+1))^2}$$

$$\theta_i(k+1) = \tan^{-1}\left(\frac{(Y_i - y_r(k+1))}{(X_i - x_r(k+1))}\right) - \varphi(k+1)$$

where

$$x_r = x(k) + a\cos(\varphi(k))$$

$$y_r = y(k) + a\sin(\varphi(k))$$

The best estimate for this observation $\mathbf{Z}(k + 1|k)$ derived from previous information can be represented in the following form;

$$\mathbf{Z}(k + 1) = \begin{bmatrix} \hat{Z}_r(k + 1|k) \\ \hat{Z}_\theta(k + 1|k) \end{bmatrix} = \begin{bmatrix} \sqrt{(X_i - \hat{x}(k + 1|k))^2 + (Y_i - \hat{y}(k + 1|k))^2} \\ \tan^{-1}\left(\frac{(Y_i - \hat{y}(k+1|k))}{(X_i - \hat{x}(k+1|k))}\right) \end{bmatrix}$$

The prediction of covariance can be obtained as

$$P(k + 1|k) = \nabla f_x(k)P(k|k)\nabla f_x^T(k) + \nabla f_w(k)\Sigma(k)\nabla f_w^T(k)$$

where ∇f_x represents the gradient of Jacobean of $f(.)$ evaluated at time k with respect to the states, ∇f_w is the Jacobean of $f(.)$ with respect to the noise sources, and $\Sigma(k)$ is the noise strength given by

$$\Sigma(k) = \begin{bmatrix} \sigma_V^2 & 0 \\ 0 & \sigma_\omega^2 \end{bmatrix}$$

$$\nabla f_x(k) = \begin{bmatrix} 1 & 0 & \Delta T V(k)\sin(\phi(k|k)) \\ 0 & 1 & \Delta T V(k)\cos(\phi(k|k)) \\ 0 & 0 & 1 \end{bmatrix}$$

$$\nabla f_w(k) = \begin{bmatrix} \Delta T \cos(\varphi(k|k)) & 0 \\ \Delta T \sin(\varphi(k|k)) & 0 \\ 0 & \Delta T \end{bmatrix}$$

The innovation (observation prediction error) covariance $S(k)$, which is used in the calculation of the Kalman gains, can be calculated by squaring the estimated observation error and taking the expectations of the measurements up to kth time step and can be written as follows

$$S(k + 1) = \nabla h_x(k + 1)P(k + 1|k)\nabla h_x^T(k + 1) + R(k + 1)$$

where $R(k + 1)$ is the observation variance (which is diagonal in most robotics applications due to the independent nature of the measurements)

$$R(k) = \begin{bmatrix} \sigma_r^2 & 0 \\ 0 & \sigma_\theta^2 \end{bmatrix}$$

$$\nabla h_x(k + 1) = \begin{bmatrix} \frac{\hat{x}(k+1|k)-X_i}{d} & \frac{\hat{y}(k+1|k)-Y_i}{d} & 0 \\ \frac{-\hat{y}(k+1|k)-Y_i}{d^2} & \frac{\hat{x}(k+1|k)-X_i}{d^2} & -1 \end{bmatrix}$$

where

$$d = \sqrt{(X_i - \hat{x}_r(k+1))^2 + (Y_i - \hat{y}_r(k+1))^2}$$

Finally, the state update equations for the EKF are given by (adapting general equations in the previous section)

$$\hat{x}(k+1/k+1) = \hat{x}(k+1|k) + W(k+1)[z(k+1)-h(\hat{x}(k+1|k))]$$

where

$$W(k+1) = P(k+1|k)\nabla h_x^T(k+1)S(k+1)^{-1}$$

is the *Kalman gain*.

The algorithm is now complete, and as the robot proceeds from time $t = 0$ observing the environment, it can be applied recursively to determine the current location. A set of map features must be first initialised. In the example in Fig. 9.11, we have used a set of retroreflective beacons spread out in the environment. The locations of these beacons were surveyed and recorded for initialisation. A SICK laser range finder was used to detect and measure the range and bearing to these beacons.

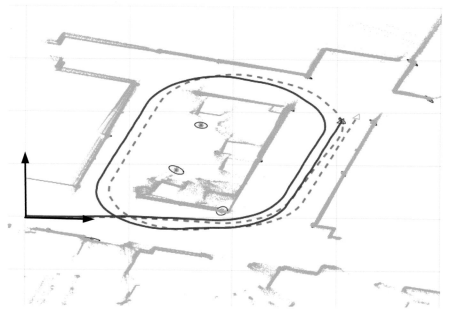

Fig. 9.11 An implementation of the EKF-based localisation algorithm. The solid blue line indicates the EKF estimate of the robot path. The red dotted line is odometry. The * denotes the locations of the surveyed beacons

9.8 Summary

This chapter looked briefly at a set of fundamental problems in robot navigation. The localisation problem answers the question 'where am I?' and the mapping problem asks the question 'how to generate a map of the robot's environment?' when the robot's location is known. The Simultaneous Localisation and Mapping (SLAM) problem involves solving both the localisation and the mapping problem concurrently. We discussed an estimation theoretic approach to solving these problems using probabilistic techniques. The extended Kalman filter was presented as an implementation of this approach using a linear approximation to the nonlinear system models. In combination, these algorithms and techniques should enable your robot never to get lost in space (or earth!).

9.9 Review Questions

- Assume a robot equipped with a sensor that can detect the state of a door. For simplicity, let us assume that the door could be in only one of two possible states, open or closed. Let us now assume the robot's sensors are noisy. The noise is characterised by the following conditional probabilities: ($\mathbf{Z}$-observation, $\mathbf{X}$-door state)

$$p(\mathbf{Z} = \mathbf{sense\_open}|\mathbf{X} = \mathbf{is\_open}) = 0.63$$

$$p(\mathbf{Z} = \mathbf{sense\_closed}|\mathbf{X} = \mathbf{is\_closed}) = 0.95$$

What is the value of the conditional probability $p(\mathbf{Z} = \mathbf{sense\_closed}|\mathbf{X} = \mathbf{is\_open})$.
- What is meant by the robot localisation problem?
- Why is data association important for successful localisation?

9.10 Further Reading

This chapter only scratched the surface of the localisation and mapping problem. Considerable research has happened since the '90s, with many successful implementations now in various production platforms operating at large-scale environments. An excellent book on the subject by (Thrun et al., 2005) provides an excellent deep dive into the subject. The essential tutorial on SLAM by Bailey and Durrant-Whyte (Bailey & Durrant-Whyte, 2006; Durrant-Whyte & Bailey, 2006) provides a great quick reference to the SLAM problem. The seminal work by (Dissanayake et al., 2001; Durrant-Whyte, & Csorba, 2001) provides proof of the existence of a solution to the SLAM problem. If you are interested in understanding the underlying probabilistic estimation techniques and theories (Bar-Shalom et al., 2001) is highly recommended.

References

Bailey, T., & Durrant-Whyte, H. (2006). Simultaneous localization and mapping (SLAM): Part II. *IEEE Robotics & Automation Magazine, 13*(3), 108–117. Retrieved from https://doi.org/10.1109/MRA.2006.1678144.

Bar-Shalom, Y., Li, X.-R., & Kirubarajan, T. (2001). *Estimation with applications to tracking and navigation*. Wiley InterScience.

Dissanayake, M. W. M. G., Newman, P., Clark, S., Durrant-Whyte, H. F., & Csorba, M. (2001). A solution to the simultaneous localization and map building (SLAM) problem. *IEEE Transactions on Robotics and Automation, 17*(3), 229-241.

Durrant-Whyte, H., & Bailey, T. (2006). Simultaneous localisation and mapping (SLAM): Part I the essential algorithms.

Elfes, A. (1989). Using occupancy grids for mobile robot perception and navigation. *Computer, 22*(6), 46–57.

Grisetti, G., Stachniss, C., & Burgard, W. (2007). Improved techniques for grid mapping with rao-blackwellized particle filters. *IEEE Transactions on Robotics, 23*(1), 34–46. Retrieved from https://doi.org/10.1109/TRO.2006.889486.

Thrun, S., Burgard, W., & Fox, D. (2005). *Probabilistic robotics cambridge*. MIT Press.

Damith Herath is an Associate Professor in Robotics and Art at the University of Canberra. Damith is a multi-award winning entrepreneur and a roboticist with extensive experience leading multidisciplinary research teams on complex robotic integration, industrial and research projects for over two decades. He founded Australia's first collaborative robotics startup in 2011 and was named one of the most innovative young tech companies in Australia in 2014. Teams he led in 2015 and 2016 consecutively became finalists and, in 2016, a top-ten category winner in the coveted Amazon Robotics Challenge—an industry-focused competition amongst the robotics research elite. In addition, Damith has chaired several international workshops on Robots and Art and is the lead editor of the book *Robots and Art: Exploring an Unlikely Symbiosis*—the first significant work to feature leading roboticists and artists together in the field of Robotic Art.

Chapter 10
How to Manipulate? Kinematics, Dynamics and Architecture of Robot Arms

Bruno Belzile and **David St-Onge**

Learning Objectives

The objective at the end of this chapter is to be able to:

- recognize the architecture and mobilities of a robot arm;
- solve the forward and inverse kinematics problem of serial and parallel manipulators;
- obtain the Jacobian relating the velocities of the joints to the end-effector;
- analyze the Jacobian to obtain the different singularities and understand their physical meaning;
- obtain the equations defining the dynamics of a robotic manipulator.

Introduction

Manipulators are not fundamentally different than any other robotic systems regarding their kinematics and dynamics. They are defined by their number of degrees-of-freedom (DoF) and their architecture, which are critical for the envisioned application. This chapter will provide you with an overview of the kinematics of robot arms, including the direct kinematics problem (DKP), the inverse kinematics problem (IKP) and the different types of singularities and how to find them. As kinematics alone is not sufficient for advanced control, you will need to understand also the dynamics of a robotic manipulator; we will cover it briefly.

B. Belzile (✉) · D. St-Onge
Department of Mechanical Engineering, ÉTS Montréal, 1100 Notre-Dame Street West,
Montreal, QC H3C 1K3, Canada
e-mail: bruno.belzile.1@ens.etsmtl.ca

D. St-Onge
e-mail: david.st-onge@etsmtl.ca

An Industry Perspective

Juxi Leitner, Co-Founder

LYRO Robotics.

My background is in computer science. I started programming computers when I was young (and there was not much else to do in my very tiny hometown in the middle of the alps).

When I was about 15-16, I realised that most of my code lives in a computer and did not really interact or change things in the real world. I started to become more and more interested in robotics and getting inspired by the movies coming out then, such as The Matrix, I robot, and Minority Report (I wanted to build those spider robots!) So, I looked for ways to learn more about it, and I enrolled in a Joint European Master Degree in Space Robotics.

I have researched robotics in academic settings for over a decade before trying to transfer the technology into real-world applications with our current startup LYRO Robotics.

Initially, I was looking at robot swarms and multi-robot coordination (for space exploration particularly), but I got lucky and was able to attend a summer programme in Lisbon to work with the then in-development iCub European Humanoid. I was fascinated by how easy certain tasks come to us, yet how hard they are for robotic systems, like detecting the world around the robot (even how to decide what to focus your "eyes"/cameras on) or how hard it is to pick up an object, even a simple one from a table in front of the robot.

That was eye-opening, and I got excited by the topic of embodiment and how to integrate perception and smarts with the physicality of the robotic system to enable physical interaction with the world! I still find it fascinating, and it is more than 17 years later :)

Another pivotal moment for me was entering and eventually winning the Amazon Robotics Challenge in 2017. There are specific things that industrial robots were designed for, and it's not picking random objects out of dynamic clutter. Building the team (we were 20+) and designing the robotic system was really just a lot of fun. The part of solving a real-world problem with fundamental tech we researched for years was particularly exciting (and frustrating at the same time ;)

The win showed that thinking about all the options from hardware to software, is important for designing robots that work. So, we started looking for real-world applications and founded LYRO in 2019 to bring robots to markets that are currently underserved due to various reasons (robots too expensive or too incapable is a big one).

Lot of the theory discussed in this chapter are relevant in the real world. For example, the iCub was inspired by the kinematics of a young child. In particular, the hand has a lot of degrees of freedom, three in the shoulder, two in the elbow, and two in the wrist. Then the hand has nine more (given it has five digits). It highlighted an interesting issue for me that the forward kinematics is pretty straightforward (if you have correct measurements), but inverse kinematics, like when I have a position of an object I want to grasp, how do I need to move my various joints, is a very hard and tricky problem with singularities and non-linearities everywhere.

During my PhD, we regularly had to fix the cables in the iCub's arm due to us running into (or over) limits and breaking things!

I work in Robotic Grasping, and the advent of machine learning 20 years ago, and deep learning ten years ago has clearly had an impact. However, while "grasping is solved" is an often-cited quote, it is still non-trivial to get a robotic arm to pick up any random object in any random configuration and perform some useful task with it.

The area is expanding, which is good, but it lacks reproducibility which is slowing down progress.

On the other hand, it is a very exciting time to enter as the whole field shifts more towards robots that perform tasks in a smart fashion rather than "simply" perform the same action over and over again.

Architectures

The physical embodiment of a robotic manipulator (we will use the term robot loosely for this chapter) is a kinematic chain composed by a set of *rigid bodies*, called *links*, connected in series together by *joints* (formally known as *kinematic pairs*). In other words, a joint constrains the motion between two bodies. There are two types of joints, namely *lower kinematic pairs* (LKP) and *higher kinematic pairs* (HKP). By definition, the former involves "a contact taking place along a surface common to the two bodies" (Angeles, 2014). You most likely encountered already the two most common joints that belong to this category: the revolute (rotation, R) and prismatic (translation, P) joints. While there are also four other types of LKP, helical (screw, H), cylindrical (C), universal (U) and spherical (S), all of them can be obtained with a combination of revolute and prismatic joints. Therefore, the content of this chapter will nearly exclusively focus on those two types of joints. While most joints commonly used in robots only have a single *degree-of-freedom* (DoF), namely the revolute and prismatic joints mentioned above, other types of joints, such as the spherical and cylindrical joints, exist, with, respectively, three and two DoFs. As it will be seen in the subsection on wrist-partitioned serial manipulators, the last three revolute joints of this type of robot are equivalent to a spherical joint.

The architectures of robotic manipulators can be classified into two main categories: *serial* and *parallel*. The former, more common in the manufacturing industry, consist of manipulators made of simple and open kinematic chains. They are known for their reach and simplicity. The Kinova Gen3 lite, shown in Fig. 10.1, falls into this category with its 6$\underline{R}$ kinematic chain,[1] i.e., an open loop of six actuated revolute joints in a serial array. The latter, parallel manipulators, are based on complex kinematic chains made of at least one loop. They are known for their structural rigidity, speed and the ability to lift a larger payload with respect to the robot mass. While for the serial manipulator, most actuators need to be moved during the robots' motion, the actuators of a parallel manipulator can all be attached rigidly to the base.

Manipulators can also be classified by their **mobility**, which include their DoFs and the type of motion they can generate. For instance, one of the most important type of robotic manipulators is the Schönflies-motion generators (SMG). These 4-DoF robots, capable of three translations and one rotation about an axis of fixed direction (usually the vertical axis), are commonly called SCARA-like robot, after one of the first and well-known SMG, the Selective-Compliance Assembly Robot Arm (SCARA), a serial robot with one prismatic and three revolute joints (Makino et al., 2007). These manipulators can have a serial or a parallel architecture. Nowadays, most industrial manipulator will have 5–7 DoFs, such as the Kinova Gen3 and Gen3 lite.

Kinematics of Serial Manipulators

Serial manipulators are considered simple kinematic chains, i.e., each link can be coupled via one or two joints, to one or two links. The first link is the *base* and the last link is the *end-effector* (EE), sometimes called tool. In the sequel, we will take a closer look to the direct and inverse kinematics of serial manipulators.

Direct Kinematics

Kinematics are used to describe the motion of a robot without considering the dynamics, namely the forces and the torques causing the motion. Therefore, kinematics

---

[1] An underline letter representing a joint means it is actuated.

Fig. 10.1 Kinova Gen3 lite, a serial 6-DoF robotic manipulator

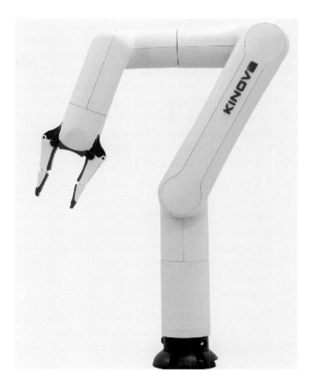

problems are geometric problems. First, we consider the direct kinematics (DK), sometimes called forward kinematics (FK), of a serial robot. The DK equations are used to **map the joint variables**, called the *posture* or *configuration* of the robot, **into the position and the orientation of the EE**, namely its *pose*. In the end, you will obtain an explicit system of nonlinear equations to compute $\mathbf{p} = [p_x \quad p_y \quad p_z]^T$, the three-dimensional vector representing the Cartesian position of the EE, as well as a 3×3 orthogonal orientation (rotation) matrix $\mathbf{Q}$ made of three unit vectors parallel to the X-, Y- and Z-axes of the EE (expressed in the base reference frame). Both $\mathbf{p}$ and $\mathbf{Q}$ can be assembled into a single 4×4 homogeneous matrix, as you will see.

Denavit-Hartenberg Convention

It is impossible to discuss the subject of direct kinematics of serial robots without bringing up the *Denavit-Hartenberg convention*. It is a powerful tool that will help you solve the forward kinematics of a serial manipulator in a systematic way. Since this method was first introduced by Hartenberg and Denavit (Hartenberg and Denavit, 1964), some variations were proposed. Here, we use Paul's notation, also known as the distal variant (Lipkin, 2008). Each link is numbered from 0 to n, 0 being the base, while n is the nth link, namely the flange of the robot to which the end-effector is attached. The ith joint is defined as the one connecting the $(i - 1)$th and ith links. While the forward kinematics of a serial robot can be solved without the use of the DH convention (or any other), it simplifies considerably the process and can be easy understood by other engineers familiar with the DH notation. Brace yourself, the following lines cover several definitions and formulas, but the procedure quickly become easy to use after trying some examples. For each link, a Cartesian frame is defined. Two such frames are shown in Fig. 10.2. You should note that the (X_i, Y_i, Z_i) axes are rigidly attached to the $(i - 1)$th link. The following convention is used:

1. Z_i is the axis of the ith kinematic pair/joint.
2. X_i is the common normal between $Z_{(i-1)}$ and Z_i. Contrary to Z_i, which does not have a prescribed direction, X_i is oriented from $Z_{(i-1)}$ toward Z_i. If they intersect, resulting in an undefined direction for X_i, the convention is to use the cross product of unit vectors parallel to $Z_{(i-1)}$ and Z_i $\mathbf{i}_i = \mathbf{k}_{(i-1)} \times \mathbf{k}_i$. In the case the former and the latter are parallel, X_i is arbitrarily chosen to complete the Cartesian frame, i.e., orthogonal to $Z_{(i-1)}$ and Z_i.
3. with the right-hand rule,[2] Y_i is defined.

With these frames and their respective axes, four parameters are defined for $i = 1 \ldots n$: $\theta_i, \alpha_i, d_i, a_i, i = 1 \ldots n$, being respectively the *joint angle*, the *link twist*, the *link offset* and the *link length*. They are defined below:

1. a_i is the distance[3] between Z_i and $Z_{(i+1)}$ along $X_{(i+1)}$.
2. d_i is the coordinate,[4] along Z_i, from the origin of the ith frame to the intersection with $X_{(i+1)}$.
3. α_i is the angle between Z_i and $Z_{(i+1)}$, measured with respect to the positive direction of $X_{(i+1)}$.
4. θ_i is the angle between X_i and $X_{(i+1)}$, measured with respect to the positive direction of Z_i.

An homogeneous transformation matrix, as defined in Chap. 6 (Section 6.4.4), is obtained from these parameters, i.e.,

$$
\mathbf{H}_{i-1}^i \equiv
\begin{bmatrix}
\cos \theta_i & -\sin \theta_i \cos \alpha_i & \sin \theta_i \sin \alpha_i & a_i \cos \theta_i \\
\sin \theta_i & \cos \theta_i \cos \theta_i & -\cos \theta_i \sin \alpha_i & a_i \sin \theta_i \\
0 & \sin \alpha_i & \cos \alpha_i & d_i \\
0 & 0 & 0 & 1
\end{bmatrix}
\tag{1}
$$

---

[2] As explained in Chap. 4 the thumb of the right hand points along the direction of the Z-axis; the curl of the fingers while closing the hand represents a motion from the X-axis toward the Y-axis.

[3] Always positive by definition.

[4] Being a signed distance, it can be negative.

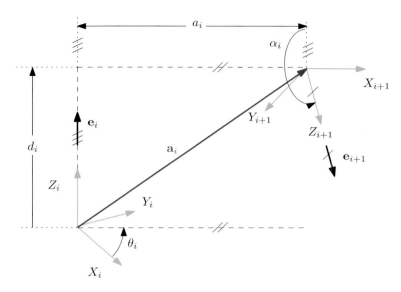

Fig. 10.2 Frames' representation in the DH convention

where subscript refers to the reference frame in which the coordinates are given, $(i-1)$ in this notation. It can also be separated into the rotation matrix $\mathbf{Q}_{i-1}^i$ and the displacement vector $\mathbf{a}_i$, i.e.,

$$\mathbf{H}_{i-1}^i = \begin{bmatrix} \mathbf{Q}_{i-1}^i & \mathbf{a}_{i-1}^i \\ \mathbf{0}^T & 1 \end{bmatrix} \tag{2}$$

The orientation and position of the EE are thus obtained by multiplying the individual transformation matrices associated with the DH parameters, giving us

$$\mathbf{Q} = \mathbf{Q}_0^1 \mathbf{Q}_1^2 \mathbf{Q}_2^3 \mathbf{Q}_3^4 \mathbf{Q}_4^5 \mathbf{Q}_5^6 \tag{3a}$$

$$\mathbf{p} = \sum_{i=1}^{6} \mathbf{a}_0^i \quad \text{or} \tag{3b}$$

$$\mathbf{p} = \mathbf{a}_0^1 + \mathbf{Q}_0^1 \mathbf{a}_1^2 + \mathbf{Q}_0^1 \mathbf{Q}_1^2 \mathbf{a}_2^3 + \mathbf{Q}_0^1 \mathbf{Q}_1^2 \mathbf{Q}_2^3 \mathbf{a}_3^4 + \mathbf{Q}_0^1 \mathbf{Q}_1^2 \mathbf{Q}_2^3 \mathbf{Q}_3^4 \mathbf{a}_4^5 + \mathbf{Q}_0^1 \mathbf{Q}_1^2 \mathbf{Q}_2^3 \mathbf{Q}_3^4 \mathbf{Q}_4^5 \mathbf{a}_5^6 \tag{3c}$$

$$\mathbf{H} = \begin{bmatrix} \mathbf{Q} & \mathbf{p} \\ \mathbf{0}^T & 1 \end{bmatrix} = \mathbf{H}_0^1 \mathbf{H}_1^2 \mathbf{H}_2^3 \mathbf{H}_3^4 \mathbf{H}_4^5 \mathbf{H}_5^6 \tag{3d}$$

where $\mathbf{H}$ is the homogeneous transformation matrix representing both the position and orientation of the EE. For the sake of brevity, in the sequel, if only a subscript is given for a rotation/transformation matrix, it is given in the previous reference frame.

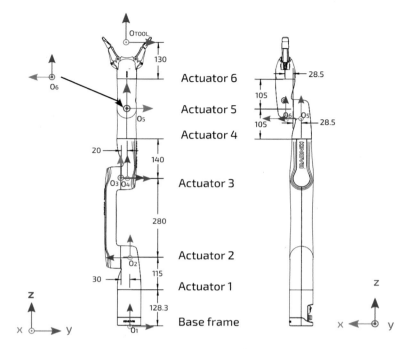

Fig. 10.3 DH frames for each joint for the Kinova Gen3 lite (extracted from the manipulator user manual)

For a joint with a single DoF, such as a revolute or a prismatic joint, only one of the four parameters $(a_i, d_i, \theta_i, \alpha_i)$ is a variable, the others are constant. As previously mentioned, a homogeneous transformation matrix is characterized by six parameters in 3D space. Here, this number is reduced to four since, with the DH convention, the location of the origin of frame i is not arbitrarily chosen. Indeed, we have two constraints for the X-axis of each subsequent frame: (1) X_i must be normal to Z_i and (2) it must also intersect it. The frame is rigidly attached to link i, but it is not necessarily located at the end of the link, as one may expect. In fact, it may lie outside the link itself. The reduced number of parameters defining the transformation matrices is one of the main assets of the DH notation.

The DH frames applied to the Kinova Gen3 lite are shown in Fig. 10.3, and the corresponding DH parameters are detailed in Table 10.1. Since the six joints of the Gen3 lite are revolute, all θ_i are unknowns. In Chap. 18, more precisely in Project 3, you will have to find the DH parameters of a 3-DoF version of this manipulator,[5] as well as compute its forward and inverse kinematics.

---

[5] Three of its six joints will be locked.

Table 10.1   DH parameters of the Kinova Gen3 lite

i	1	2	3	4	5	6
a_i	0	a_2	0	0	0	0
d_i	d_1	d_2	d_3	d_4	d_5	d_6
α_i	$\pi/2$	π	$\pi/2$	$\pi/2$	$\pi/2$	0

Inverse Kinematics

As mentioned at the beginning of this chapter, solving the IKP allows the engineer to obtain the set of joint coordinates, namely the posture of the robot, from a position and orientation of the end-effector, namely the pose. Contrary to the DKP, which give only one EE pose from a set of joint coordinates, there may be **more than one solution** to the IKP, i.e., more than one posture that corresponds to a position/orientation of the EE. However, an analytical (symbolic and exact) solution to the inverse kinematics is not necessarily always obtainable, depending on the architecture of the robotic manipulator. In some cases, a numerical approach is preferable. Numerical approaches are also better fit for simulator compatible with various manipulator architectures. The different solutions to the IKP are called configuration types. Usually, while moving, a manipulator will keep the same configuration type, as alternating from one configuration type to another requires large joint angle variations to obtain, in the end, the same EE coordinates. Switching configuration can also risk passing through a singularity, which we will discuss later. The controller of commercially available manipulators takes these elements into account while computing the positions and velocities of the joints.

To solve the IKP symbolically for the explicit equations, we start with the same equations used above, i.e., the ones defined by the 4×4 homogeneous transformation matrix, i.e., **H**. Since the last line is always [0 0 0 1], we thus have 12 nonlinear equations, but only six unknowns in the case of a non-redundant[6] spatial manipulator. Of course, if the robot has additional joints, for example, to reach a target within a cluttered workspace (ex. welding operations), the number of potential solutions increases. Within this chapter, only non-redundant manipulators are considered.

As previously mentioned, while we have nine equations for the orientation of the EE, only thhree are independent, giving us a system of six equations with six unknowns (three for the orientation, three for the position). Solving the IKP for a general serial manipulator is thus a challenging mathematical problem considering the nonlinearity of the equations. However, you will find that most commercially

---

[6] A spatial serial redundant manipulator has more than six joints. Notwithstanding the mechanical limits of the joints, the limits of the reachable workspace and singularities, only six joints are needed to reach any point with any orientation of the EE. You should be careful if you come across the term "redundant," as it can have different meanings depending on the context. A parallel robot can be redundantly actuated, i.e., more actuators than DoFs, and any manipulator can be *kinematically redundant* with respect to its task, for example, pointing tasks, which only require two DoFs.

available manipulators fall in the special category of *wrist-partitioned*, greatly simplifying the problem, as we will show below.

Wrist-Partitioned Manipulators

The architecture of decoupled serial manipulator (wrist-partitioned) makes it possible to separate the orientation problem from the position problem. Therefore, we obtain *explicit* equations, avoiding the need for a numerical method to solve the IKP. The problem is thus split into the *inverse position kinematics* and the *inverse orientation kinematics*. By definition, the axes of the last three joints of decoupled manipulators intersect. This point is known as the *wrist center*. Looking back to the DH parameters, this means that $a_4 = a_5 = a_6 = 0$. This also means that the last three DH frames share the same origin. The coordinates of the latter are given by vector $\mathbf{p}_w$ in frame 0, i.e.,

$$\mathbf{p}_w = \mathbf{a}_1 + \mathbf{Q}_1\mathbf{a}_2 + \mathbf{Q}_1\mathbf{Q}_2\mathbf{a}_3 + \mathbf{Q}_1\mathbf{Q}_2\mathbf{Q}_3\mathbf{a}_4 \qquad (4)$$

Since $a_4 = 0$, $\mathbf{a}_4$ is not a function of θ_4, as the equation of $\mathbf{p}_w$ above. With Eq. (3b), we can rewrite the above equation as

$$\mathbf{p}_w = \mathbf{p} - \mathbf{Q}_1\mathbf{Q}_2\mathbf{Q}_3\mathbf{Q}_4\mathbf{a}_5 - \mathbf{Q}_1\mathbf{Q}_2\mathbf{Q}_3\mathbf{Q}_4\mathbf{Q}_5\mathbf{a}_6 \qquad (5)$$

which can be simplify, knowing that with a decoupled wrist, $\mathbf{a}_5 = \mathbf{0}$, as

$$\mathbf{p}_w = \mathbf{p} - \mathbf{Q}\mathbf{Q}_6^T\mathbf{a}_6 \qquad (6)$$

This equation is solely function of constant DH parameters and the target position and orientation coordinates of the EE in the case of an IKP. Therefore, the location of the wrist, $\mathbf{p}_w$, can be computed in the base frame without the joint coordinates, decoupling the position from the orientation.

In short, we solve the position problem by first computing the location of the wrist with Eq. (6), then by isolating the first three joint coordinates in Eq. (4), which is a simpler 3-DoF problem with three equations and three unknowns.

Example: 3-DoF Serial Manipulator

As an example, we can solve 3-DoF inverse position problem for a generic serial manipulator with three revolute joints. It should be noted that the procedure below may need to be slightly adapted in certain special cases (null DH parameters, certain angles, division by zero, etc.). First, we need to rewrite Eq. (4):

$$\mathbf{Q}_1^T(\mathbf{p}_w - \mathbf{a}_1) = \mathbf{a}_2 + \mathbf{Q}_2\mathbf{a}_3 + \mathbf{Q}_2\mathbf{Q}_3\mathbf{a}_4 \qquad (7)$$

This can be done because rotation matrices are orthogonal, thus $\mathbf{Q}_i^{-1} = \mathbf{Q}_i^T$. Developing the above equation in terms of its components, we have

$$A\cos\theta_2 + B\sin\theta_2 = x_w\cos\theta_1 + y_w\sin\theta_1 - a_1 \qquad (8a)$$

$$A\sin\theta_2 - B\cos\theta_2 = \cos\alpha_1(y_w\cos\theta_1 - x_w\sin\theta_1) + (z_w - b_1)\sin\alpha_1 \qquad (8b)$$

$$C = \sin\alpha_1(x_w\sin\theta_1 - y_w\cos\theta_1) + (z_w - b_1)\cos\alpha_1 \qquad (8c)$$

with

$$A = a_2 + a_3 \cos \theta_3 + b_4 \sin \alpha_3 \sin \theta_3 \tag{8d}$$

$$B = -a_3 \cos \alpha_2 \sin \theta_3 + b_3 \sin \alpha_2 + b_4 \cos \alpha_2 \sin \alpha_3 \cos \theta_3 + b_4 \sin \alpha_2 \cos \alpha_3 \tag{8e}$$

$$C = b_2 + a_3 \sin \alpha_2 \sin \theta_3 + b_3 \cos \alpha_2 - b_4 \sin \alpha_2 \sin \alpha_3 \cos \theta_3 + b_4 \cos \alpha_2 \cos \alpha_3 \tag{8f}$$

We can see that the right-hand side of Eqs (8a–8c) is only function of θ_1, the position of the wrist and the DH parameters. Let

$$D = x_w \cos \theta_1 + y_w \sin \theta_1 - a_1 \tag{9}$$

$$E = \cos \alpha_1 (y_w \cos \theta_1 - x_w \sin \theta_1) + (z_w - b_1) \sin \alpha_1 \tag{10}$$

we can cast Eq. (8a–8b) in matrix form, i.e.,

$$\begin{bmatrix} A & -B \\ -B & A \end{bmatrix} \begin{bmatrix} \cos \theta_2 \\ \sin \theta_2 \end{bmatrix} = \begin{bmatrix} D \\ E \end{bmatrix} \tag{11}$$

We are now able to compute explicit functions of $\sin \theta_2$ and $\cos \theta_2$:

$$\cos \theta_2 = (AD - BE)/(A^2 + B^2) \tag{12}$$

$$\sin \theta_2 = (BD - AE)/(A^2 + B^2) \tag{13}$$

which leads to

$$\theta_2 = \arctan2(\sin \theta_2, \cos \theta_2) \tag{14}$$

Obviously, θ_2 cannot be computed right away since the values of the other two joint angles are needed. To this aim, we need to make θ_2 disappear. This is done by calculating the sum of squares of each side of Eq. (8a–8c). Knowing $\sin^2 \theta_2 + \cos^2 \theta_2 = 1$, we obtain

$$A^2 + B^2 + C^2 = x_w^2 + y^2 + (z_w - b_1)^2 + a_1^2 - 2a_1 x_w \cos \theta_1 - 2a_1 y_w \sin \theta_1 \tag{15}$$

The left-hand side of the above equation is only a function of DH parameters and θ_3, while the right-hand side is only dependent on DH parameters and θ_1. Moreover, Eq. (15) is linear in $\sin \theta_1$, $\sin \theta_3$, $\cos \theta_1$ and $\cos \theta_3$. Computing the sum of the squares of Eq. (8a–8b) would not have been useful, here, to eliminate θ_2, as the resulting equation would not have been linear in the terms mentioned above, which is necessary for the following steps. Therefore, Eq. (15) is rewritten as

$$F_1 \cos \theta_1 + G_1 \sin \theta_1 + H_1 \cos \theta_3 + I_1 \sin \theta_3 + J_1 = 0 \tag{16}$$

where F_1, G_1, H_1, I_1 and J_1 are only functions of DH parameters and the position of the wrist, all these terms being known at this stage. Then, Eq. (8c) is rewritten in a similar form, i.e.,

$$F_2 \cos\theta_1 + G_2 \sin\theta_1 + H_2 \cos\theta_3 + I_2 \sin\theta_3 + J_2 = 0 \tag{17}$$

Again, F_2, G_2, H_2, I_2 and J_2 are only functions of DH parameters and the position of the wrist. Having two linear equations and four unknowns, the next step is obtaining explicit expressions of $\cos\theta_1$ and $\sin\theta_1$, as we did with θ_2. Thus, we obtain

$$\cos\theta_1 = \frac{-G_2(H_1 \cos\theta_3 + I_1 \sin\theta_3 + J_1) + G_1(H_2 \cos\theta_3 + I_2 \sin\theta_3 + J_2)}{F_1 G_2 - F_2 G_1} \tag{18}$$

$$\sin\theta_1 = \frac{F_2(H_1 \cos\theta_3 + I_1 \sin\theta_3 + J_1) - F_1(H_2 \cos\theta_3 + I_2 \sin\theta_3 + J_2)}{F_1 G_2 - F_2 G_1} \tag{19}$$

$$\theta_1 = \arctan2(\sin\theta_1, \cos\theta_1) \tag{20}$$

Finally, we eliminate θ_1 by computing the sum of the $\sin^2\theta_1$ and $\cos^2\theta_1$, which results in

$$K \cos^2\theta_3 + L \sin^2\theta_3 + M \cos\theta_3 \sin\theta_3 + N \cos\theta_3 + P \sin\theta_3 + Q = 0 \tag{21}$$

where the coefficients in front of the trigonometric functions of θ_3 are functions of F_i, G_i, H_i, I_i and $J_2 i$, for $i = 1, 2$, which are in turn functions of DH parameters and the position of the wrist. We, therefore, have a nonlinear equation with known coefficients where the only unknown is θ_3. To solve this implicit equation, we use a well-known identity in the field of kinematics, the Weierstrass substitution (also known as the tangent half-angle substitution):

$$\cos\theta_3 \equiv \frac{1 - T_3^2}{1 + T_3^2}, \quad \sin\theta_3 \equiv \frac{2T_3}{1 + T_3^2}, \quad T_3 = \tan(\theta_3/2) \tag{22}$$

With this substitution, Eq. (21) is rewritten as an equation of degree four in T_3:

$$RT_3^4 + ST_3^4 + UT_3^2 + VT_3 + W = 0 \tag{23}$$

All four possible values for T_3 are thus obtained by computing the roots of the above equation. These values are then used to calculate the solutions for θ_3 with

$$\theta_3 = 2\arctan T_3 \tag{24}$$

The values for the remaining joint coordinates are then computed with first Eq. (20) then Eq. (14), for θ_1 and θ_2, respectively. Therefore, we have solve the inverse position problem for a 3-DoF serial manipulator, obtaining four sets of joint coordinates. If we replace the revolute joints with prismatic joints, the problem becomes less challenging to solve, as two prismatic joints (and one revolute) lead to a maximum of two solutions to the inverse position problem and three prismatic joints lead to only one solution to the inverse position problem. The position of the wrist now known; the next step is to find the solutions for the remaining three joints.

Spherical Wrist

The first three rotation matrices $\mathbf{Q}_1$, $\mathbf{Q}_2$, $\mathbf{Q}_3$ now fully known; the next step is to compute the solutions for the last three transformation matrices, which are function of the last three joint coordinates. First, we recall Eq. (3a) and rewrite it with everything known on the right, i.e.,

$$\mathbf{Q}_4\mathbf{Q}_5\mathbf{Q}_6 = \mathbf{R} \tag{25a}$$

$$\mathbf{R} = \mathbf{Q}_3^T\mathbf{Q}_2^T\mathbf{Q}_1^T\mathbf{Q}_d = \begin{bmatrix} r_{11} & r_{12} & r_{13} \\ r_{21} & r_{22} & r_{23} \\ r_{31} & r_{32} & r_{33} \end{bmatrix} \tag{25b}$$

Now you should remember that according to the DH notation, the angle between the axes Z_5 and Z_6 is α_5. These two axes are defined by the unit vectors $\mathbf{e}_5$ and $\mathbf{e}_6$. Therefore, according to the dot product, we have

$$\mathbf{e}_5 \cdot \mathbf{e}_6 = \cos \alpha_5 \tag{26}$$

We need to express these two vectors in one single reference frame. The DH frame 4 is chosen since it simplifies the equations. In this frame, $\mathbf{e}_5$ is simply the last column of $\mathbf{Q}_4$. As for $\mathbf{e}_6$ is the last column of $\mathbf{Q}_4\mathbf{Q}_5$. To avoid introducing more than one unknown in the equation, we use the fact that

$$\mathbf{Q}_4\mathbf{Q}_5 = \mathbf{R}\mathbf{Q}_6^T \tag{27}$$

We thus obtain an equation where θ_4 is the only unknown variable:

$$X \cos \theta_4 + Y \sin \theta_4 = Z \tag{28a}$$

where

$$X = -\sin \alpha_4 (r_{22} \sin \alpha_6 + r_{23} \cos \alpha_6) \tag{28b}$$
$$Y = \sin \alpha_4 (r_{12} \sin \alpha_6 + r_{13} \cos \alpha_6) \tag{28c}$$
$$Z = -\cos \alpha_4 (r_{32} \sin \alpha_6 + r_{33} \cos \alpha_6) + \cos \alpha_5 \tag{28d}$$

Using the Wieirstrass substitution introduced previously, the above equation is then transformed into a quadratic equation in T_4, where the roots are computed and substituted in $\theta_4 = 2 \arctan T_4$. To find the possible values for the remaining to joint angles, we need to go back to Eq. (25a). We keep only the unknown terms on the lefthand side by premultiplying by $\mathbf{Q}_4^T$, resulting in

$$\mathbf{Q}_5\mathbf{Q}_6 = \mathbf{Q}_4^T\mathbf{R} \tag{29}$$

By developing the components of the above equation and by simple inspection, we find

$$\cos\theta_6 = \frac{r_{12}\sin\alpha_4\sin\theta_4 - r_{22}\sin\alpha_4\cos\theta_4 + r_{32}\cos\alpha_4 - \cos\alpha_5\sin\alpha_6}{\sin\alpha_5\cos\alpha_6} \qquad (30a)$$

$$\sin\theta_6 = \frac{r_{11}\sin\alpha_4\sin\theta_4 - r_{21}\sin\alpha_4\cos\theta_4 + r_{31}\cos\alpha_4}{\sin\alpha_5} \qquad (30b)$$

As previously done, we put both values into

$$\theta_6 = \arctan2(\sin\theta_6, \cos\theta_6) \qquad (31)$$

Finally, θ_5 is found in a similar fashion but with Eq. (27) instead. By inspection, we find

$$\cos\theta_5 = \frac{\cos\alpha_4\cos\alpha_5 - r_{32}\sin\alpha_6 - r_{33}\cos\alpha_6}{\sin\alpha_4\sin\alpha_5} \qquad (32a)$$

$$\sin\theta_5 = \frac{r_{31}\cos\theta_6 - r_{32}\cos\alpha_6\sin\theta_6 + r_{33}\sin\alpha_6\cos\theta_6}{\sin\alpha_4} \qquad (32b)$$

and we compute

$$\theta_5 = \arctan2(\sin\theta_5, \cos\theta_5) \qquad (33)$$

Other Manipulators

In the case of a serial manipulator without a decoupled wrist, there is no simple recipe to solve the IKP. In some case, a numerical solver is necessary to obtain the joint coordinates from a set of EE coordinates. In other cases, explicit equations can be obtained, for instance, the Kinova Gen3 lite, but they are unique to the robots with the same architecture. However, while the solutions are different, the approach to solve the IKP of non-wrist-partitioned manipulators is generally similar, which is reducing the number of unknowns to only one to obtain the roots of a univariate polynomial equation to compute the values for one joint coordinate, then computing those for the other joints by backsubstitution, as we did with the inverse position problem of the wrist-partitioned manipulator. Indeed, this approach relies mostly on trigonometric identities, e.g.:

- $\sin^2\theta + \cos^2\theta = 1$
- $\sin\alpha\sin\beta + \cos\alpha\cos\beta = \cos(\alpha - \beta)$
- $\cos\alpha\cos\beta - \sin\alpha\sin\beta = \cos(\alpha + \beta)$
- $\sin\alpha\cos\beta + \cos\alpha\sin\beta = \sin(\alpha + \beta)$
- $\sin\alpha\cos\beta - \cos\alpha\sin\beta = \sin(\alpha - \beta)$

and the concept of *dyalitic elimination*. The latter is used to reduce the number of unknowns in a system of non-homogeneous equations. The procedure consists of four steps:

1. Rewrite the equations as polynomial expressions where one of the variables is included into the coefficients; this variable is dubbed the *eliminated variable*.
2. As many equations as the number of unknowns is needed; therefore, we may need to generate a new one by multiplying one of the equations by one of the unknowns, for instance, the equations are then casted into matrix form $\mathbf{Ax} = \mathbf{0}$, where $\mathbf{A}$ is a function of powers of the eliminated variable only, and $\mathbf{x}$ of the other unknowns; it should be noted that the last component of $\mathbf{x}$ is equal to 1.
3. Since one component of $\mathbf{x}$ is not equal to zero by definition, $\mathbf{A}$ must be singular; thus, its determinant has to be equal to zero; the next step is thus to compute the roots of $\det(\mathbf{A}) = 0$ to find the possible values of the eliminated variable.
4. The last step is to compute the null space of $\mathbf{A}$; knowing the last component must be equal to 1, we simply need to scale the obtain vector to make sure its last component is equal to 1.

Example: IKP of the Kinova Gen3 lite
The inverse kinematics problem of the Kinova Gen3 lite can be solved without the use of a numerical approach. Considering the number of joints, a large set of solutions are obtained for each feasible position and orientation of the end-effector. The methodology to solve the IKP of the Kinova Jaco manipulator, which shares an architecture similar to the Gen3 Lite, can be found in the literature (Gosselin and Liu, 2014). The feasible solutions for an arbitrarily chosen pose of the EE are shown in Fig. 10.4. Four are shown here, but more solutions could have been obtained if we did not take into account the joint rotational limitations. One unique solution can be chosen with a particular criterion, for instance, to minimize the joint rotations, to minimize the torque generated by joint actuator to lift a payload, to simply avoid obstacle, etc. While the topic of the optimal solution to the IKP will not be covered in this chapter, numerous criteria can be found in the literature.

Numerical Approach to the IKP
The method presented above to find the symbolic solution to the IKP is not necessarily adequate to all practical use cases. For instance, computing the roots of a high-degree polynomial, which is often the case with manipulators with several DoFs, may lead to numerical instabilities; thus, imprecision on the values of the joint coordinates obtained. The analytical approach may not be fast enough as well. Therefore, to avoid numerical instabilities and finding the symbolic solution to a challenging IKP, the numerical approach is often used in the industry. To this regards, we introduce the Newton-Gauss algorithm, but other avenues are possible. You first need to use the orientation and position of the end-effector to obtain a system of nonlinear equations that can be written as

$$\mathbf{f}(\mathbf{x}) = \mathbf{0} \tag{34}$$

Let the desired orientation matrix (defined for instance by Euler angles) and desired position vector

$$\mathbf{Q}_d = \begin{bmatrix} \mathbf{q}_{d,1} & \mathbf{q}_{d,2} & \mathbf{q}_{d,3} \end{bmatrix}, \quad \mathbf{p}_d = \begin{bmatrix} p_{x,d} & p_{y,d} & p_{z,d} \end{bmatrix}^T \tag{35}$$

Fig. 10.4 Possible postures
for the same EE pose

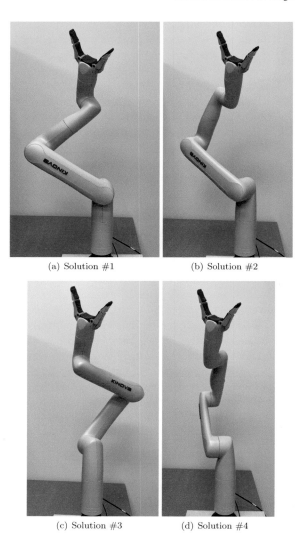

(a) Solution #1   (b) Solution #2

(c) Solution #3   (d) Solution #4

and the solution to the forward kinematics defined in Eq. (3a–3b). The former can
also be shown in a format similar to $\mathbf{Q}_d$ and $\mathbf{p}_d$:

$$\mathbf{Q} = \begin{bmatrix} \mathbf{q}_1 \ \mathbf{q}_2 \ \mathbf{q}_3 \end{bmatrix}, \quad \mathbf{p} = \begin{bmatrix} p_x \ p_y \ p_z \end{bmatrix}^T \tag{36}$$

For a generic 6-DoF serial manipulator, we thus have a system of 12 equations:

$$f = \begin{bmatrix} q_1 - q_{d,1} \\ q_2 - q_{d,2} \\ q_3 - q_{d,3} \\ p - p_d \end{bmatrix} = 0 \tag{37}$$

where f is a 12-dimensional vector, 0 is the null vector of the same dimension and the six unknowns are the joint coordinates we are looking for. The Newton-Gauss algorithm can now be applied to find x. Through this process, we will find a sequence of approximations of x, denoted $x_1, x_2, \ldots, x_k$ converging toward the solution of the IKP. The next estimation is denoted x_{k+1}. This algorithm is based on the Taylor series of the first degree; therefore, we have

$$x_{k+1} = x_k + \Delta x_k \tag{38a}$$

and

$$f(x_{k+1}) = f(x_k + \Delta x_k) = f(x_k) + J_f(x_k)\Delta x_k = 0 \tag{38b}$$

where $J_f(x_k)$ is the mathematical Jacobian of f with respect to x (Section 6.6.2), i.e., $J_f = \partial f/x$), evaluated at x_k. It should not be confused with the Jacobian(s) of the manipulator, which will be introduced later in this chapter. Equation (38b) can be rewritten as

$$J_f(x_k)\Delta x_k = -f(x_k) \tag{39}$$

To be able to compute the next increment Δx_k to obtain Δx_{k+1}, we thus need to solve the overdetermined system of equation defined by the above equation (J_f being a tall matrix, i.e., more rows than columns). Since you nearly never have an exact solution for an overdetermined system, we will find the solution minimizing the least squares of the error, known as the least square solution. This is done with the left Moore-Penrose generalized inverse J_f^L (Section 6.3.3), i.e.,

$$J_f^L = (J_f^T J_f)^{-1} J_f^T \tag{40a}$$

$$\Delta x_k = -J_f^L(x_k)f(x_k) \tag{40b}$$

You should **not** compute the generalized inverse *per se* with the equation above, since it is known to generate numerical issues (the condition number of $J_f^T J_f$ is, roughly, the square of that of matrix J_f itself, resulting into a badly conditioned system (Forsythe, 1970)). Instead, algorithms such as the QR decomposition and the householder reflections are used, achieving the same results while minimizing potential numerical issues.[7] Depending on the value of x_1, the algorithm will converge toward one feasible solution (if any). To obtain at least some of the other potential solutions (thus different configuration types), several starting points x_1 must be tested.

---

[7] Section 6.10.

Jacobian

The forward and inverse kinematics derived in the previous sections relate the joints coordinates to the position and orientation of the end-effector and vice-versa. Now, we consider the velocity of the EE and the joint rates. Mathematically, the relationship between both is the *Jacobian* of the function defining the FKP. The Jacobian is useful to plan smooth trajectory, to compute the *wrench* applied by the EE, to determine singular postures, etc. For your understanding, a wrench is the six-dimensional vector representation of forces and moments. Similarly, a *twist* is the six-dimensional vector representation of linear and angular velocities. The expressions of the twist and the wrench are, respectively,

$$\mathbf{t} \equiv \begin{bmatrix} \boldsymbol{\omega} \\ \dot{\mathbf{p}} \end{bmatrix}, \quad \mathbf{w} \equiv \begin{bmatrix} \mathbf{n} \\ \mathbf{f} \end{bmatrix} \tag{41}$$

where $\dot{\mathbf{p}}$, $\boldsymbol{\omega}$, $\mathbf{f}$ and $\mathbf{n}$ are the 3-dimensional linear velocity, angular velocity, force and moment, respectively.

The Jacobian for a *n*-link serial manipulator is a $(6 \times n)$ matrix mapping the n joint velocities into the six-dimensional vector consisting of the linear and angular velocities of the EE, i.e., the twist mentioned above. Let uss assume only revolute joints for now. Given the angular velocity vector of each link

$$\boldsymbol{\omega}_0 = \mathbf{0} \tag{42a}$$

$$\boldsymbol{\omega}_1 = \dot{q}_1 \mathbf{e}_1 \tag{42b}$$

$$\boldsymbol{\omega}_2 = \dot{q}_2 \mathbf{e}_2 + \boldsymbol{\omega}_1 \tag{42c}$$

$$\boldsymbol{\omega}_3 = \dot{q}_3 \mathbf{e}_3 + \boldsymbol{\omega}_2 \tag{42d}$$

$$\cdots$$

$$\boldsymbol{\omega}_n = \dot{q}_n \mathbf{e}_n + \boldsymbol{\omega}_{(n-1)} \tag{42e}$$

where $\dot{q}_i$ is the velocity of the *i*th joint, $\mathbf{e}_i$ is a unit vector parallel to the axis of the *i*th joint, namely the Z_i-axis of the *i*th DH frame, and $\mathbf{0}$ is the three-dimensional null vector. The angular velocity of the end-effector, $\boldsymbol{\omega}$, is simply equal to $\boldsymbol{\omega}_n$. As previously mentioned, the position of the EE is

$$\mathbf{p} = \sum_{i=1}^{n} \mathbf{a}_i \tag{43}$$

Differentiating the above equation with respect to time, we obtain

$$\dot{\mathbf{p}} = \sum_{i=1}^{n} \dot{\mathbf{a}}_i, \text{ where } \dot{\mathbf{a}}_i = \boldsymbol{\omega}_i \times \mathbf{a}_i, \quad i = 1, \ldots, n \tag{44}$$

Substituing Eqs. (42a) into (44), and with some manipulation, we obtain

$$\dot{\mathbf{p}} = \sum_{i=1}^{n} \dot{q}_i \mathbf{e}_i \times \mathbf{r}_i, \quad \mathbf{r}_i \equiv \sum_{j=i}^{n} \mathbf{a}_j \tag{45}$$

where $\mathbf{r}_i$ is defined as the vector from the ith DH frame to the last DH frame attached to the EE. We can rewrite the previous equations in a more compact matrix form:

$$\boldsymbol{\omega} = \mathbf{A}\dot{\mathbf{q}}, \quad \dot{\mathbf{p}} = \mathbf{B}\dot{\mathbf{q}} \tag{46}$$

with

$$\mathbf{A} \equiv \begin{bmatrix} \mathbf{e}_1 & \mathbf{e}_2 & \dots & \mathbf{e}_n \end{bmatrix}, \quad \mathbf{B} \equiv \begin{bmatrix} \mathbf{e}_1 \times \mathbf{r}_1 & \mathbf{e}_2 \times \mathbf{r}_2 & \dots & \mathbf{e}_n \times \mathbf{r}_n \end{bmatrix} \tag{47}$$

Therefore, the Jacobian mapping $\dot{\mathbf{q}}$ into $\mathbf{t}$ is

$$\mathbf{J} = \begin{bmatrix} \mathbf{A} \\ \mathbf{B} \end{bmatrix} = \begin{bmatrix} \mathbf{j}_1 & \mathbf{j}_2 & \dots & \mathbf{j}_n \end{bmatrix}, \quad \mathbf{j}_i = \begin{bmatrix} \mathbf{e}_i \\ \mathbf{e}_i \times \mathbf{r}_i \end{bmatrix} \tag{48}$$

where (3×6) submatrices $\mathbf{A}$ and $\mathbf{B}$ are, respectively, known as the orientation and position Jacobians.

Earlier in this section, we assumed only revolute joints to compute the Jacobian of a serial manipulator. If a ith joint is prismatic instead, the angular and linear velocities of the ith link are written as

$$\boldsymbol{\omega}_i = \boldsymbol{\omega}_{i-1}, \quad \dot{\mathbf{a}}_i = \boldsymbol{\omega}_{i-1} \times \mathbf{a}_i + \dot{d}_i \mathbf{e}_i \tag{49}$$

We can then prove that the contributing member of the ith joint to the Jacobian, i.e., the ith column, is expressed as

$$\mathbf{j}_i = \begin{bmatrix} \mathbf{0} \\ \mathbf{e}_i \end{bmatrix} \tag{50}$$

Example: Jacobian of a 6-DoF Wrist-Partitioned Serial Manipulator
Since the axes of the last joints of a wrist-partitioned serial manipulator intersect at one point, known as the spherical wrist, its Jacobian matrix is simplified, resulting in

$$\mathbf{J} = \begin{bmatrix} \mathbf{J}_{11} & \mathbf{J}_{12} \\ \mathbf{J}_{21} & \mathbf{0} \end{bmatrix} \tag{51}$$

where $\mathbf{0}$, $\mathbf{J}_{11}$, $\mathbf{J}_{12}$ and $\mathbf{J}_{21}$ are (3×3) matrices. You should note that to simplify the equations, the Jacobian matrix given here maps the joint rates into the twist of P_w, namely the location of the intersection of the axes of the last three joints. Therefore, we have

$$\mathbf{t}_w = \mathbf{J}\dot{\mathbf{q}} \tag{52}$$

where $\mathbf{t}_w = [\boldsymbol{\omega}^T \quad \dot{\mathbf{p}}_w^T]^T$. As you can see, the angular velocity vector $\boldsymbol{\omega}$ is not a function of the location of P_w. The linear velocity of P_w, which is only a function of the first three joint velocities, is computed with the following equation, i.e.,

$$\dot{\mathbf{p}}_w = \dot{q}_1\mathbf{e}_1 \times \mathbf{r}_1 + \dot{q}_2\mathbf{e}_2 \times \mathbf{r}_2 + \dot{q}_3\mathbf{e}_3 \times \mathbf{r}_3 \tag{53}$$

where $\mathbf{r}_i$ is defined as the vector from the ith DH frame to the P_w and $\mathbf{e}_i$ is the unit vector parallel to the axis of the ith joint, as mentioned above. The angular velocity of the EE is computed with the formula given earlier in this section, i.e.,

$$\boldsymbol{\omega} = \dot{q}_1\mathbf{e}_1 + \dot{q}_2\mathbf{e}_2 + \dot{q}_3\mathbf{e}_3 + \dot{q}_4\mathbf{e}_4 + \dot{q}_5\mathbf{e}_5 + \dot{q}_6\mathbf{e}_6 \tag{54}$$

Therefore, we can determine that the submatrices included in expression (51) are

$$\mathbf{J}_{11} = \begin{bmatrix} \mathbf{e}_1 & \mathbf{e}_2 & \mathbf{e}_3 \end{bmatrix} \tag{55a}$$

$$\mathbf{J}_{12} = \begin{bmatrix} \mathbf{e}_4 & \mathbf{e}_5 & \mathbf{e}_6 \end{bmatrix} \tag{55b}$$

$$\mathbf{J}_{21} = \begin{bmatrix} \mathbf{e}_1 \times \mathbf{r}_1 & \mathbf{e}_2 \times \mathbf{r}_2 & \mathbf{e}_3 \times \mathbf{r}_3 \end{bmatrix} \tag{55c}$$

Singularities

In robotics, when a manipulator is in a singular posture, or simply in a singularity, it cannot displace its EE along at least one direction. Mathematically, this corresponds to a **singular Jacobian matrix** use to compute joint velocities. We assumed previously this matrix was non-singular, i.e., for a robot with six DoFs, its Jacobian is inversible and its determinant is not equal to zero (Section 6.4). It might not be the case for certain configurations. Beyond the numerical issue of inverting a singular matrix, the corresponding posture of the robot also has a physical meaning related to the **limits of the workspace** of the robot or a **loss of mobility**, as mentioned above. Moreover, if we refer back to the configuration types discussed earlier in this chapter, the singularities correspond to **boundaries** between these entities within the workspace of the robot.

A posture close to a singularity is also problematic for a manipulator and a robot in general, as the determinant of its Jacobian matrix will be close to zero, yielding a division by a number close to zero. This will result in significantly high joint velocities, which raises safety concerns and reduces the trajectory-tracking accuracy. Let

$$\mathbf{t} = \mathbf{J}(\mathbf{q})\dot{\mathbf{q}} \tag{56}$$

where $\dot{\mathbf{q}}$, $\mathbf{t}$ and $\mathbf{J}(\mathbf{q})$ are, respectively, the n-dimensional joint-rate vector, the six-dimensional EE twist and the $6 \times n$ Jacobian matrix, where n is the number of joints. It is thus trivial to see that any given feasible EE twist, namely its linear and angular velocity, as defined in Sect. 10.4.4, is a linear combination of the joint velocities. To be able to achieve any arbitrary value of $\mathbf{t}$, the rank of $\mathbf{J}$, which is a function of the posture of the robot, i.e. $\mathbf{q}$, must be equal to six for a robot in 3D space. If it is the case, any given twist of the EE is feasible. However, it should be noted that since the Jacobian is posture-dependent, it is not always the case. If the rank($\mathbf{J}$) becomes lower than six, this is call a singular posture, or, for brevity, a singularity. Depending on which part of the Jacobian matrix generates a singularity, we can have a position or an orientation singularity, each having a different physical interpretation.

Singularity of the Position Jacobian

For a 6-DoF wrist-partitioned serial manipulator, a singularity of the submatrix $\mathbf{J}_{21}$ causes a position singularity, corresponding to the impossibility of computing the joint rates for this location. This occurs when the determinant of $\mathbf{J}_{21}$ is equal to zero. Considering Eq. (51), the determinant can be written as

$$\det(\mathbf{J}_{21}) = (\mathbf{e}_1 \times \mathbf{r}_1) \times (\mathbf{e}_2 \times \mathbf{r}_2) \times (\mathbf{e}_3 \times \mathbf{r}_3) = 0 \tag{57}$$

This situation occurs in two situations. First, you will find this type of singularity when one column of $\mathbf{J}_{21}$ is equal to zero, for instance, when $\mathbf{e}_i$ and $\mathbf{r}_1$ are parallel, which is commonly called a *shoulder singularity*. This particular case corresponds physically to the wrist center being located on the first joint axis, resulting in the instantaneous loss of one DoF. It can also be true for the second or third joint (wrist center being located on the ith joint axis), but this is usually avoided by carefully designing the manipulator.

Otherwise, we can also have $\det(\mathbf{J}_{21}) = 0$ when two columns of $\mathbf{J}_{21}$ become coplanar, resulting in a rank-deficiency. Multiple postures/configurations of the robot can lead to this, notably, but not only, a fully extended arm at the limit of the reachable workspace. This includes *elbow singularities*, which occurs for vertically articulated[8] manipulators such as the Meca500 sold by Mecademics.[9] when the wrist center lies on the plane passing through the second and third axes. This can also happen in theory with the manipulator folded on itself, but mechanical limits normally prevents this situation from occurring.

Singularity of the Orientation Jacobian

In the case of a wrist-partitioned manipulator, an orientation singularity occurs when $\det(\mathbf{J}_{12}) = 0$. This can only happen when $\mathbf{e}_4$, $\mathbf{e}_5$ and $\mathbf{e}_6$ are **coplanar**. In this configuration, only angular velocity vector on the plane generated by the three vectors mentioned above are possible at the EE. Considering the typical kinematic chain of a serial wrist-partitioned manipulator, it generally occurs when the axes of the fourth and sixth revolute joints are coincident. This type of singularity is sometimes called a **wrist singularity**.

Singularities with a Non-Wrist-Partitioned Manipulators

We now have seen the different singularities within the workspace of a serial wrist-partitioned manipulator thought an analysis of its Jacobian. Mathematically, you

---

[8] A vertically articulated architecture is common for commercially available wrist-partitioned six-axis serial manipulators: the axes of the second and third joints are parallel, the axes of the first and fourth joints are orthogonal to the axes of the second and third joints and the axis of fifth joint is orthogonal to the axes of fourth and sixth joints.

[9] https://www.mecademic.com/en/what-are-singularities-in-a-six-axis-robot-arm.

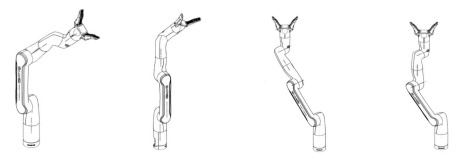

Fig. 10.5 Singular postures of the Kinova Gen3 lite (extracted from user manual)

can apply the same process to find singularities in the workspace of a non-wrist-partitioned manipulator. However, we will look at the full Jacobian matrix in this case, since we do not have decoupled kinematics for the orientation and position. To this aim, we will use the Kinova Gen3 lite previously mentioned to illustrate the process. Potential singular postures are shown in Fig. 10.5.

In this figure, from left to right, we have four different configurations corresponding to singularities of Jacobian matrix that differ from a fully extended arm, another singular configuration. We have, from left to right (all axes mentioned are illustrated in red in the figure),

1. The axis of the first joint and the X-axis of the third DH frame, i.e., the common perpendicular between the axes of joints 2 and 3, are parallel; the axes of joints 4 and 6 are also parallel.
2. The axes of the first and fourth joints are both parallel to the common perpendicular between the axes of joints 2 and 3.
3. The axes of the third and fifth joints are parallel; the fourth joint is also parallel to the common perpendicular between the axes of joints 2 and 3.
4. The axis of the third joint is parallel with the fifth joint axis and the fourth joint axis is parallel with the sixth joint axis.

All four cases illustrated above involve a double alignment in the posture of the Gen3 Lite, which loses a DoF momentarily. For example, in the second case, the EE cannot move in the direction of the fourth joint axis. In the third and fourth cases, motion is impossible in the direction of the axis of the third joint.

Kinematics of Parallel Manipulators

As we mentioned at the beginning of this chapter, parallel manipulators are known for their structural rigidity, speed and the ability to lift a larger payload compared to serial manipulators with similar mass and size. While their architecture is composed of at least one loop, they commonly have more. Among the well-known parallel architectures, the three-limb *Delta* (sometimes with a telescopic Cardan shaft to add a fourth DoF) (Clavel, 1990) as well as the four-limb *Par4* (Pierrot et al., 2003)

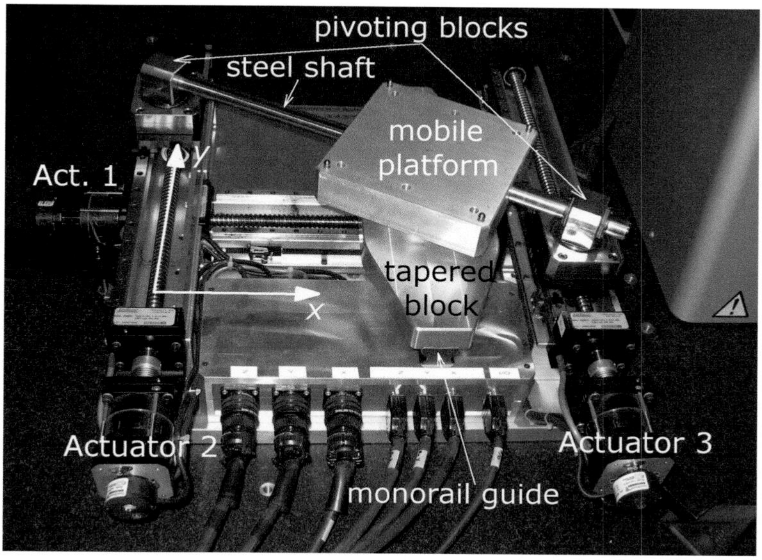

Fig. 10.6 PPR-2PRP parallel robot, from (Joubair et al., 2012)

(*Adept Quattro*) have been patented and commercialized. Before starting with the kinematics of parallel manipulators, you should know that the EE of a parallel robot is commonly called the mobile (or moving) platform, considering it is attached to the base with several limbs.

Direct and Inverse Kinematics

While solving the forward kinematics of a serial kinematic chain is generally a simple task, it is not the case with parallel robots. Indeed, the tool we used in Sect. 10.4.2, the Denavit-Hartenberg convention, is not appropriate for parallel manipulators, as it only accepts a maximum of two joints for each link. In general, it is not possible to obtain an explicit function of the Cartesian coordinates of the EE with respect to the joint coordinates, even for a simple parallel robot. Therefore, iterative methods are commonly used for this purpose.

Contrary to the forward kinematics, solving the IKP of a parallel robot is usually less challenging than with a serial robot. We will obtain an implicit function equal to zero where $\mathbf{q}$ and $\mathbf{p}$ are the variables, i.e.,

$$\mathbf{f}(\mathbf{q}, \mathbf{p}) = \mathbf{0} \tag{58}$$

Example: Kinematics of a PPR-2PRP Parallel Robot

Here is a planar parallel robot with three prismatic actuated joints connected to three limbs attached to the mobile platform, shown in Figs. 10.6 and 10.7. One is a PPR

Fig. 10.7 Geometry of a
PPR-2PRP parallel robot,
from (Joubair et al., 2012)

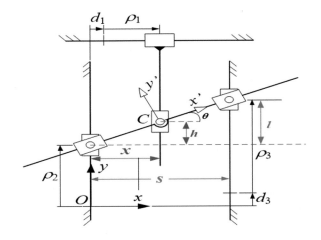

Fig. 10.7 Geometry of a
PPR-2PRP parallel robot,
from (Joubair et al., 2012)

chain, while the other two are PRP chains. The mobile platform's coordinates are
(x, y, θ), and the joint coordinates are (ρ_1, ρ_2, ρ_3). We thus need to find expressions
of the former as a function of the latter. Using simple geometric relationships, we
have:

$$\theta = \arctan\left(\frac{\rho_3 + d_3 - \rho_2}{s}\right) \tag{59a}$$

$$x = \rho_1 + d_1 \tag{59b}$$

For the last Cartesian coordinate, knowing

$$\frac{h}{x} = \frac{l}{s} \tag{60}$$

we can compute

$$y = \rho_2 + (\rho_1 + d_1)\frac{\rho_3 + d_3 - \rho_2}{s} \tag{61}$$

These three expressions above represent the solution to the FKP. The solution to
the IKP is straightforward from this point:

$$\rho_1 = x - d_1 \tag{62a}$$

$$\rho_2 = y - x\tan\theta \tag{62b}$$

$$\rho_3 = y + (s - x)\tan\theta - d_3 \tag{62c}$$

Jacobians

As mentioned above, the kinematics model of a parallel robot is generally expressed as an implicit function, namely Eq. (58). By differentiating it with respect to time, we have

$$\mathbf{J}\dot{\mathbf{p}} = \mathbf{K}\dot{\mathbf{q}} \qquad (63)$$

where both $\mathbf{J}$ and $\mathbf{K}$ are Jacobian matrices.

Singularities

From these two Jacobian matrices, we can define three types of singularities:

1. Type I: When $\mathbf{K}$ is singular, i.e., $\det(\mathbf{K}) = 0$. This usually corresponds to a limit of the reachable workspace or an internal limit of the workspace where two branches of solutions to the IKP meet. Therefore, certain Cartesian velocities at the EE will not be possible to generate.
2. Type II: When $\mathbf{J}$ is singular, i.e., $\det(\mathbf{J}) = 0$. These singularities occur at locations within the reachable workspace where two branches of solutions to the FKP meet. Therefore, even for a fixed joint coordinates, an infinitesimal motion of the end-effector is possible. This also means that the robot cannot balance certain external wrenches applied to the EE, thus resulting in a loss of control, which must be absolutely avoided.
3. Type III: A combination of both types above, thus when $\det(\mathbf{J}) = \det(\mathbf{K}) = 0$. In this case, Eq. (58) degenerates, resulting in an unusable EE. This kind of singularity only exists for certain architectures.

Figure 10.8 depicts singular postures of a pantograph, a common five-bar mechanism that can be used as a planar parallel manipulator. The EE is on the middle revolute joint and the two revolute joints attached to the base are actuated. As can be seen in this figure, the EE cannot move further up since the mechanism is fully extended for the illustrated type-I singularity. In the case of the type II singular posture depicted, it is impossible to control the vertical motion of the EE. With a small perturbation, the EE could move up or down for the same velocities of the actuated base joints.

Dynamics

According to the Merriam-Webster dictionary, dynamics is "a branch of mechanics that deals with forces and their relation primarily to the motion but sometimes also to the equilibrium of bodies."[10] Forces can be linear, but also rotational, namely torque.

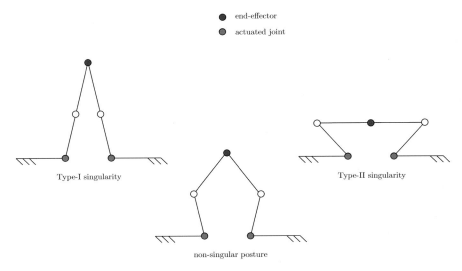

Fig. 10.8 Pantograph, a 2-DoF planar parallel manipulator

The second Newton's law is particularly significant when it comes to the quantitative analysis of the dynamics of a system, as it states that "the time rate of change of the momentum of a body is equal in both magnitude and direction to the force imposed on it."[11] Similarly to kinematics, we can define two different problem:

- forward dynamics, from the actuators to the motion, useful for simulations;
- inverse dynamics, from the motion to the actuators, essential for control.

In this chapter, a brief overview of two approaches to compute the dynamics model of a robot is given, namely the Euler-Lagrange and the Newton-Euler methods.

Euler-Lagrange

The Euler-Lagrange method is based on energy. The *Lagrangian* is defined as

$$\mathcal{L} = \mathcal{T} - \mathcal{V} \tag{64}$$

where T and V are, respectively, the total kinetic and potential energies in the system. From the Lagrangian, the dynamics equations defining the robot's motion are computed with

$$\frac{\mathrm{d}}{\mathrm{d}t}\left(\frac{\partial L}{\partial \dot{q}_i}\right) - \frac{\partial L}{\partial q_i} = \tau_i \tag{65}$$

---

[10] www.merriam-webster.com/dictionary/dynamics.

[11] Definition from www.britannica.com/science/Newtons-laws-of-motion.

Fig. 10.9 Geometry of a
2-DoF serial robot

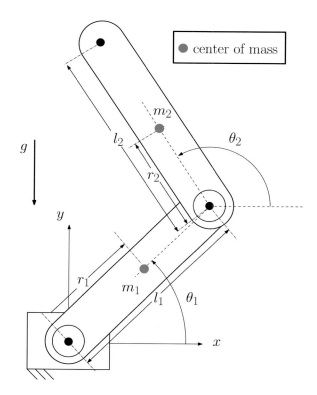

where the individual q_i and τ_i are, respectively, the generalized joint coordinates and torque (or force for a prismatic joint).

Example: Euler-Lagrange Applied to a 2-DoF Planar Manipulator

A simple 2-DoF serial planar manipulator is illustrated in Fig. 10.9. For the purpose of this example, only the mass of each link is considered and not their moment of inertia. The expression of total kinetic is

$$K = K_1 + K_2 = \frac{1}{2}m_1 v_1^2 + \frac{1}{2}m_2 v_2^2 \tag{66}$$

where v_1 and v_2 are the magnitude of the linear velocity of masses m_1 and m_2, respectively. We know, considering the geometry, that

$$v_1^2 = \dot{x}_1^2 + \dot{y}_1^2 = r_1^2 \dot{\theta}_1^2 \tag{67a}$$

$$v_2^2 = \dot{x}_2^2 + \dot{y}_2^2 \tag{67b}$$

$$\dot{x}_2 = (-l_1 \sin\theta_1 - r_2 \sin(\theta_1 + \theta_2))\dot{\theta}_1 - r_2 \sin(\theta_1 + \theta_2)\dot{\theta}_2 \tag{67c}$$

$$\dot{y}_2 = (l_1 \cos\theta_1 + r_2 \cos(\theta_1 + \theta_2))\dot{\theta}_1 + r_2 \cos(\theta_1 + \theta_2)\dot{\theta}_2 \tag{67d}$$

where m_1, r_1, l_1, l_2, r_2, m_2 and g are, respectively, the masses, distances between the origin of each link and its CoM and lengths of the first and second links and the gravitational acceleration. The total kinetic energy is thus

$$K = \frac{1}{2}m_1 r_1^2 \dot{\theta}_1^2 + \frac{1}{2}m_2 \left((l_1^2 + 2l_1 l_2 \cos\theta_2 + l_2^2)\dot{\theta}_1^2 + 2(l_2^2 + l_1 l_2 \cos\theta_2)\dot{\theta}_1\dot{\theta}_2 + l_2^2\dot{\theta}_2^2 \right)$$
(68)

Finally, again considering the geometry, the total potential energy is

$$T = T_1 + T_2 = m_1 g l_1 \sin\theta_1 + m_2 g \left(l_1 \sin\theta_1 + l_2 \sin(\theta_1 + \theta_2) \right)$$
(69)

You can complete the procedure as an exercise.

Newton-Euler

The Newton-Euler approach is a recursive method. You first compute the angular and linear velocities and accelerations of each link individually in the inertial frame, starting from the base. Then, the forces and torques applied by each link on the previous one are computed, starting from the end-effector. It is used here to solve the inverse dynamics of serial manipulators.

Velocities and Accelerations
First, it should be noted that in this procedure, it is the velocity and acceleration of the center of mass (CoM) of each body, not frame, that you need to compute. The velocities and accelerations are obtained with the Algorithm 1. In this table, the components of vectors $\mathbf{a}_i$ and $\mathbf{e}_i$ (cf. Fig. 10.2) in frame $(i + 1)$ are

$$[\mathbf{e}_i]_{i+1} = \begin{bmatrix} 0 & \sin\alpha_i & \cos\alpha_i \end{bmatrix}^T$$
(70a)

$$[\mathbf{a}_i]_{i+1} = \begin{bmatrix} a_i & b_i \sin\alpha_i & b_i \cos\alpha_i \end{bmatrix}^T$$
(70b)

Forces and Moments
The next step is to compute the forces and moments on each link, starting with the EE. The wrench applied by the $(i - 1)$th link on the ith link is defined as

$$\mathbf{w}_i = \begin{bmatrix} \mathbf{n}_i^T & \mathbf{f}_i^T \end{bmatrix}^T$$
(71)

where the three-dimensional vectors $\mathbf{n}_i^T$ and $\mathbf{f}_i^T$, are, respectively, the force and moment associated to this wrench. One component of each wrench is the actuation associated to the corresponding joint, namely the third component for a revolute joint and the sixth joint for a prismatic joint. The remaining components are the reaction force and moment between the two links. The procedure to compute the

---

Algorithm 1 Velocities and accelerations

Require: $[\boldsymbol{\omega}_0]_1$, $[\dot{\mathbf{c}}_0]_1$, $[\dot{\boldsymbol{\omega}}_0]_1$ and $[\ddot{\mathbf{c}}_0]_1$

  for $i = 1$ to n **do**

    if ith joint is revolute **then**

      $[\boldsymbol{\omega}_i]_{i+1} \leftarrow \mathbf{Q}_i^T [\boldsymbol{\omega}_{i-1}]_i + \dot{\theta}_i [\mathbf{e}_i]_{i+1}$

      $[\dot{\mathbf{c}}_i]_{i+1} \leftarrow \mathbf{Q}_i^T [\dot{\mathbf{c}}_{i-1}]_i + [\boldsymbol{\omega}_i]_{i+1} \times [(\mathbf{a}_i + \mathbf{s}_i)]_{i+1} - \mathbf{Q}_i^T [\boldsymbol{\omega}_{i-1} \times \mathbf{s}_{i-1}]_i$

      $[\dot{\boldsymbol{\omega}}_i]_{i+1} \leftarrow \mathbf{Q}_i^T [\dot{\boldsymbol{\omega}}_{i-1}]_i + \ddot{\theta}_i [\mathbf{e}_i]_{i+1} + \dot{\theta}_i (\mathbf{Q}_i^T [\boldsymbol{\omega}_{i-1}]_i) \times [\mathbf{e}_i]_{i+1}$

      $[\ddot{\mathbf{c}}_i]_{i+1} \leftarrow \mathbf{Q}_i^T [\ddot{\mathbf{c}}_{i-1}]_i + [\dot{\boldsymbol{\omega}}_i]_{i+1} \times [(\mathbf{a}_i + \mathbf{s}_i)]_{i+1} + [\boldsymbol{\omega}_i]_{i+1} \times [\boldsymbol{\omega}_i]_{i+1} \times [(\mathbf{a}_i + \mathbf{s}_i)]_{i+1} -$
$\mathbf{Q}_i^T [\dot{\boldsymbol{\omega}}_{i-1} \times (\boldsymbol{\omega}_{i-1} \times \mathbf{s}_{i-1})]_i$

    else if ith joint is prismatic **then**

      $[\boldsymbol{\omega}_i]_{i+1} \leftarrow \mathbf{Q}_i^T [\boldsymbol{\omega}_{i-1}]_i$

      $[\dot{\mathbf{c}}_i]_{i+1} \leftarrow \mathbf{Q}_i^T [\dot{\mathbf{c}}_{i-1}]_i + [\boldsymbol{\omega}_i]_{i+1} \times [(\mathbf{a}_i + \mathbf{s}_i)]_{i+1} + \dot{d}_i [\mathbf{e}_i]_{i+1} - \mathbf{Q}_i^T [\boldsymbol{\omega}_{i-1} \times \mathbf{s}_{i-1}]_i$

      $[\dot{\boldsymbol{\omega}}_i]_{i+1} \leftarrow \mathbf{Q}_i^T [\dot{\boldsymbol{\omega}}_{i-1}]_i$

      $[\ddot{\mathbf{c}}_i]_{i+1} \leftarrow \mathbf{Q}_i^T [\ddot{\mathbf{c}}_{i-1}]_i + [\dot{\boldsymbol{\omega}}_i]_{i+1} \times [(\mathbf{a}_i + \mathbf{s}_i)]_{i+1} + [\boldsymbol{\omega}_i]_{i+1} \times [\boldsymbol{\omega}_i]_{i+1} \times [(\mathbf{a}_i + \mathbf{s}_i)]_{i+1} -$
$\mathbf{Q}_i^T [\dot{\boldsymbol{\omega}}_{i-1} \times (\boldsymbol{\omega}_{i-1} \times \mathbf{s}_{i-1})]_i + 2[\boldsymbol{\omega}_i]_{i+1} \times \dot{b}_i [\mathbf{e}_i]_{i+1} + \ddot{b}_i [\mathbf{e}_i]_{i+1}$

    end if

  end for

---

wrench on each link is detailed in Algorithm 2. You may wonder where the effect of gravity appears in the algorithm. To simplify the procedure while still obtaining an equivalent solution, we use a simple trick. Here, we suppose a virtual acceleration $-\mathbf{g}$ at the base of the robot, namely the first link. Therefore, even though the base is fixed and not moving, we have

$$[\ddot{\mathbf{c}}_0]_1 = [-\mathbf{g}]_1 \tag{72}$$

where $-\mathbf{g}$ is the gravitational acceleration.

Algorithm 2 Wrench on each link

$[\mathbf{f}_n]_n \leftarrow \mathbf{Q}_n [m_n \ddot{\mathbf{c}}_n - f]_{n+1}$

$[\mathbf{n}_n]_n \leftarrow \mathbf{Q}_n [\mathbf{I}_n \dot{\boldsymbol{\omega}}_n + \boldsymbol{\omega}_n \times \mathbf{I}_n \boldsymbol{\omega}_n - \mathbf{n} + (\mathbf{a}_n + \mathbf{s}_n) \times \mathbf{f}_n]_{n+1}$

for $i = n - 1$ to 1 **do**

  $[\mathbf{f}_i]_i \leftarrow \mathbf{Q}_i [m_i \ddot{\mathbf{c}}_i + f_{i+1}]_{i+1}$

  $[\mathbf{f}_i]_{i+1} \leftarrow \mathbf{Q}_i [\mathbf{f}_i]_i$

  $[\mathbf{n}_i]_i \leftarrow \mathbf{Q}_i [\mathbf{I}_i \dot{\boldsymbol{\omega}}_i + \boldsymbol{\omega}_i \times \mathbf{I}_i \boldsymbol{\omega}_i + \mathbf{n}_{i+1} + (\mathbf{a}_i + \mathbf{s}_i) \times \mathbf{f}_i - \mathbf{s}_i \times \mathbf{f}_{i+1}]_{i+1}$

end for

We now have all forces $\mathbf{f}_i$ and moments $\mathbf{n}_i$; the final step is thus to compute what we were looking for at the beginning, the actuation torques for revolute joints and actuation forces for prismatic joint. This is done with the following two equations:

$$\tau_i = \mathbf{e}_i^T \mathbf{n}_i, \quad \text{for revolute joints} \tag{73a}$$

$$\tau_i = \mathbf{e}_i^T \mathbf{f}_i, \quad \text{for prismatic joints} \tag{73b}$$

Chapter Summary

In this chapter, an introduction to the fundamentals of robotics manipulators, from the mechanics point-of-view, was given. We first introduced the typical architectures, serial and parallel, their pros and cons, as well as notable characteristics such as their DoFs and the type of motion they can generate. Then, we focused on the kinematics of both categories, from the joints to the end-effector (direct kinematics) and the other way around (inverse kinematics). While standard approaches exist for both the forward and inverse kinematics of a serial manipulator, notably if it is wrist-partitioned, it is not the case for their parallel counterparts. However, the IKP of a parallel robot can generally be solved more easily. We have studied the relations between the joint velocities and the twist of the EE, which includes its angular and linear velocities. These equations can be put together to obtain the Jacobian matrix, an useful tool in the analysis of serial and parallel robots. Indeed, from this matrix, singular postures of the robot can be found: these configurations must be avoided, because they may cause safety and control issues. Finally, we did a brief overview of the dynamics of robotic manipulators, namely two common approaches, Euler-Lagrange and Newton-Euler.

Revision Questions

Question #1
Which equations are valid (there could be more than one or none)?

1. $\mathbf{H}_{tool}^{workshop} = \mathbf{H}_{workshop}^{0} \mathbf{H}_1^0 \mathbf{H}_2^1 \mathbf{H}_3^2 \mathbf{H}_{tool}^3$
2. $\mathbf{H}_{tool}^{workshop} = \mathbf{H}_{workshop}^{0} \mathbf{H}_1^0 \mathbf{H}_2^1 \mathbf{H}_3^2 \mathbf{H}_0^{workshop}$
3. $\mathbf{H}_3^0 = \mathbf{H}_1^0 \mathbf{H}_2^1 \mathbf{H}_3^2$
4. $\mathbf{H}_0^3 = \mathbf{H}_1^0 \mathbf{H}_2^1 \mathbf{H}_3^2$
5. $\mathbf{H}_{EE}^{workshop} = \mathbf{H}_{workshop}^{0} \mathbf{H}_0^1 \mathbf{H}_1^2 \mathbf{H}_2^3 \mathbf{H}_3^{EE}$

Question #2
Inverse kinematics makes it possible to obtain. . .

Please choose an answer:

1. the pose of the robot effector, based on its parameters and joint coordinates;
2. the values of the joint coordinates of the robot, from the pose of the effector and the parameters of the robot;
3. the position of the robot effector, based on its parameters and joint coordinates.

Question #3
The kinematic chains of parallel robots are made of:

1. passive and active joints;
2. only passive joints;
3. only active joints.

Table 10.2   DH parameters of a wrist-partitioned 6R manipulator

i	1	2	3	4	5	6
a_i	0	135 mm	38 mm	0	0	0
d_i	135 mm	0	0	120 mm	0	70 mm
α_i	$-\pi/2$	0	$-\pi/2$	$\pi/2$	$\pi/2$	0
θ_i	q_1	$q_2 - \pi/2$	q_3	q_4	$q_5 + \pi$	q_6

Question #4
Regarding the computation of the Jacobian matrix of a serial manipulator, the vector $\mathbf{e}_{i-1}^{workshop}$ represents:

1. the unit vector parallel to the X-axis of the $(i-1)$th frame with respect to the workshop;
2. the unit vector parallel to the Y-axis of the $(i-1)$th frame with respect to the workshop;
3. the unit vector parallel to the Z-axis of the $(i-1)$th frame with respect to the workshop.

Question #5
The DH parameters of a wrist-partitioned manipulator are given in Table 10.2. First, compute the six homogeneous transformations matrices. Then, compute the solution(s) to the inverse kinematics for a Cartesian position of the wrist of $(250, 0, 150)$ mm and an orientation with the Euler angles of $(0°, 90°, 0°)$ according to the *XYZ* mobile convention.

Further Reading

This chapter only gave you a short summary on the mechanics of robotic manipulators. If you want to learn more, you can first take a look into the original DH notation and its variants. To this aim, you can refer to a paper published by Harvey Lipkin (Lipkin, 2008). Moreover, given the fact that we did not go into the details of the dynamics of robots, extensive literature can be found on this topic. Notably, you can look into the Kane's equations, similar to Lagrangian approach. Also, (Angeles, 2014) introduced an alternative method to solve the inverse dynamics of a robotic system, the *natural orthogonal complement* (NOC). Regarding mathematical tools useful for the analysis of the mobility, kinematics and dynamics of robotic systems and mechanisms, you can take a look into group theory (Angeles, 2014), screw theory (Davidson, 2004; Müller, 2017), where the twist and wrench concept originate, and dual-numbers algebra, useful to combine a translation and a rotation into one single variable. Finally, you can also look into the concept of *constraint singularities* for parallel mechanisms (Zlatanov et al., 2002).

References

Angeles, J. (2014). *Fundamentals of Robotic Mechanical Systems, Mechanical Engineering Series* (Vol. 124). Springer International Publishing. https://doi.org/10.1007/978-3-319-01851-5

Clavel, R. (1990). *Device for the movement and positioning of an element in space*

Davidson, J. K. (2004). *Robots and screw theory?: Applications of kinematics and statics to robotics.* Oxford University Press.

Forsythe, G. E. (1970). Pitfalls in Computation, or Why a Math Book isn't Enough. *The American Mathematical Monthly, 77*(9), 931–956. https://doi.org/10.1080/00029890.1970.11992636

Gosselin, C., & Liu, H. (2014). Polynomial Inverse Kinematic Solution of the Jaco Robot. In *ASME International Design Engineering Technical Conferences and Computers and Information in Engineering Conference, ASME*, Buffalo, NY, p. V05BT08A055. https://doi.org/10.1115/detc2014-34152

Hartenberg, R., & Denavit, J. (1964). *Kinematic synthesis of linkages.* McGraw-Hill Book Company.

Joubair, A., Slamani, M., & Bonev, I. A. (2012). A novel XY-Theta precision table and a geometric procedure for its kinematic calibration. *Robotics and Computer-Integrated Manufacturing, 28*(1), 57–65. https://doi.org/10.1016/J.RCIM.2011.06.006

Lipkin, H. (2008). A note on Denavit-Hartenberg notation in robotics. In *Proceedings of the ASME International Design Engineering Technical Conferences and Computers and Information in Engineering Conference*, DETC2005, Vol.7B, pp. 921–926. https://doi.org/10.1115/DETC2005-85460

Makino, H., Kato, A., & Yamazaki, Y. (2007). Research and commercialization of SCARA Robot: The case of industry-university joint research and development. *International Journal of Automation Technology, 1*, 61–67.

Müller, A. (2017). Screw theory: A forgotten Tool in Multibody Dynamics. *PAMM, 17*(1), 809–810. https://doi.org/10.1002/PAMM.201710372

Pierrot, S., Morita Pierrot, F., Shibukawa, T., & Morita, K. (2003). *Four-degree-of-freedom parallel robot*

Zlatanov, D., Bonev, I. A., & Gosselin, C. M. (2002). Constraint singularities of parallel mechanisms. *Proceedings—IEEE International Conference on Robotics and Automation, 1*, 496–502. https://doi.org/10.1109/ROBOT.2002.1013408

Bruno Belzile is a postdoctoral fellow at the INIT Robots Lab. of ÉTS Montréal in Canada. He holds a B.Eng. degree and Ph.D. in mechanical engineering from Polytechnique Montréal. His thesis focused on underactuated robotic grippers and proprioceptive tactile sensing. He then worked at the Center for Intelligent Machines at McGill University, where his main areas of research were kinematics, dynamics and control of parallel robots. At ÉTS Montréal, he aims at creating spherical mobile robots for planetary exploration, from the conceptual design to the prototype.

David St-Onge (Ph.D., Mech. Eng.) is an Associate Professor in the Mechanical Engineering Department at the École de technologie supérieure and director of the INIT Robots Lab (initrobots.ca). David's research focuses on human-swarm collaboration more specifically with respect to operators' cognitive load and motion-based interactions. He has over 10 years' experience in the field of interactive media (structure, automatization and sensing) as workshop production director and as R&D engineer. He is an active member of national clusters centered on human-robot interaction (REPARTI) and art-science collaborations (Hexagram). He participates in national training programs for highly qualified personnel for drone services (UTILI), as well as for the deployment of industrial cobots (CoRoM). He led the team effort to present the first large-scale symbiotic integration of robotic art at the IEEE International Conference on Robotics and Automation (ICRA 2019).

Chapter 11
Get Together! Multi-robot Systems: Bio-Inspired Concepts and Deployment Challenges

Vivek Shankar Varadharajan and Giovanni Beltrame

11.1 Objectives of the Chapter

At the end of this chapter, you will:

- understand the different types of multi-robot systems,
- be aware of the task allocation problem,
- be able to point out the different types of swarm programming techniques,
- be familiar with the fundamentals of swarm programming,
- understand the real-world deployment challenges with robot swarm.

11.2 Introduction

Swarm robotics is a branch of robotics that focuses on multi-robot systems that coordinate to perform complex tasks through simple behavioral rules. Swarm robotics combines multi-robot systems with swarm intelligence (Bonabeau et al., 1999), a field that studies how complex behaviors emerge from simple and local interactions (Dorigo et al., 2021) in natural systems like schools of fish, flocks of birds and colonies of insects (see Fig. 11.1). These natural systems are of high interest because they exhibit efficiency, robustness, parallelism and adaptivity. Ant colonies are an excellent model for swarm intelligence, as ants work in parallel and use incredibly low amounts of energy to perform tasks (efficiency), the loss of several ants does not compromise the colony (robustness), and they can overcome complex environmental challenges: as an example, fire ants can form rafts with their bodies to carry the colony to safety in case of floods (adaptivity). Swarm robotics research started out as an use case to swarm intelligence on virtual and physical agents. Swarm intelligence is a property of groups of simple individuals whose collective behavior exhibit

V. S. Varadharajan (✉) · G. Beltrame
Department of Computer and Software Engineering, Polytechnique Montréal, Montreal, Canada
e-mail: vivek-shankar.varadharajan@polymtl.ca

G. Beltrame
e-mail: giovanni.beltrame@polymtl.ca

D. Herath and D. St-Onge (eds.), *Foundations of Robotics*,
https://doi.org/10.1007/978-981-19-1983-1_11

Fig. 11.1 Some examples of natural swarms are a flock of birds, colony of bees, schools of fish and swarms of ants. *Credits* Bee colony—flickr.com/Sy, Fish school—iStock.com/armiblue, Army ants—flickr.com/Axel Rouvin, Ant raft—wikimedia.org/TheCoz and Starling swarm—wikimedia.org/Walter Baxter

capabilities that are beyond the capacity of a single individual. The phenomenon of having many simple things performing complex activities when working as a group is known as *emergence*.

Swarm intelligence was initially applied to virtual agents as an approach to solve optimization problems that are otherwise considered very hard. Some examples of such computational algorithms are ant colony optimization (Dorigo et al., 2006) and particle swarm optimization (Kennedy & Eberhart, 1995). Ant colony optimization applies the foraging behavior observed in ants to optimization: a group of simulated agents move randomly in the search space (i.e., the space of possible parameters), locate optimal solutions and lay virtual pheromones (analogue to the chemical traces left by real ants) to direct other agents. Similarly, particle swarm optimization uses a group of agents moving in a search space. These techniques have been very successful in a wide range of domains like antenna design (Chang et al., 2012), vehicle routing (Bell & McMullen, 2004), and scheduling problems (Xing et al., 2010).

Applying swarm intelligence to multi-robot systems in the real world is not as straightforward as for virtual agent based optimization algorithms: robots need to perceive their environment, determine their position, interact with other robots and the (potentially unstructured) environment itself. Performing all these activities in a single complex robot is already a daunting challenge, and having them emerge from the interaction of many simple robots requires novel approaches to design and synthesize robotic systems. This additional complexity means that only a very limited number of works have demonstrated out-of-the-laboratory operation capability and there is no real-world application to date that directly uses swarm robotics design principles (Dorigo et al., 2021). However, swarm robotics is rapidly finding new application domains (logistics, agriculture, space exploration, and many others) in which it can provide a definite advantage, and swarm-based real-world applications are bound to happen in the near future.

In this chapter, we will provide a brief introduction to multi-robot and swarm system design approaches, swarm programming concepts and finally outline some challenges to be addressed in realizing a real-world swarm system.

An Industry Perspective

Patrick Edwards-Daugherty

Spiri Robotics

My formal education was in mathematics, applied to theoretical physics. I began programming at a young age, among my other interests in chess, music, and science fiction. As a child, I was inspired by the positive and hopeful thinking of imaginative writers and scientists. About a year after graduating in 1998, I started a tech company. I pivoted to robotics in 2012. The use of robots for space exploration had always been interesting to me from a distance. But that year, when I saw early displays of small drones able to maneuver without human control, I became convinced of a tangible possibility to create truly autonomous robots that could improve the human condition.

When my company's robotics team was at the first major public exhibition of our work, at the most embarrassing moment, our batteries caught fire in their recharging cradles. For the rest of the conference, a security guard with a fire extinguisher was stationed next to our display. He was very pleasant and supportive. In my journey with robotics, I have found the biggest cliffs are the ones right between the "completion" of a design and algorithm on the board, and the first field test that works out. As a result, at my company, we try to fail fast and often (and as much as possible, inexpensively) as part of our method. Ensemble action by autonomous agents, sometimes called swarming or flocking, first needed a basic method for group communications and consensus. The way a flock of starlings or a school of anchovy can move as one is an inspiration. The communications part has come a long way in the past decade. The next challenge, which will remain a challenge for a long time, is to figure out what actions are useful for the robotic ensembles to engage in, and what, specifically, are the desired outcomes, so these can be programmed and optimized. The communication part is the underlying first step, and each action can be thought of as analogous to a group behavior of an animal species. There are many, and they are very particular to the need and context.

11.3 Types of Multi-robot Systems

Robot swarms are a special type of multi-robot systems that rely on three guiding principles: (a) control is *decentralized* (i.e., there are no external controlling entities); (b) there are no leaders or predefined roles; and (c) robots make decisions based on local interaction with other robots. To better understand robot swarms, we must introduce a taxonomy of multi-robot systems, clarifying the differences between decentralized approaches (such as robot swarms) and other types of multi-robot systems. Multi-robot systems (MRS) are generally considered to have two or more robots that coordinate to perform a task. The robots in an MRS can be simple, as the actual potential of the system can lie within group's emergent behavior. Consider the task of collaborative transport (as seen in natural ant colonies): robots need to lift and move an object that would be too heavy for a single robot. In this case, a single robot is incapable of performing the task, but several robots can, although requiring a high degree of coordination. In general, multi-robot systems are preferred for large, spatially distributed tasks which benefit from the inherent parallelism of using multiple robots.

An MRS can be homogeneous or heterogeneous: a homogeneous MRS is composed of identical robots (same sensors, computing resources and actuators), while a heterogeneous MRS contains robots that are fundamentally different (in sensors, computing resources and/or actuators). Homogeneous MRSs are the most common type of MRS because they are relatively simple to design and manage, whereas heterogeneous MRS design needs sophisticated task planning to determine the appropriate type of robot to perform each task.

MRSs can be further classified into centralized, distributed and decentralized based on the decision making strategy that they use, as illustrated in Fig. 11.2. In Fig. 11.2: (a) centralized system with each robot connecting to a central server, the centralized server performs the decision making. (b) One of the robots is elected to perform decision making in a distributed system. (c) All the robots in a decentralized system perform decision making on-board by collecting information from other robots.

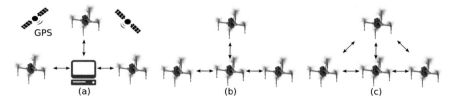

Fig. 11.2 Decision making architecture classes: **a** centralized, **b** distributed and **c** decentralized

Similar to other fields of research, decision making in an MRS can be considered as a process of analyzing a sequence of alternatives to determine the best choice of action to perform, using the available information. Most general forms of MRS design brakes down the global problem into smaller sub-problems that can be assigned to individual robots. Individual robots take up one or more of these sub-problems and work toward solving them. Task allocation (TA) is a process of optimally assigning tasks to a robot that will maximize the overall system performance and it can be considered as an example of decision making. TA in a MRS is commonly referred to as Multi-Robot Task Allocation (MRTA), where a set of tasks are assigned optimally to a set of robots to maximize the overall performance of the system.

An MRTA is generally modeled as a combinatorial optimization problem: consider N_t to be the number of sub-tasks that need to be assigned to N_r robots to minimize the global combined cost, or maximize the reward. The cost function (a metric to define the quality of global task performance) and the customized constraints on the optimization for each of the robots depend on the specific task performed by the robots. The goal of the optimization problem is to obtain a specific sub-task assignment for the robots, which is generally defined by the tuple (r_i, t_i), where $t_i \in 1, \ldots, N_t$ and $r_i \in 1, \ldots, N_r$. A generalized MRTA problem is of the following form:

$$\max \sum_{r_i=1}^{N_R} \sum_{t_i=1}^{N_T} b_{it} x_{it} \tag{11.1}$$

subject to

$$\sum_{r_i=1}^{N_R} x_{it} \leq 1 \qquad \forall t_i \in 1, \ldots, N_T$$

$$\sum_{t_i=1}^{N_T} x_{it} \leq L_T \qquad \forall r_i \in 1, \ldots, N_R$$

$$x_{it} \in \{0, 1\} \qquad \forall r_i \in 1, \ldots, N_R \quad \forall t_i \in 1, \ldots, N_T$$

where b_{it} is the reward accumulated by assigning the task t_i to robot r_i. $x_{it} \in \{0, 1\}$ is a binary variable indicating whether robot r_i is assigned to task t_i. The constraint $\sum_{r_i=1}^{N_R} x_{it} \leq 1, \forall t_i \in 1, \ldots, N_T$ restricts that assignment of one single task to one robot. L_T indicates the maximum number of tasks that can be assigned to each robot; when $L_T = 1$, it is referred to as single-assignment problem, where every single robot only performs one task.

11.3.1 Centralized Multi-robot System

Centralized MRS generally have a single hub, either a server or a robot, which gathers the sensory data from all the robots and aggregates a global view to then perform task allocation. A centralized MRS effectively is one large system with a global view of the environment and the states of all the robots, and hence, this system has the ability to produce globally optimal task assignment and plans. There exists a wide variety of centralized decision systems (Luna & Bekris, 2011; Wurm et al., 2008; Yan et al., 2010). Some of them rely heavily on centralized localization (like Global Positioning System, GPS) and a few other approaches (McLurkin, 2009) suited for indoor applications use motion capture systems or ceiling-mounted camera. An interesting example of a centralized system is the Intel Shooting Star drones, which have been used in several light shows. These aerial robots form a large pack that operate synchronously to create 3D visual effects in the night sky. These drones have a centralized coordination stations to plan predetermined GPS trajectories and perform role-specific behaviors that are pre-scripted.

While these pre-scripted displays are impressive, the ability of centralized approaches to handle dynamic environments is limited and does not scale efficiently for larger numbers of robots. Centralized approaches also have other drawbacks—they are not robust to robot failure and are vulnerable to security threats due to their single point of failure: if the centralized hub malfunctions or compromised, the system is rendered useless.

11.3.2 Distributed Multi-robot System

Distributed MRS uses opportunistic centralization, where one robot in the system (referred to as the "master") is elected to act as a centralized hub that receives task-related information from all robots for TA. The term opportunistic centralization is used mainly because centralization is performed only for the time being until the TA is performed; for the next round of TA, a different robot or the same robot is used. Distributed MRS is used in a wide range of application domains, for instance, in formation control (Michael et al., 2008), exploration control (Sheng et al., 2006) and navigation control (Fan et al., 2020). Distributed MRS is comparable to a distributed computing cluster (Hwang et al., 2013). The main difference with a distributed computing cluster and a distributed MRS is that the nodes rely on a static topology with reliable communication, failures are rare (nodes operate in safe server rooms), and state of the system is completely controllable. The election of a master compute node and task assignment in a distributed MRS is generally done through auction, voting and assignment. In an assignment, the master node aggregates all the information of other nodes and makes a decision on the task assignment without any feedback from the other nodes in the system. In contrast to assignment, both auction and voting involves receiving a resource estimate (bid) or a preference for performing a certain task from each robot in the system to perform the TA.

Auction

An auction can be generally considered an activity in which a seller presents an item for sale to a set of buyers. For instance, in a distributed MRS, the auctions are for assigning a particular sub-task to a given robot in the system, with the seller being the central robot and the buyers being all other robots in the system. An auction is a preferred routine when the sellers do not have a good estimate on the buyers true value of an item. Here, we will briefly discuss the common types of auctions, and for a more detailed comparison, you can refer to Chap. 9 of Easley et al. (2012).

- English auctions: This type of auctions is also known as the ascending-bid auction. In this type of auction, the seller raises the price of an item and the bidders drop out of the auction gradually until there remains only one final buyer, who is declared as the winner.
- Dutch auction: This type of auctions is also known as the descending-bid auction. In this type of auction, the bidders gradually decrease the price of the item until one of the bidder accepts the current price. The bidder that accepts the current price is declared as the winner.
- Japanese auction: In this type of auction, the value of the item starts with a zero price. The bidders gradually increase the price, bidder leave auction when the price becomes too high and the last bidder standing is declared as the winner.
- First-price sealed-bid: In this type of auction, the bidders submit closed bids that are unknown to other bidders. The highest bidder is declared as the winner.
- Second-price sealed-bid: This type of auctions are also known as the Vickery auctions, named after the Noble price winner William Vickrey. In this type of auction, the bidders submit closed bids to the seller; the highest bidder is declared as the winner and will pay the second-highest bid value.

The auctions can be further classified into sequential, parallel and combinatorial based on the order the tasks are sold by the seller. In a sequential auctions, the seller sells the items one at a time until all the items are sold; the auction lasts several rounds until all the items are sold. Parallel auction requires the seller to sell all the items at once and the buyers bid on it in parallel; the auction of all the items is performed in one single round. In a combinatorial auction, the seller sells a combination of different items, and the bids are cast on packages of items; the seller sells the items based on an assignment that maximize the revenue.

In a MRS, the bids are generally determined by the sellers' cost (resources required) in performing the task. For example, if the tasks (item) correspond to spatial goals the robots have to reach, the cost of reaching the goal (distance) is fixed. First price sealed bid is one of the most commonly used type of auction in MRS (Otte et al., 2020), mainly because of the nature of the tasks involved and the auctions can be performed in a single round rather than multiple rounds (as in English, Dutch and Japanese auction). However, there are type of tasks that are favorable for multi-round auctions that use other type of auctions. We refer the reader to Otte et al. (2020) for more information on types of auctions that are used in MRS for different types of tasks.

Voting

The voting is generally considered to be an activity in which a group of individuals express their preference over a sequence of alternatives which are aggregated to obtain the preference of the whole group. Voting and auctions are both used to aggregate information across a group; thus, it might be hard to completely distinguish between the two. However, the circumstances under which voting and auction are used can be clearly distinguished. Voting is applied when the group is trying to reach a single distinct decision that defines the group preference using individual preferences, whereas auction is applied when an estimate on the choices (bids) can be used to aggregate the preferences.

In voting, there exists a set of alternatives m that needs be ranked by each individual as strictly dominating $A \prec_i B$ or as weakly dominating $A \preceq_i B$, with A and B being the two alternatives. $A \prec_i B$ means that individual i strictly prefers alternative A over alternative B and $A \preceq_i B$ means A is preferred weakly by individual i over B. With these individual preferences in hand, different types of rules are applied in various voting systems to aggregate the individual preference.

- Plurality: This type of voting is also called the majority rule and considered the most natural way of voting on alternatives. In this type of voting, each alternative receives a score when it is ranked first and the group preference is a ranking that is produced with the aggregated score.
- Borda Count: An alternative receives a score of m when it is ranked first by an individual, receives m-1 when it is ranked second and 0 when it is ranked last. A summation of all the individual ranking scores are produced to obtain the aggregated group preference.
- Copeland Count: Elections are conducted pairwise between individuals; the alternatives that win a pairwise election receive a score of two, a score of one for a tie and no points for a defeat. An aggregated score of all the alternatives are produced to obtain the group ranking. The alternative that wins the most pairwise election is ranked first.
- Bucklin: The alternative that receives a first ranking from more than half of the voters is placed as the first group ranking alternative; if there is no alternative that is preferred first, then the second alternative preferred by more than half of the voters is ranked first. This process of selecting the alternative is iteratively performed to obtain the group ranking.
- STV: The alternative that receives the least votes is removed in each round of voting and the last standing alternative is ranked first.
- Slater: A combined ranking is produced with alternatives that is consistent with the majority of the pairwise elections.
- Kemeny: A group ranking of alternatives is produced based on as few disagreements as possible. For each disagreement, an alternative is pushed behind in ranking, and a final ranking is produced with this final disagreement ranking of alternatives.

These voting procedures are reasonable and produce desirable properties, but it might be difficult to clearly distinguish the advantages between these procedures (Kacprzyk et al., 2020). A well founded set of evaluation criterion might be required to evaluate the different advantages of these systems. In the context of MRS, plurality or majority count is commonly used, since it is the most natural form of voting and applies to the type of tasks dealt in MRS (Karpov et al., 2016). There are some application scenarios like formation selection (Iocchi et al., 2003), where voting procedures like Bucklin is applied to select a formation that is preferred by more than half of robots.

Many of these voting systems produce different group preferences based on the order the voting is conducted; this gives rise to two important properties: Unanimity and independence of irrelevant alternatives (IIA). Unanimity states that when $A \prec_i B$ is the preference of every individual in the group, then the group ranking should reflect this preference. Whereas, IIA requires that the group ranking between A and B should only depend on the individual preferences between A and B and not on the preference of other alternatives. Using these two properties, we can now state Arrow's impossibility theorem: *A voting system that satisfies both unanimity and IIA must correspond to a dictatorship by one individual, when there are three alternatives or more.*

Arrows impossibility theorem essentially means the voting system that satisfies both Unanimity and IIA will not suffer from the drawback of the order in which voting is conducted. However, there is no voting system that will satisfy both these properties for more than two alternatives.

11.3.3 Decentralized Multi-robot System

Robot swarms being a subset of decentralized multi-robot systems, arise from the intersection of two domains: collective robots and swarm intelligence.

The key design principle that is followed in the design of decentralized systems are:

- Control should be decentralized: All robots in the system are considered to equip independent decision making capability.
- No leaders: There should be no master node that coordinates and manages the agents in the system.
- No predefined agent roles: There should be no fixed role for agents in the system.
- Simple, local interactions: All the interactions with the agents should be simple and should happen only on a local scale (within the communication range).

These rules directly apply to a concept generally referred to as *emergence*. Emergence is a property that a system exhibits which the individual parts of the system are incapable of exhibiting on their own, a behavior that demonstrate emergence is called emergent behavior. In the context of multi-robot systems, emergent behavior can be thought of as collective behavior that is exhibited by the system as a whole when they aggregate together. These kind of collective behaviors are widely found in natural swarms. Consider, a school of fish that exhibits a circling behavior as a measure to protect itself from predators (Fig. 11.1 shows one such circling behavior). The system that demonstrates emergence is in general very attractive because they exhibit some inherent capability to produce the following properties: scalability, efficiency, robustness, parallelism and adaptivity. Swarm robotics is the field of engineering that study emergence in robots.

Swarm robotics design problem

The problem of the design of swarm systems (see Fig. 11.3) can be defined as: given a set of high-level requirements for a swarm, how can these requirements be translated into a set of robot rules. For instance, consider the task of cleaning a room, the high-level requirements is cleaning the room and the designer task is to derive a set of robot commands to satisfy the requirements. In formal terms, the design problem is to drive the states of all the robots from an initial state to a desired final state. Consider the state of the swarm $S_0 = \{s_1, s_2, \ldots, s_n\}$, with s_i being the initial state of robot i; the goal is to derive a function $f : S_0 \rightarrow S_T$ that will drive the system to a desired final state $S_T = \{s_{1,T}, s_{2,T}, \ldots, s_{n,T}\}$ within the swarm state space $\mathbb{S}$. Before we delve into the methods available to design these rules, it might be useful to first understand what are the states of the system and how could one model a swarm system.

Swarm states

The swarm state space ($\mathbb{S}$) contains all the possible configurations for all the robots in the swarm, and each of these configurations is called a swarm state, i.e., a combination of all the individual robot states. Each individual robot has a different perception of the environment, communicates with different neighbors and hence has a potentially unique internal state. Figure 11.4 shows the individual robot states that are combined to form the overall swarm state. The individual robot state in the swarm can be broken down into:

- environmental state s_e: the state of the robot surroundings,

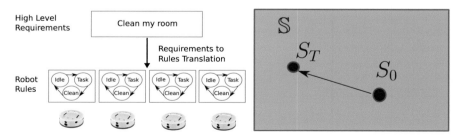

Fig. 11.3 The swarm robotics design problem: how do we translate swarm-level instructions to commands for each robot? More formally, how do we change the system state $S_0 \in \mathbb{S}$ to a desired S_T?

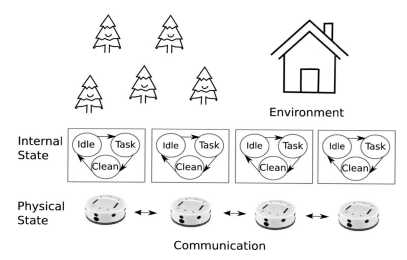

Fig. 11.4 States of the robots in a swarm are a combination of the environmental state, internal state, physical state and communication state of the robot

- internal state s_i: e.g., battery level, memory use, etc.,
- physical state s_p: the state of the sensors, actuators, and other mechanical parts of the robot,
- communication state s_c: the internal state resulting from communication with neighbors.

Consider, $s_i = \{s_e, s_i, s_p, s_c\}$ the state of the robot i, the state of the swarm would be $S = \{s_i\}, i \in (1, n)\}$ with n the number of robots in the swarm. As one can observe, the swarm state contains the state of all robots, making the formal modeling of swarms rather challenging.

A swarm of robots is generally considered as a single machine with evolving state, and it is generally called an open machine, since only a parts of its state is controllable. The environmental state s_e around the robot is dynamic, and it can only be partially modeled because the sensors of the robot are only capable of capturing a subset of the environmental state with some amount of uncertainty. Furthermore, the environment around the robot keeps evolving as the robot is performing its task and can only be considered partially controllable. Similarly, the physical state s_p of the robot is also only partially controllable (e.g., the battery level cannot be controlled). Another reason for s_p being partially controllable is due to the presence of a non-zero probability for a hardware failure. On the contrary, the internal state of the robot s_i is considered to be fully controllable by the robots through programming.

The communication state s_c depends on the underlying communication topology created by the robots and the state of the communication medium. When more and more robots are sending information, the chances of collisions and packet drops increase. Every robot in a swarm is assumed to have a limited communication range

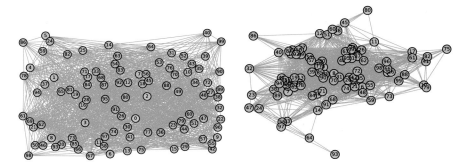

Fig. 11.5 Communication topology created by a swarm of 100 robots: on the left the robots form a cluster topology and on the right a scale-free topology

based on its underlying communication hardware. The ability of the robots to communicate can be represented by a graph structure called the communication topology. In a communication topology graph, the nodes represent the robots and the edges represent the communication links between them. Figure 11.5 shows some types of communication topology, scale-free, and cluster topology. The communication topology of a MRS is continuously affected by the movement of the robots, and the communication affects movements; hence, only a part of this state is controllable. The problem of connectivity maintenance (i.e., maintaining a desired communication topology) is usually formulated as a dual problem that addresses both movement and communication simultaneously.

Design of swarm robotic systems

The task of programming a swarm robotic system starts by the definition of the requirements that define what the swarm is required to accomplish. The requirements are then translated into a set of robot rules defining the behavior of each of the robot in the swarm. The task of the programmer is to create these robot rules from the requirements, which can be generally referred to as the control software design process. The design of control software for robot swarms can be manual or automatic.

Automatic methods

The task of control software design is formulated as an optimization problem where the parameters for optimal robot behavior are found via search (see Fig. 11.6). The search in automatic methods generally involves a robot simulator and a set of template alternative robot control architectures. The performance of a given alternative on the swarm is then evaluated using a performance metric. The performance metric ideally captures the efficiency and effectiveness of the swarm when completing the given task. The configuration space that contains all possible alternatives to the template control architecture is referred to as the design space. The candidate solutions from the design space are drawn with some search rules to identify the optimal solution that maximize the performance metric.

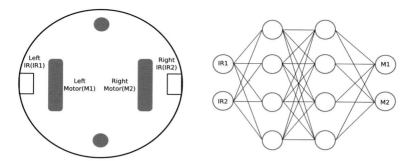

Fig. 11.6 An illustration of automatic design in its basic form, the sensory input of the robots are considered to be two binary IR sensors (IR1 and IR2), and the actuators of the robot are two motors (M1 and M2) taking real value inputs. In this scenario, an evolutionary algorithm could be used to obtain the weights of the neural network that performs the task of aggregation (similar the behavior observed in ants while forming rafts, as seen in Fig. 11.1)

The combinatorial nature of the design space demands the use of metaheuristics like evolutionary algorithm to search the design space using the performance metric as an objective function to evaluate the quality of the solutions. The most common form of automatic design is to use an evolutionary algorithm (EA) to search the design space paired with an artificial neural network (see Chap. 15 for more information on artificial neural networks) to act as the template control architecture. EA is a population-based optimization procedure inspired from biological evolution. In EA metaheuristics, a virtual population containing best performing candidate solutions are maintained; in each round of optimization, two candidate solutions are selected to be combined and produce an offspring (via procedures like crossover or recombination), and this new offspring is mutated to introduce some variance and novelty in the population. The artificial neural network maps the sensory inputs to actuation commands, and the design space contains all possible combinations of neural network weights. Some initial studies (Nol & Floreano, 2000) have proven that automatic control software design is a viable option for design of decentralized robotic systems.

Some other types of automatic design focus on designing methods that promote and search for novel behaviors in the design space, such as novelty search (Gomes et al., 2013). Novelty search promotes diversity over performance: this type of automatic design procedures are known to not suffer from problems like immature convergence or stagnation of solutions around local minima. Apart from using a neural network, automatic methods can be applied to other control architectures like parametric finite state automate (Hecker et al., 2012) or behavioral trees (Kuckling et al., 2018). Based on the type of system used to evaluate the candidate solutions, automatic methods can be further classified as online and offline.

Offline methods

The offline design process involves generating the control software before the deployment of the swarm. During the design phase, simulation is typically used to evaluating a large number of possible settings from the design space and generate and appropriate control software. The use of simulation offers the benefit of being faster than real robot evaluations and avoids damage to the physical hardware caused by low-quality candidate solutions. The most common characteristics of offline methods are:

- The behaviors produced are usually for homogeneous swarms, executing an identical version of the control software.
- The objective function (also known as the performance metric) is evaluated in a centralized manner for the whole swarm rather than evaluating the performance of the individual robots.
- A typical evolutionary approach evaluates the populations of up to 200 robots using the control software settings altered through evolutionary procedures (elitism, recombination and mutation).
- The general performance metric used is based on the spatial change of the robots relative to other robots, with an evaluation across 10–30 runs to take unknown stochastic variables into account.

Online methods

Online methods perform directly on the real deployment environment, and the performance is evaluated directly on the physical hardware. The most natural benefit of using online methods is that they can benefit from the feedback received by deploying the robots directly in the operational environment. Online methods generally produce mission specific control software rather than generalized control software that can be applied to a wide range of missions. The optimization being performed on deployed robots, only a limited number of alternatives can be evaluated due to resource limitations, and potentially robot harmful behavior has to be filtered out before evaluation. In addition, the optimization has to be distributed (with opportunistic centralization), since the swarm cannot rely on a centralized node to compute the performance metrics and guide the design space search. The limitation of using a performance indicator that can only be evaluated in a local and distributed manner makes online approaches less effective in comparison to offline design methods. However, use of hybrid approaches that combine online and offline methods is an effective way of reaping the benefit of both worlds and is under active research by the community. Some notable characteristics of online design methods are:

- The robots are asynchronously used to explore a portion of the design space, with each robot evaluating a sub-population of the evolutionary instance of the control software.
- the robots continuously exchange the best performing instance of the control software allowing other robots to include this information in its local population for further search.

- The behaviors executed by the robots are usually heterogeneous in nature; each robot executes a different control software instance. However, there is a possibility that the robots will eventually reach a point in the search, where they execute a similar version of the control software.
- As the robots are completely decentralized, the performance metric used has to be computed locally, using the information available on the robots. This severely limits the type of tasks that can apply online approaches.

Manual methods

Manual methods involve the design of the control software for the robots either by hand using a trial-and-error approach, or using the designer's expertise. The general procedure is to use a state machine to model and encode the robot control software. The state machine allows the robots to decompose the overall goal into elementary tasks. Some state transitions are performed by the swarm as a whole to ensure consensus among the individual robots in the swarm.

The designer picks a tool that best fits the task at hand and devises a set of rules that will allow the robots in a swarm to produce a self-organizing behavior. Some notable self-organizing behaviors are aggregation, circling and pattern formation. The main advantage of using manual methods is that the programmer has complete control over the design software and can customize them to best fit the robotic mission. One of the downsides of this approach is that it is very hard to manually design a decentralized behavior for the robots since only a part of the state is controllable and known to the programmer.

11.4 Swarm Programming

Swarm programming is the process of writing code to describe swarm behaviors. A swarm programming language is a *domain-specific language* that can be used for describing control software for robot swarms. Like other domains in computer science, a swarm programming language can be compiled into machine code containing a set of instructions that can be executed by each robot. The basic requirements of a swarm programming language are to provide a rich feature to allow arbitrary missions and to provide support for most robotic hardware. Other desirable properties of a swarm programming language are:

1. Composability: The control software should be able to work in parts and as well as, work as a single control software, when various parts of the code are put together. For instance, a programmer can design a particular part of the code separately as a function and test it, when putting together such similar functions, it should work as a whole behavior.
2. Predictability: When looking at a piece of code, the designer should be able to reason the behavioral outcome that will be observed on the robots.

3. Heterogeneous hardware support: The programming language should provide support for designing swarms that contain various types of robotic hardware.
4. Hardware agnostic: The programming language should produce invariable behavioral outcome across various robotic hardware. A given piece of control software designed using the programming language should be compatible to be deployed on a wide range of robotic hardware.

11.4.1 Swarm Programming Languages

Over a decade of research in the field of swarm robotics have produced a wide variety of methods that are used in programming the control software for robot swarms. In this section, we will discuss some of the notable programming languages and paradigms that are used in design of robot swarms.

Robot oriented

The main focus in robot-oriented programming is to provide the designer with as precise control as possible to program every single robot in the group. In robot-oriented programming, the designer focuses on designing an individual robot behavior that will work synchronously to realize a desired group behavior, this type of swarm programming is also known as bottom-up approach. One of the most common tool used for robot-oriented programming is the robot operating system (ROS) (Quigley et al., 2009). ROS is considered to be one of the widely used tools in programming both single robot and multi-robot systems. ROS being programming language flexible allows a designer to design a control software using various programming languages (Python, C/C++ or Java). Chap. 5 introduces the fundamentals of ROS and can be used as a reference to ROS. One of the main advantages of using ROS is the availability of several robustified packages and drivers that can be readily used to program every single robot in the group. On the downside of programming swarms with ROS, the programmer has to take into account each of the ROS node interactions and its details (not limited to the naming used to connect ROS nodes). The complexity of managing the ROS specific details increases exponentially with the number of robots in the system.

Spatial computing

Spatial computing focuses on providing programming tools for programming the swarm as a whole rather than considering the individual robot's behavior. Spatial computing can be used when the programmer is not interested in programming individual robots but would like to design group level behavior; this kind of approach is called the top-down approach. The robots in spatial computing are considered to be a collection of communicating compute devices that are distributed in an arbitrary operational space, capable of performing a local computational task. The frameworks in spatial computing abstract the individual robot and provide swarm specific primitives that will allow design of global behaviors. Some examples of spatial computing are

Proto (Beal & Bachrach, 2006) and Protelis (Pianini et al., 2015). In proto, the robots are assumed to be deployed on a manifold of space called amorphous medium with a physical and computational state. The robot program defines the way they interact with the neighbors and the environment to perform a location specific behavior in the amorphous medium. Spatial computing being a powerful tool for designing swarm behaviors still lose the robot individuality and the capability to program each robot in the swarm. Programming of heterogeneous robots with spatial computing is not possible.

Goal oriented and task oriented
Goal-oriented programming is considered a bottom-up programming approach where the individual robots are assigned spatial goals. The global task is broken down into elementary spatial goals and assigned to robots; the robots coordinate and reach these spatial goals in parallel. The main focus in goal-oriented programmer is placed on decomposing the global requirements into spatial goal rather than the logic used to perform the task. Some example of goal-oriented programming languages are SWARMORPH (O'Grady et al., 2012) and Termes (Petersen et al., 2011). Goal-oriented programming is more suited when the mission requires spatial organization among the robots and has minimal to no robot failures, since the approaches do not have contingent mechanisms for robot failures.

In task-oriented programming, the global task is broken down into a set of sub-tasks (such as spatial goals) that can be performed by a single robot and optimally assigned to robots. The robotic swarm is considered as a system with parallel machines that can be scheduled jobs using a scheduler (a system that assigns resource to a specific task). These type of systems are referred to as deterministic parallel machines in sequencing and scheduling theory (Pinedo, 2012). The task of control software design in goal-oriented programming is to formulate the global problem as a scheduling problem and design a scheduling system that will assign jobs (sub-goals) to the robots in a swarm. Task-oriented programming is considered to be a bottom-up approach since the individual robots are assigned tasks separately. Karma (Dantu et al., 2011) and Voltron (Mottola et al., 2014) are some examples to task-oriented programming. Task-oriented oriented programming is more suited for missions that can be decomposed into a set of sub-tasks that can be performed on a single robot. Task-oriented programming cannot be used in missions that require active inter-robot coordination (e.g., when a sub-task require two or more robots to complete it).

11.4.2 Programming in Buzz

Buzz (Pinciroli & Beltrame, 2016) is considered to be a hybrid domain-specific programming language that provide programming primitives similar to robot-oriented programming, spatial computing and goal-oriented programming. It allows programmers to maintain desirable levels of abstraction while programming, the swarm can be programmed as a whole (top-down) or individual robot behaviors can be designed (bottom-up) at the same time in a single control software. For instance, the language provides support for both setting the actuation commands (bottom-up) and support for neighbors management to consider the swarm as a whole and perform operations in the neighborhoods (top-down). A pure bottom-up approach suffers from scalability issues and conversely; a top-down approach suffers from inability to fine-tune individual robot behaviors. A concurrent design used in Buzz allows the designer to pick the right amount of abstraction required at the various stages of the mission.

Buzz satisfies most of the desirable properties of a swarm programming language: the code can be organized as functions and classes (composable), language syntax is intuitive with similarities to Python and Lua (predictable), swarm programming constructs allows concurrent use of heterogeneous robots in a swarm (heterogeneous hardware support) and unified Buzz virtual machine (BVM) for use with various hardware platform (hardware agnostic). These properties make Buzz a promising approach to design control software for robot swarms and hence, in this chapter, we will provide a detailed introduction to programming in Buzz.

Communication and execution model

The reference communication model used in Buzz is situated communication; it is a communication paradigm introduced by Stroy et al. (2001) and commonly used in swarm robotics. In situated communication, the receiver of a message knows the positional information (distance and bearing) of the sender using a specialized communication device. The robots using Buzz either equip such a communication device or simulate situated communication through other sensory measurements. The measurements in a situated communication device are obtained as a positional and payload pair. As illustrated in Fig. 11.7 left, the positional data includes the relative range (distance) and bearing (angle) of the sender in the receivers' coordinate frame. The payload part of the message includes a serialized messages from the internal behavior programmed on the robots. The robots in a swarm, broadcast messages, the robots in communication range receive these messages (often assumed to be line-of-sight, a requirement for situated communication devices) and process these messages. The information flow in the swarm happens in a gossip based communication (i.e., from one neighborhood of robots to another) until all the robots in the swarm have similar information.

The execution of the control software follows a discretized step wise execution phase with each step denoting one control loop (illustrated in Fig. 11.7 right). During each control loop, the robots perform the following actions in order: reading the sensors, processing the input messages, performing a loop of the code, sending messages and updating the actuation commands.

Buzz Virtual Machine

Buzz considers the swarm as a collection of devices that uses a virtual machine called Buzz Virtual Machine[1] (BVM). The BVM contains an interpreter[2] to execute the control software designed for the robots (a script called Buzz script). BVM is

[1] https://github.com/buzz-lang/Buzz.
[2] A program used to execute code, for example, Python interpreter.

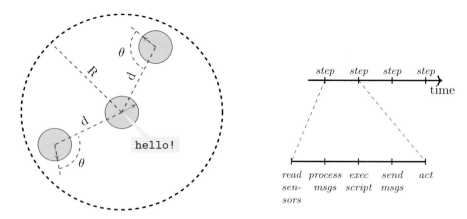

Fig. 11.7 On the left, the reference communication model performing situated communication, the sender robot in the center broadcast messages within its communication range (R) and the two receiver robots on the top/bottom, measure the distance (d) and angle (θ) of the sender in their coordinate frame. On the right, the reference execution model containing the discretized step-wise execution

written entirely in C and uses a stack (datatype providing a collection of elements)-based operations to execute the control software. Figure 11.8 illustrates the internal structure of the BVM. For more details on the BVM, we refer the reader to Pinciroli and Beltrame (2016). BVM is designed to be compact in size (about 12 kB) providing the possibility to deploy it on most of the robots used in swarm robotics; there exists a compact BVM optimized for microcontroller called BittyBuzz.[3] The internal datatype used to store information inside the BVM is key hashable tables (referred to as data holders). The reference execution architecture discussed earlier directly translates into BVM operations performed at each step: latest sensor readings and input messages update the respective data holders, the values from the data holders are used to perform a code step resulting in updating the data holders and the values from the data holders are used to update the actuation commands and output messages.

In practice, a designer writes his code in Buzz, which gets compiled into a bytecode and the bytecode is executed by the BVM. Buzz offers command line tools like *bzzc* to compile the buzz script into a BVM interpretable buzz code. The compilation is generally performed on the programmers machine and the corresponding byte code is then uploaded onto the robots for execution. Buzz is an extensible language allowing programmers to attach custom C/C++ functions as closures that can be called from the Buzz script. For instance, consider, the *take_off* routine that can be implemented for flying robots and *set_wheels* for ground robots. In the compilation phase, these custom closures are set as symbolic references and referenced during execution phase.

---

[3] https://github.com/buzz-lang/BittyBuzz.

Fig. 11.8 Internal structure
of the buzz virtual machine,
figure obtained from
Pinciroli and Beltrame
(2016)

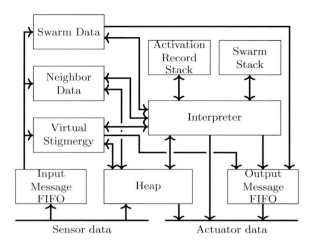

Deploying Buzz on robots

Deployment of Buzz on robots requires an adapter called the Buzz controller. The main purpose of a Buzz controller is to connect the robot sensors and actuators to the BVM data holders that store sensor and actuator information. Buzz controller also serves the purpose of connecting the communication hardware with the BVM, updating the in and out message queue inside the BVM. Buzz controllers are comparable to a hardware abstraction layer (HAL) that abstracts the robot specific sensor/actuator communication to the BVM. There exists several buzz controllers that can be readily used for robot deployments: 1. ARGoS Buzz controller, a controller that is available with the BVM implementation and can be used with ARGoS3 simulator (Pinciroli et al., 2012), 2. BzzKh4,[4] a controller for KheperaIV[5] robots and 3. ROSBuzz,[6] a controller that can be used with ROS compatible robots. Buzz controllers can be considered more than a HAL wrapper to BVM because some controllers leverage the extensible nature of the language (using custom C/C++ function-based primitives) to provide additional features. For instance, ROSBuzz provides features to Geo-Fence robots (limit the operational space for robots), compute veronoi tessellation for robot groups (a method used to partition the space into sub-groups), exploration primitives (methods to plan an exploration path in unknown spaces), etc.

Programming primitives

The programming primitives are pre-built software packages and constructs that can be used to create a more sophisticated control software for the robots. As mentioned earlier, buzz offers constructs for both bottom-up (programming operations performed on individual robots) and top-down (programming operations performed with groups of robots) programming. Robot-wise operations available in Buzz are:

---

[4] https://github.com/MISTLab/BuzzKH4.

[5] http://www.k-team.com/khepera-iv.

[6] https://github.com/MISTLab/ROSBuzz.

assignment of variables, loops, branching and function definitions. The use of robot-wise operations is analogous to other scripting languages (like Python). As for the top-down programming primitives Buzz offer: Neighbor management, swarm management and virtual Stigmergy. Each of this programming primitive takes inspiration from natural swarm and virtually replicates a phenomenon from natural swarm intelligence. The basic data types available in Buzz are: nil, Int, float, string, table, closure, swarm and virtual stigmergy. Data types nil, Int, float and string are analogous to other scripting languages. Whereas, tables are the only structured datatype available in Buzz that can be either used as tables or dictionaries. Closures correspond to function pointers that can be stored as global variables and referenced at the execution time. Swarm and virtual stigmergy are primitives for top-down programming, and we will discuss them in the following.

Neighbor management

The neighbor management in Buzz is used either for performing operations with the positional information or communicating information within the robots' neighborhood. Figure 11.9 shows a comparison of a behavior observed in nature (flocking) and artificial behavior (boids rule) performing a similar behavior. Neighbors construct simplifies this implementation by using the function *neighbours.foreach* that loops through all the neighbors of a robot and apply a function. The function applied for each neighbor could compute vectors for all three components (separation, cohesion and alignment) for this neighbor. The result of this operation would be one aggregated vector for each component (separation, cohesion and alignment) that can be averaged to obtain the common heading of the robot. There are also other functions in Buzz that could be leveraged in a neighbor based operation: map, reduce and filter. Communication functions like *neighbours.broadcast* and *neighbours.listen* can be used to broadcast messages in the robots neighborhood. For instance, *neighbours.broadcast* can be used to broadcast the value of a Buzz datatype under a topic and *neighbours.listen* can be used to register a callback function to execute when a message is received from a topic.

11.4.2.1   Swarm Management

Programming heterogeneous swarms are a challenging task, locomotion and sensing used by the different types of robot can be fundamentally different. Consider flying, legged robots and rolling robots, each of these robot types need different kinds of sensors and actuators to realize locomotion. With different kinds of sensors and actuators comes different types of programming constructs to operate the robots. Buzz offers swarm construct that can take into account heterogeneity while programming the robots. Figure 11.10 illustrates swarm construct with a heterogeneous swarm containing flying and ground robots. Sub-groups within the robots can be created to perform robot specific operations; these virtually tagged robot groups are called a swarm. From a programming perspective, *swarm.create()* function can be used to create a virtual group (swarm) and functions like *swarm.select()* and *swarm.join()*

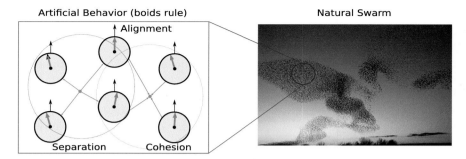

Fig. 11.9 Illustration of flocking: on the right, a starling swarm flocking and on the left, artificial swarm intelligence performing the equivalent behavior using boids rule (separation, alignment and cohesion). This behavior require looping through a robot's neighbors to compute the current movement, neighbors construct in Buzz provides *neighbors.foreach* function to loop through all neighbors and compute the current heading vectors. *Credits* Starling swarm—wikimedia.org/Walter Baxter

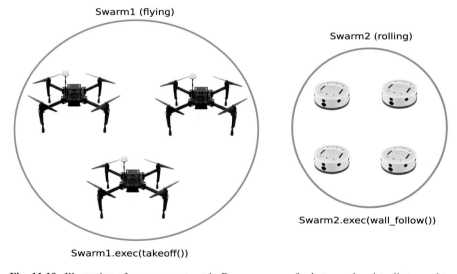

Fig. 11.10 Illustration of swarm construct in Buzz, a group of robots can be virtually tagged to assign group specific behavior, flying robots are assigned takeoff task and ground robots are assigned a wall following behavior

can be applied to join a swarm. As in Fig. 11.10, the function *swarm.exec()* can be used to assign a group specific function to execute for a given swarm.

Virtual stigmergy

Virtual stigmergy is a programming construct derived from natural swarm intelligence called stigmergy. Stigmergy is widely found in insect swarms, consider termites, they change the structure of the mold they build to communicate with other termites, in this case the information flow is environment mediated. Another exam-

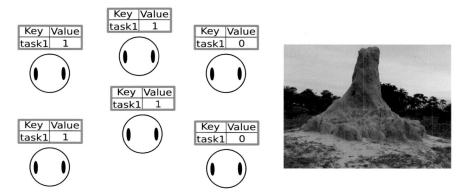

Fig. 11.11 Stigmergy in termites modifying the mold structure to communicate with other termites (right), a virtual implementation of this phenomenon (virtual stigmergy) provide (key, value) tuples to propagate information (left). *Credits* Termite mound—flickr.com/Justin Hall

ple to stigmergy can be found in ants; they spray pheromones to communicate the shortest path to a food source. The environment acts as a medium to relay information to other insects. Virtual stigmergy is a programming construct that allows programmers to replicate this phenomenon in a virtual manner on the robots. Unlike stigmergy in insects, the robots using virtual stigmergy make use of data structures to store and propagate information. From a usage point of view, virtual stigmergy is comparable to shared memory and distributed ledgers, acting as a black board for writing information from one robot and reading it on other robots. Figure 11.11 illustrates virtual stigmergy on robots by making comparison to termite swarms. In programming robots, virtual stigmergy table can be created using *stigmergy.create()*, *stigmergy.put()* can be used to add or modify entries and *stigmergy.get()* can be used to read the latest value. The internal implementation of virtual stigmergy optimizes the information to be broadcast to achieve a guaranteed network wide propagation. The information flow happens in a gossip-based fashion from one robots' neighborhood to another until a unified information is present in the whole group.

11.5 Deployment of Real-World Swarm Systems

Swarm robotics is a very young field of robotics that has received an increasing attention over the past decade due to its inherent benefits. However, the field has not matured enough to have robust real-world deployments, mainly due to the fact that some of the underlying engineering concepts are not completely clear for full autonomy. These challenges have given rise to creation of technologies that can allow humans to supervise and manage the system once deployed.

11.5.1 Human Swarm Interaction

Rapid advances in artificial intelligence are driving the adoption of robotics and automation in transport and logistics, providing new solutions to highway systems (Shladover, 2018), passenger transport (Pavone, 2015), last-mile delivery (Grippa et al., 2019), and automated warehouses (Enright & Wurman, 2011). For the foreseeable future, humans will remain indispensable to supervise and manage such fleets because we are transitioning from systems that are generally already in use; technology gaps prevent us from performing all of the required functions autonomously; and particularly in visible, safety-critical applications, society's trust in decentralized technology will be earned gradually. However, integrating increasingly sophisticated AI techniques leads to increasingly opaque robot control programs. Furthermore, human supervisors' cognitive capacities are challenged (and eventually exceeded) as the size of autonomous fleets grows. The difficulty of ensuring operational performance is compounded when incoming information is scattered, delayed, asynchronous or unreliable. These factors lead to increased pressure on human supervisors' cognitive resources and their ability to maintain situational awareness, detect problems and make successful decisions. There are some methodologies and approaches (St-Onge et al., 2019b) for the supervision of AI-driven swarm systems, deployed across domains such as transportation and logistics.

Given that the operator is indispensable in a robotic fleet to solve complex tasks and communicate with the swarm, the major focus in the field of human swarm interaction (HSI) are the following: 1. Operator cognitive complexity, 2. Communication with the swarm, 3. Control architectures, 4. Level of autonomy and 5. Methods to interact with operator. All of these modules that the field focus on are tightly coupled with one another, for instance, the level of communication of robot states depends on the level of autonomy, which in turn affects the operator effort to control the swarm. A detailed consideration to the concepts in HSI can be found in Kolling et al. (2015).

Operator cognitive complexity
In the field of computer science, the term computational complexity is defined as the resources (such as time and memory) required to solve an algorithmic problem. The required resources are generally considered to be a function of the size of the input. Computational complexity is used to classify the solvable computational algorithm from the unsolvable ones. Higher computational complexity algorithm might work reasonably for smaller number of inputs and fail for larger number of inputs. In HSI, a similar concept exists called the cognitive complexity for the robot control task; instead of the algorithm, an operator is replaced. The main task of an operator in a swarm system is to supervise and manage the robots by performing a sequence of actions on observation of a robot status. Operators cognitive load can be defined as the complexity of actions to perform by the operator on observation of a status. The analogy between computational complexity and cognitive complexity was first drawn in Lewis (2013).

Consider a group of aerial robots are performing a search operation in the forest to locate human survivors, when the robots are managed individually by the operator (checking each of the robot camera feed individually for a human and sending commands to further explore) then the cognitive complexity of this mission is $O(n)$. Conversely, when the operator deploys the robots and selects a search area, the robots subdivide the tasks autonomously, run a human detection algorithm internally using the camera feeds and send the operator of only a possible human detection for verification then the complexity here is $O(1)$, which is the minimal possible cognitive complexity. Another term that relates to cognitive complexity is the negligence tolerance, the time required by the robots to show performance degradation when left unattended. For optimal operation of the fleet, the operator has to attend to the robots before negligence tolerance time. In reality, the cognitive complexity of the system lie between $O(1)$ and $O(n)$ could also be sometimes worse than $O(n)$, when the operator has to deal with a cascade of tasks for a given robot in the swarm.

Communication with the swarm
An operator communicating with the swarm is an essential routine in real-world missions, the current level of robot autonomy demands an operator to be present in the system. An operator generally use a specialized device called the base station to communicate with the swarm. There are two types of communication that might be necessary between a base station and the operator: 1. The operator has to relay high-level goals to the system (commands) and 2. Operator has to obtain situational awareness on the robot fleet (states). Realizing both the goals require a reliable communication infrastructure within the system. Maintaining a reliable communication among the robots in the fleet is a challenging problem. The robots need to move to perform their mission, the movement in turn results in the change of communication topology, the swarm needs to realize the communication topology change for information propagation. One common approach to communication in robot swarms is to design a connectivity maintenance algorithm that will maintain a desired level of connectivity in the swarm allowing a base station to connect to the swarm.

Control architecture
The control architecture used in the swarm system defines the possible controls the operator can have over the system. Control architecture used in the system might influence the operator cognitive complexity, since it limits the possible controls the operator can have over the system. The desired cognitive complexity of controlling a swarm system is $O(1)$, where the operator treats the swarm as a whole, as if it were a single robot with complex dynamics. The current types of control architectures available require more fine grained interactions than swarm-level interaction, demand the operator to interact with sub-groups of robots. Some of the common control architectures used in robot swarms are:

1. Behavior library: Where a set of behaviors are implemented for the robots and the operator selects an appropriate behavior from the behavioral set based on the current situational awareness.
2. Parameter adaptation: A generic behavior is implemented initially and the operator is left to adapt the parameters of the system to control the robotic swarm.

3. Environment mediated control: The operator is made a part of the swarm and interacts with the swarm through the environmental medium (with modalities like gesture control).
4. Leader based control: A selective set of robots are assigned a leader role and the rest of the robots follow the leader robots, the operator continuously interacts with leader robot to control the swarm system. For example: the operator could teleop the leader robots to control the swarm.

Level of autonomy
The level of autonomy (LOA) of a swarm system can be defined as the degree to which the swarm system can make decisions on its own without external support (like an operator). LOA is generally defined through a 10 point scale, initially proposed by Sheridan and Verplank (1978). A scale of 1 defines the swarm to take absolutely no decisions and actions; the operator must perform all the tasks in the system. Conversely, a scale of 10 denotes the system completely disregards the human and performs all actions exclusively autonomously. It is commonly referred (Kolling et al., 2015) that the swarm system lie somewhere in or above a scale of 7, which means the system performs actions autonomously and informs the humans of the choices. The level of autonomy has no influence on the amount of situational awareness an operator acquires to interact with the swarm.

Methods to interact with operator
The method of interaction with the operator is an important factor to consider in system with operators and highly influence the cognitive complexity of the system. There are two methods to operator swarm interaction: 1. Remote interaction and 2. Proximal interactions. In remote interactions, the operator is considered to be monitoring a remote control node called the ground station. The ground station is a specialized computer that is used to obtain situational awareness on the robots mission and send commands back to the system to provide them with directives. Proximal control is another paradigm that considers the operator to be a physical part of the swarm as a special swarm member and these specialized swarm members provide directives to the swarm. Some approaches to proximal control are using gesture control, voice control and expressive motion. In these approaches, the user performs a certain gesture or voice command, which in turn creates a local interaction with the swarm to perform a task. However, the level of control that can be achieved with proximal control is minimal, since the swarm is controlled as a whole.

Human swarm interaction in Buzz
Within the framework of Buzz, several HSI approaches (St-Onge et al., 2019a, 2019b) have been designed to facilitate the operator interaction with the swarm. These works generally use ROSBuzz executing a Buzz script to realize the operator interactions. The operator is considered to be a virtual swarm member and the system uses a ground station to communicate with the operator (remote interaction). The ground station being a virtual swarm member, deploy a similar buzz script and receiving the same states that every other robot in the swarm is receiving. This system has been tested with two types of ground stations: 1. A traditional computer node and

2. A tangible robot fleet interface. A traditional computer node in this setup use a specialized visualization software to visualize and command the swarm. In tangible robot fleet interface, the operator uses a table top map with miniature robots indicating the status of the robots in the swarm. These miniature robots are used to interact with the swarm deployed in the field. The idea behind the use of tangible interface is that, the operator modifies a replica of the swarm, which in turn applies the changes to the actual swarm. The infrastructure in St-Onge et al. (2019a) was used within the framework of Pangaea-X in Lanssorate Spain, where a group of astronauts used the above elaborated interface (computer node and tangible interface) to control the swarm, while the cognitive load on the astronauts was evaluated.

11.5.2 Data Management, Communication and Mobility

In general, multi-robot systems need to collect large amounts of data from their environment, and often these data need to be aggregated, shared and distributed. Consider the task of distributed map merging (Mangelson et al., 2018) and inter-robot loop-closure detection in simultaneous localization and mapping (SLAM) (Lajoie et al., 2020), where robots need to exchange large amounts of data in the form of map fragments and/or pose graphs along with certain key-frame images. Many multi-robot systems are designed to share state information and commands, but their communication infrastructure is often too limited for significant data transfers. A mechanism called SOUL (Varadharajan et al., 2020a) allows members of a fully distributed system to share data with their peers. SOUL leverage a BitTorrent-like strategy to share data in smaller chunks, or datagrams, with policies that minimize reconstruction time. The main challenges addressed in this approach are: 1—cope with dynamic network topologies, 2—optimize the data fragmentation and reconstruction, and 3—optimize the distribution of the datagrams (chunks of injected data). Since peer-to-peer (P2P) file sharing mechanisms are well established in literature, with ample research to demonstrate their robustness and scalability (Reid, 2015), this method leverages some of their strategies (e.g., with the use of distributed hash tables) and integrates additional concepts from decentralized robotic systems. There are few other methods like Swarm mesh (Majcherczyk & Pinciroli, 2020) that provide location based data storage, referred to as spatial consensus to allow robot in a swarm to leverage the storage space on all robots.

The key principle that needs to be addressed for real-world deployments is addressing the perception-action-communication loop in robot swarm. Real-world robot deployments need to perform the following cascading action loop: to perceive the environment, estimate its state, perform an action, communicate its state to its neighbors. This cascading sequence of actions affects the other robots in the swarm and hence is a tightly coupled state that affects each other. A control software designer must consider the presence of perception-action-communication loop at design time.

The ability of a swarm to coordinate and exchange information depends largely on the underlying communication graph. A reliable communication infrastructure allows the robots to exchange information at any time. However, real deployments include many potential sources of failures (environmental factors, mobility, wear and tear, etc.) that can break connectivity and compromise the mission. The underlying assumption taken by several works (St-Onge et al., 2017) includes the robots ability to exchange information. There are two general approaches to connectivity maintenance in multi-robot systems: strict end-to-end connectivity (Stephan et al., 2017) or relaxed intermittent connectivity (Kantaros et al., 2019). Many of these approaches are either computationally intensive or cannot integrate the presence of an operator. There are some alternatives that use lightweight algorithm (Varadharajan et al., 2020b) allowing a heterogeneous group of robots to navigate to a target in complex 3D environments while maintaining connectivity with a ground station by building a chain of robots. The fully decentralized algorithm is robust to robot failures, can heal broken communication links and exploits heterogeneous swarms: when a target is unreachable by ground robots, the chain is extended with flying robots.

11.5.3   Fault Handling

When multi-robot systems are deployed in real-world scenarios, there is an increasing concern regarding the safety and reliability of the system. Robots that are faulty could potential harm humans or infrastructure. The robot control designed for the robots needs to explicitly design mechanisms that can tolerate some common malfunctions at the minimum. Faults in robotic hardware are inevitable, reliable mechanism incorporation within the control software could minimize the risks caused by faulty hardware. There are generally two kinds of robot failures: 1. Endogenous and 2. Exogenous faults. Endogenous faults are generally faults that occurs within the robotic hardware and exogenous faults are the faults that occur as a result of factors in the environment, and the robots interaction with the environment.

There are two kinds of approaches to detect faults in robot swarms: 1. Introspection and 2. Extrospection. In introspection, the robots run some kind of internal diagnostics to determine if the hardware is faulty. Extrospection is using the diagnostics of the neighboring robots to determine if a given robot is faulty. Some kinds of faults can be addressed using introspection and others require extrospection, currently resolved using an operator in the loop. Extrospection is an interesting solution, when dealing with multi-robot systems, for instance, a deviation from normal operation of a robot can be detected by other robots. Some existing approaches to fault detection are:

- Communication based (Christensen et al., 2009; Ozkan et al., 2010): Robots inability to communicate is detected by using periodic ping messages between the robots. These method can only detect completely failed robots or robots with communication issues.
- Model based (Millard et al., 2014): Robots use a model to compare their behavior to determine normal operation.

- Task effort based (Lau et al., 2011): Robots compute their contribution to the fulfilment of the task and estimate if they are contributing to the global task to determine their fault state.
- Online methods (Tarapore et al., 2017): An online classification model is learnt to distinguish between faulty and normal behavior of the robots; robots evaluate their behavior with the neighbor to determine faultiness. There are some methods that use a immunology inspired models to predict faulty robots (Tarapore et al., 2015).

11.6   Chapter Summary

This chapter provides an introduction to the different types of multi-robot systems and introduces the task allocation problem used in assigning tasks to different robots in a multi-robot system. A particular concentration is given to decentralized systems and various methods available to design decentralized control software. Fundamentals of programming a robotic swarm is discussed using the Buzz programming language. Toward the end of the chapter, a discussion is made regarding the necessity of an operator in a swarm system and the challenges toward the real-world deployment of swarm systems.

11.7   Chapter Revision

Question #1
What are the fundamental differences between centralized, distributed and decentralized systems?

Question #2
What are the design rules followed in the design of decentralized system?

Question #3
What are different methods to design control software for robot swarms?

Question #4
What are the desirable properties expected from a swarm programming language?

Question #5
What is the reference communication and execution model used in Buzz?

Question #6
When a operator needs to send individual commands to each robot in a swarm, what is the operator cognitive complexity in this situation?

Question #7
What are the types of information that needs to be exchanged between an operator and a robot swarm?

Question #8
How many point scales are generally used to identify the level of autonomy in a swarm robotic system?

Question #9
What are the common methods used for operator interaction with the swarm?

Question #10
Why is it important to maintain a desired communication topology in robot swarms?

Question #11
What are the types of faults that arise in robot swarms?

11.8 Further Reading

For further information on distributed multi-robot methods such as auction and voting, we refer the reader to Chaps. 9 and 23 of Easley et al. (2012), respectively. For more information on decentralized control software design, we refer the reader to Francesca and Birattari (2016). For more information on the Buzz programming, we refer the reader to Beltrame (2016). As a further reading on human swarm integration, we refer the reader further reading on human swarm integration, we refer the reader to Kolling et al. (2015).

References

Beal, J., & Bachrach, J. (2006). Infrastructure for engineered emergence on sensor/actuator networks. *IEEE Intelligent Systems, 21*(2), 10–19.

Bell, J. E., & McMullen, P. R. (2004). Ant colony optimization techniques for the vehicle routing problem. *Advanced Engineering Informatics, 18*(1), 41–48.

Bonabeau, E., Theraulaz, G., & Dorigo, M. (1999). *Swarm intelligence*. Springer.

Chang, L., Liao, C., Lin, W., Chen, L. L., & Zheng, X. (2012). A hybrid method based on differential evolution and continuous ant colony optimization and its application on wideband antenna design. *Progress in Electromagnetics Research, 122*, 105–118.

Christensen, A. L., OGrady, R., & Dorigo, M. (2009). From fireflies to fault-tolerant swarms of robots. *IEEE Transactions on Evolutionary Computation, 13*(4), 754–766.

Dantu, K., Kate, B., Waterman, B., Bailis, P., & Welsh, M. (2011). Programming micro-aerial vehicle swarms with karma. In *Proceedings of the 9th ACM Conference on Embedded Networked Sensor Systems* (pp. 121–134).

Dorigo, M., Birattari, M., & Stutzle, T. (2006). Ant colony optimization. *IEEE Computational Intelligence Magazine, 1*(4), 28–39. https://doi.org/10.1109/MCI.2006.329691

Dorigo, M., Theraulaz, G., & Trianni, V. (2021). Swarm robotics: Past, present, and future. *Proceedings of the IEEE, 109*(7), 1152–1165.

Easley, D., & Kleinberg, J. (2012). *Networks, crowds, and markets*. Cambridge Books.

Enright, J. J., & Wurman, P. R. (2011). Optimization and coordinated autonomy in mobile fulfillment systems. In *Workshops at the Twenty-Fifth AAAI Conference on Artificial Intelligence*. Citeseer.

Fan, T., Long, P., Liu, W., & Pan, J. (2020). Distributed multi-robot collision avoidance via deep reinforcement learning for navigation in complex scenarios. *The International Journal of Robotics Research, 39*(7), 856–892.

Francesca, G., & Birattari, M. (2016). Automatic design of robot swarms: Achievements and challenges. *Frontiers in Robotics and AI, 3*, 29.

Gomes, J., Urbano, P., & Christensen, A. L. (2013). Evolution of swarm robotics systems with novelty search. *Swarm Intelligence, 7*(2), 115–144.

Grippa, P., Behrens, D. A., Wall, F., & Bettstetter, C. (2019). Drone delivery systems: Job assignment and dimensioning. *Autonomous Robots, 43*(2), 261–274.

Hecker, J. P., Letendre, K., Stolleis, K., Washington, D., & Moses, M. E. (2012). Formica ex machina: Ant swarm foraging from physical to virtual and back again. In *International Conference on Swarm Intelligence* (pp. 252–259). Springer.

Hwang, K., Dongarra, J., & Fox, G. C. (2013). *Distributed and cloud computing: From parallel processing to the internet of things*. Morgan Kaufmann.

Iocchi, L., Nardi, D., Piaggio, M., & Sgorbissa, A. (2003). Distributed coordination in heterogeneous multi-robot systems. *Autonomous Robots, 15*(2), 155–168.

Kacprzyk, J., Merigó, J. M., Nurmi, H., & Zadrozny, S. (2020). Multi-agent systems and voting: How similar are voting procedures. In *International Conference on Information Processing and Management of Uncertainty in Knowledge-Based Systems* (pp. 172–184). Springer.

Kantaros, Y., Guo, M., & Zavlanos, M. M. (2019). Temporal logic task planning and intermittent connectivity control of mobile robot networks. *IEEE Transactions on Automatic Control, 64*(10), 4105–4120. https://doi.org/10.1109/tac.2019.2893161

Karpov, V., Migalev, A., Moscowsky, A., Rovbo, M., & Vorobiev, V. (2016). Multi-robot exploration and mapping based on the subdefinite models. In *International Conference on Interactive Collaborative Robotics* (pp. 143–152). Springer.

Kennedy, J., & Eberhart, R. (1995). Particle swarm optimization. In *Proceedings of ICNN'95—International Conference on Neural Networks* (Vol. 4, pp. 1942–1948). https://doi.org/10.1109/ICNN.1995.488968

Kolling, A., Walker, P., Chakraborty, N., Sycara, K., & Lewis, M. (2015). Human interaction with robot swarms: A survey. *IEEE Transactions on Human-Machine Systems, 46*(1), 9–26.

Kuckling, J., Ligot, A., Bozhinoski, D., & Birattari, M. (2018). Behavior trees as a control architecture in the automatic modular design of robot swarms. In *International Conference on Swarm Intelligence* (pp. 30–43). Springer.

Lajoie, P. Y., Ramtoula, B., Chang, Y., Carlone, L., & Beltrame, G. (2020). Door-slam: Distributed, online, and outlier resilient slam for robotic teams. *IEEE Robotics and Automation Letters, 5*(2), 1656–1663.

Lau, H., Bate, I., Cairns, P., & Timmis, J. (2011). Adaptive data-driven error detection in swarm robotics with statistical classifiers. *Robotics and Autonomous Systems, 59*(12), 1021–1035.

Lewis, M. (2013). Human interaction with multiple remote robots. *Reviews of Human Factors and Ergonomics, 9*(1), 131–174.

Luna, R., & Bekris, K. E. (2011). Efficient and complete centralized multi-robot path planning. In *2011 IEEE/RSJ International Conference on Intelligent Robots and Systems* (pp. 3268–3275). IEEE.

Majcherczyk, N., & Pinciroli, C. (2020). SwarmMesh: A distributed data structure for cooperative multi-robot applications. In *2020 IEEE International Conference on Robotics and Automation (ICRA)* (pp. 4059–4065). IEEE.

Mangelson, J. G., Dominic, D., Eustice, R. M., & Vasudevan, R. (2018). Pairwise consistent measurement set maximization for robust multi-robot map merging. In *2018 IEEE International Conference on Robotics and Automation (ICRA)* (pp. 2916–2923). https://doi.org/10.1109/ICRA. 2018.8460217

McLurkin, J. (2009). Experiment design for large multi-robot systems. In *Robotics: Science and Systems, Workshop on Good Experimental Methodology in Robotics*, Seattle, WA.

Michael, N., Zavlanos, M. M., Kumar, V., & Pappas, G. J. (2008). Distributed multi-robot task assignment and formation control. In: *2008 IEEE International Conference on Robotics and Automation* (pp. 128–133). IEEE.

Millard, A. G., Timmis, J., & Winfield, A. F. (2014). Run-time detection of faults in autonomous mobile robots based on the comparison of simulated and real robot behaviour. In *2014 IEEE/RSJ International Conference on Intelligent Robots and Systems* (pp. 3720–3725). IEEE.

Mottola, L., Moretta, M., Whitehouse, K., & Ghezzi, C. (2014). Team-level programming of drone sensor networks. In *Proceedings of the 12th ACM Conference on Embedded Network Sensor Systems* (pp. 177–190).

Nolfi, S., & Floreano, D. (2000). *Evolutionary robotics: The biology, intelligence, and technology of self-organizing machines*. MIT Press.

Otte, M., Kuhlman, M. J., & Sofge, D. (2020). Auctions for multi-robot task allocation in communication limited environments. *Autonomous Robots, 44*(3), 547–584.

Ozkan, M., Kirlik, G., Parlaktuna, O., Yufka, A., & Yazici, A. (2010). A multi-robot control architecture for fault-tolerant sensor-based coverage. *International Journal of Advanced Robotic Systems, 7*(1), 4.

O'Grady, R., Christensen, A. L., & Dorigo, M. (2012). SWARMORPH: Morphogenesis with self-assembling robots. In *Morphogenetic engineering* (pp. 27–60). Springer.

Pavone, M. (2015). Autonomous mobility-on-demand systems for future urban mobility. In *Autonomes Fahren* (pp. 399–416). Springer.

Petersen, K. H., Nagpal, R., & Werfel, J. K. (2011) TERMES: An autonomous robotic system for three-dimensional collective construction. *Robotics: Science and Systems, VII.*

Pianini, D., Viroli, M., & Beal, J. (2015). Protelis: Practical aggregate programming. In *Proceedings of the 30th Annual ACM Symposium on Applied Computing* (pp. 1846–1853).

Pinciroli, C., & Beltrame, G. (2016). Buzz: An extensible programming language for heterogeneous swarm robotics. In *2016 IEEE/RSJ International Conference on Intelligent Robots and Systems (IROS)* (pp. 3794–3800). IEEE.

Pinciroli, C., Trianni, V., O'Grady, R., Pini, G., Brutschy, A., Brambilla, M., Mathews, N., Ferrante, E., Di Caro, G., Ducatelle, F., Birattari, M., Gambardella, L. M., & Dorigo, M. (2012). ARGoS: A modular, parallel, multi-engine simulator for multi-robot systems. *Swarm Intelligence, 6*(4), 271–295.

Pinedo, M. (2012). *Scheduling* (Vol. 29). Springer.

Quigley, M., Conley, K., Gerkey, B., Faust, J., Foote, T., Leibs, J., Wheeler, R., & Ng, A. Y. (2009). ROS: An open-source robot operating system. In *ICRA Workshop on Open Source Software*, Kobe, Japan (Vol. 3, p. 5).

Reid, N. (2015). *Literature review: Purely decentralized P2P file sharing systems and usability* (Technical report). Rhodes University, Grahamstown.

Schwager, M., Dames, P., Rus, D., & Kumar, V. (2017). A multi-robot control policy for information gathering in the presence of unknown hazards. In *Robotics research* (pp. 455–472). Springer.

Sheng, W., Yang, Q., Tan, J., & Xi, N. (2006). Distributed multi-robot coordination in area exploration. *Robotics and Autonomous Systems, 54*(12), 945–955.

Sheridan, T. B., & Verplank, W. L. (1978). *Human and computer control of undersea teleoperators* (Technical report). Massachusetts Institute of Technology Cambridge Man-Machine Systems Lab.

Shladover, S. E. (2018). Connected and automated vehicle systems: Introduction and overview. *Journal of Intelligent Transportation Systems, 22*(3), 190–200.

St-Onge, D., Kaufmann, M., Panerati, J., Ramtoula, B., Cao, Y., Coey, E. B., & Beltrame, G. (2019a). Planetary exploration with robot teams: Implementing higher autonomy with swarm intelligence. *IEEE Robotics & Automation Magazine, 27*(2), 159–168.

St-Onge, D., Varadharajan, V. S., & Beltrame, G. (2019b). Tangible robotic fleet control. In *Proceedings of the 18th International Conference on Autonomous Agents and MultiAgent Systems* (pp. 2387–2389).

Stephan, J., Fink, J., Kumar, V., & Ribeiro, A. (2017). Concurrent control of mobility and communication in multirobot systems. *IEEE Transactions on Robotics, 33*(5), 1248–1254. https://doi.org/10.1109/TRO.2017.2705119

Styø, K. (2001). Using situated communication in distributed autonomous mobile robotics. *SCAI, Citeseer, 1*, 44–52.

Tarapore, D., Lima, P. U., Carneiro, J., & Christensen, A. L. (2015). To err is robotic, to tolerate immunological: Fault detection in multirobot systems. *Bioinspiration & Biomimetics, 10*(1), 016014.

Tarapore, D., Christensen, A. L., & Timmis, J. (2017). Generic, scalable and decentralized fault detection for robot swarms. *PLoS ONE, 12*(8), e0182058.

Varadharajan, V. S., St-Onge, D., Adams, B., & Beltrame, G. (2020a). Soul: Data sharing for robot swarms. *Autonomous Robots, 44*(3), 377–394.

Varadharajan, V. S., St-Onge, D., Adams, B., & Beltrame, G. (2020b). Swarm relays: Distributed self-healing ground-and-air connectivity chains. *IEEE Robotics and Automation Letters, 5*(4), 5347–5354. https://doi.org/10.1109/LRA.2020.3006793

Wurm, K. M., Stachniss, C., & Burgard, W. (2008). Coordinated multi-robot exploration using a segmentation of the environment. In *2008 IEEE/RSJ International Conference on Intelligent Robots and Systems* (pp. 1160–1165). IEEE.

Xing, L. N., Chen, Y. W., Wang, P., Zhao, Q. S., & Xiong, J. (2010). A knowledge-based ant colony optimization for exible job shop scheduling problems. *Applied Soft Computing, 10*(3), 888–896.

Yan, Z., Jouandeau, N., & Cherif, A. A. (2010). Sampling-based multi-robot exploration. In *ISR 2010 (41st International Symposium on Robotics) and ROBOTIK 2010 (6th German Conference on Robotics)*, VDE (pp. 1–6).

Vivek Shankar Varadharajan is a Ph.D. candidate in the Department of Computer Engineering and Software Engineering at École Polytechnique de Montréa. Varadharajan obtained his M.Sc. degree in Automation and Robotics from Technical University of Dortmund in 2015. He is a full-stack robotics developer and has vast experience in equipping robotic platforms with SLAM algorithms, navigation/traversability algorithms and robotic behaviors. His research interests include distributed robotics, multi-robot systems, machine learning, artificial intelligence and Cyber-Physical systems. During his study, he has won several prizes at technical contents, hackathon, poster presentation and demonstrations. He was a team member of CoStar that took part in the DARPA subterranean challenge along with members from NASA Jet Propulsion Laboratory. He has supervised over 5 interns during his Ph.D. He is a recipient of a BSFD student scholarship from École Polytechnique de Montréal.

Giovanni Beltrame is a full time professor in the Department of Computer Engineering and Software Engineering at École Polytechnique de Montréal. Beltrame obtained his Ph.D. in Computer Engineering from Politecnico di Milano, in 2006 after which he worked as microelectronics engineer at the European Space Agency on a number of projects spanning from radiation-tolerant systems to computer-aided design. In 2010 he moved to Montreal, Canada where he is currently Professor at Polytechnique Montreal with the Computer and Software Engineering Department. He was also Visiting Professor at the University of Tübingen in 2017/2018. Dr. Beltrame directs the MIST Lab, with more than 20 students and postdocs under his supervision. He has completed several projects in collaboration with industry and government agencies in the area of robotics, disaster response, and space exploration. His research interests include modeling and design of embedded systems, artificial intelligence, and robotics, on which he has published research in top journals and conferences.

Chapter 12
The Embedded Design Process: CAD/CAM and Prototyping

Eddi Pianca

12.1 Learning Objectives

This chapter will provide students with the following knowledge:

1. An understand of the design process and where CAD is applied within the process.
2. A general understanding of the different CAD systems, their application and file structures.
3. An insight into the use of CADD as a design tool.
4. An awareness for the various digital prototyping and visualisation tool.

12.2 Introduction

This section provides a brief introduction to the design process, how CAD/CAM fits into this process and a brief overview of the various CAD/CAM technologies. For a more detailed understanding of the topics covered in this section, references are included.

12.3 The Design Process and CAD

The design process is a method, used by designers and engineers, to divide a project into manageable steps to find a solution to a problem. Typically, the design

E. Pianca (✉)
University of Canberra, Canberra, Australia
e-mail: eddi.pianca@canberra.edu.au

© The Author(s) 2022
D. Herath and D. St-Onge (eds.), *Foundations of Robotics*,
https://doi.org/10.1007/978-981-19-1983-1_12

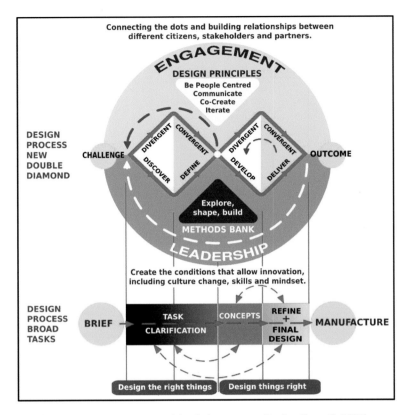

Diagram 1 Double-diamond diagram of the design process (Design Council, 2021)

process consists of five steps: 1. Briefing and task clarification, 2. Concepts generation, 3. Refinement and prototyping, 4. Final design and documentation and 5. Manufacturing (Bonollo, 2016, p. 17–18), refer Diagram 1.

Importantly, the design process is not a linear process because at any step during the design process a previous step can be revisited to gather more information or to re-consider the design, etc. Below is a brief overview of the design process steps and the application of CAD.

Briefing and Task Clarification involves understanding the client brief, defining the problem and the user, gathering information (researching the market, ergonomics, anthropometrics, standards, etc.), quantities, cost estimates, materials, processes, preparing design guidelines for the next steps of the design process, brainstorming possible ideas and preparing a project timeline. Identifying suitable materials and manufacturing processes is an important consideration as these will greatly influence the design to ensure it can be manufactured within the expected retail costs. The technologies needed to undertake the project, such as suitable CAD packages, 3D printing, etc., will also need to be identified to ensure the viability of the project.

Concept Generation is where a variety of alternative ideas are generated as possible solutions to the design problem. As many concept ideas as possible are generated, typically five or more that are originated from all different directions and fields via an out-of-the-box, free-flowing cognitive process known as divergent thinking. Typically, these concepts are hand generated using pencils and paper or via multi-touch displays such as Wacom tablets. However, with both techniques a certain level of free-hand sketching skills is required to adequately communicate the design intent. At this stage rudimentary prototypes are also sometimes produced to evaluate ideas. Today CAD is being used more often at this early stage of the design process to generate quick models for early testing and verification. This is due to advancements in CAD that have enable 2D CAD to be transferred to 3D CAD and other features such polygon modelling to generate quick complex surfaces that can then be converted to solid geometry. An example of polygon modelling is the Freestyle tool in CREO CAD software where a simple starting geometry such as a sphere, cube or flat surface can be easily pulled and pushed into the desired shape and then converted into a solid.

Refinement and Prototyping involve narrowing down the ideas to one final design through a cognitive process referred to as convergent thinking. Again, this commonly involves same freehand skills and or the use of Wacom tablets as in the previous step. Generally, a hand-generated workshop prototype is also made to verify design elements such as ergonomics, proportions, size, functionality and aesthetics. As per the concept generation step, CAD is being used more often at this step to model, test and 3D print the design to verify its design integrity.

Final Design and Documentation. This is where the CAD model is developed to be ready for manufacture. Here the CAD model can be numerically analysed via a range of CAD analysis tools such as Finite Element Analysis (FEA) both static and dynamic, etc., to ensure the structural integrity of the design, to confirm that all the parts fit together as intended, and 3D printed for further verification.

This step also involve the preparation of technical drawing and documentation for manufacturing requirements.

Once the design meets all the requirements of the brief, it can then proceed to the next step.

Manufacturing. This entails several phases as follows.

Phase 1. Source suitable manufactures and find out what file format(s) they require for the part files.

Phase 2. Send the part files, technical drawings and manufacturing requirements to the manufacturer(s) to check the parts suitability for manufacture, provide quotes to manufacture each part and recommend any changes to improve manufacturing.

Phase 3. Any recommended changes from the manufacturer(s) will require further edits to the CAD model, technical drawings and the manufacturing requirements.

Phase 4. Once the design is considered ready for manufacture a working prototype is produced to test and verify the integrity of the design in its working environment.

Phase 5. If changes are required, the CAD model is updated, and the manufacturing phases are repeated as required.

Phase 6. Design is manufactured.

12.4   The Design Process Versus Design Thinking

Whereas the design process is a way of dividing a problem into manageable chunks, design thinking (as discussed in Chap. 3) can be applied at any point along the design process to find solutions to ill-defined problems that emerge (Dam & Siang, 2018). Said another way, the design process is a series of steps while design thinking is a mindset that enables innovative solutions to emerge (Prud'homme van Reine, 2017, p.64, 70).

12.5   Cad Systems

Today there are numerous advanced computer-aided design (CAD) software packages for modelling 3D objects. However, all these packages are based on two systems, the first is parametric 3D modelling, the second is direct 3D modelling.

Parametric 3D Modelling: With this system, each entity in a model is constructed of features that are listed in a feature history tree. A feature can be a cut, extrude, revolve, datum plane, datum axis, datum point, sketch, hole, etc. Each feature, entity and all their relationships, parameters (length, width, height, radius, diameter, force, etc.), and their position within the assembly are easily tracked via the feature history tree. The parametric 3D modelling innovation emerged in 1987 when Parametric Technology Corporation (PTC) released Pro/Engineer (ProE) CAD software. Since its first release, PTC has made numerous enhancements to ProE and in 2011 renamed it to CREO. Today the parametric concept is used in most CAD programs as seen in Table 12.1 (Tornincasa & Di Monaco, 2010, p II-9 & II-17; Ault & Phillips, 2016).

In Table 12.1, level refers to the capability of the package. A high-level package offers more features and better features to generate complex geometries more easily and quickly than mid-level packages. However, high-level packages take longer to learn.

A benefit of the parametric concept is that the model is fully associative so that if a parameter in a part, assembly or drawing is changed then the related parts, assemblies and drawings will automatically update to reflect the change. For example, take a shaft that fits into a hole with a set clearance between them. If the diameter of the shaft (parameter 1) is changed, then the diameter of the hole will automatically update to maintain the set clearance (parameter 2). This ensures that the integrity of the model and the design intent is maintained.

However, on the negative side parametric 3D modelling systems requires specialised operators with expert skills and knowledge that takes considerable time and training to acquire (Tornincasa & Di Monaco, 2010, p II-5 to II-7), (Alba, 2018, March). Furthermore, considerable planning is required to ensure that the model,

Table 12.1 Comparison of some of the different systems and levels

CAD software	Company	System	Level
CREO	PTC	Parametric and direct (Hybrid)	Mid-level/High-level
Catia	Dassault systemes	Parametric with a 'Declarative modelling' direct system (Hybrid)	High-level
Fusion 360	Autodesk	Direct with a parametric component- Cloud based (Hybrid)	Mid-level
Inventor	Autodesk	Direct with a parametric component (Hybrid)	Mid-level
KeyCreator	Kubotek3D	Direct 3D modelling	Mid-level
NX	Siemens	Parametric and direct (Hybrid)	High-level
Solid edge	Siemens	Parametric and direct (Hybrid)	Mid-level
Solidworks	Dassault systemes	Parametric feature based with direct edited capability (Hybrid)	Mid-level
Sketchup	Trimble navigation	Direct modeller. Surfaces modeller. All surfaces are planer. Curved surfaces are faceted	Mid-level

features and files are properly structured to avoid problems as the model evolves. As seen in Image 1, CREO has a feature history tree displayed on the left-hand side of the window that can be used to manage and edit the model.

Direct Modelling: With this system, there are no parametric associations. Therefore, model features can be quickly and easily edited by moving, rotating, deleting,

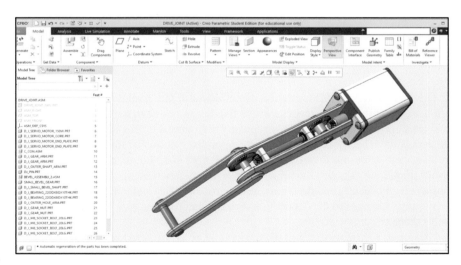

Image 1 3D model of a powered joint. Creo (version 7, 2020) parametric 3D modelling system from parametric technology corporation (PTC)

etc. features such as faces and edges without having to consider the model history (Tornincasa & Di Monaco, 2010, p II-7). This also makes it easier for a designer to work on a model that was started by someone else. Direct modelling systems can be very effective in the early stages of the design process as models can be quickly and easily generated. Another advantage is that direct modelling systems only take a short time to learn.

The disadvantage of direct modelling is that editing operations can lead to parts that are no longer dependable. For example, if the centreline of a rotating shaft has been accidently moved it can cause the shaft to interfere with other parts. Or if a change has inadvertently altered the tolerances between two parts it could lead to excessive wear when in operation.

Many of the CAD systems today offer a hybrid approach that merges both parametric 3D modelling and direct modelling (Tornincasa & Di Monaco, 2010, p II-8). These systems preserve the integrity of the model while gaining the flexibility and freedom of direct modelling.

12.6   CAD File Types

There are generally three main types of files that are used when building 3D parametric CAD models that are going to be manufactured: **assembly files, part files and drawing files**. The purpose of assembly files is to hold all the part files together, whereas part files represent the individual components that in real life cannot be broken down any further. Finally, drawing files are used for generating 2D technical drawings (also known as engineering drawings) of the model and its parts.

12.7   CAD Parametric Modelling—Assembly and Part Files

As previously noted, CAD is commonly employed at the 'final design and documentation stage' of the design process where the design has been narrowed down to one final concept. The reason CAD is typically introduced at this stage is because all the details have been decided and most importantly no major fundamental changes are foreseen. Here the details refer to the: shape, size, number of parts and their layout, materials, weight, manufacturing process, functionality, ergonomics, environmental and social considerations, aesthetics, etc. The fundamental changers refer to sweeping changes to the details that would affect the structure of the part files and their parametric associations.

Using parametric modelling, the design can be constructed in an assembly with multiple parts built to very high accuracies. This enables the CAD model to be digitally tested for correct fit between parts, structural integrity using finite element analysis (FEA), proper articulation of parts, manufacturability, aesthetics by generating realistically renders, physical verification through 3D printing, etc. Importantly, at any point during the CAD modelling process the model can be tweaked as needed.

Accordingly, CAD has greatly reduced the time needed to design products and release them onto the market.

Assembly and Part File Structure and Referencing: Before starting a 3D CAD parametric model, the structure of the files in the assembly requires considerable planning. This is important to ensure the model can be easily edited at any time by carefully considering how features are referenced to avoid circular references that can be formed between files in an assembly. The problem with circular reference is that they can cause the assembly to stop regenerating or even become unstable. Circular references are typically formed when editing a part that was created earlier in the assembly (the parent 'A') and referencing it to its child (part 'B' that was built later in the assembly). The reference sequence for this circular reference looks like A < B < A. Most CAD systems will alert you if any circular references have been created; however, this generally only happens when the whole assembly is regenerated. Therefore, it is important to regularly regenerate the whole assembly to check for circular references before they become too imbedded in the reference scheme. The best practice to avoid circular references is to only take references from files created before the one you are working in. In the feature history tree, these are files above the one you are worked in. In a 3D parametric model, an example of a feature history tree(or model tree) can be seen in Image 1, on the left side of the window. The feature history tree also represents the file structure where in Image 1 it starts with the assembly file (DRIVE_JOINT.ASM that was created first) followed by all the part files.

In the feature history tree, each part file can also be expanded to list all the features that are used to model the part geometry. This can be seen in Image 2 where the feature history tree (Model Tree) for the 'DRIVE_JOINT_SKEL.PRT' file has been expanded to show all its features.

With 3D CAD parametric modelling, panning is also required when constructing each part (Bodein et al., 2014, pp 136). Within each part file, the choice of features, how they are organised and the choice of references are important to maintain performance of the software and hardware and ensure that the part can be easily modified if needed. A poorly constructed CAD model can take considerably more time to build, edit and regenerate than a well-constructed model. Therefore, before starting a CAD model there are some essential modelling guidelines, for structuring and building assembly and part files, that operators need to know.

General CAD Modelling Guidelines forAssembly Files: When designing a product, designers typically start by sketching and drawing the product in its assembled form. This allows the designer to develop the design with all its parts as the design evolves. Developing the design from the top down also applied to modelling and designing a product on CAD. It must be noted that although CREO is used in this chapter to illustrate how CAD works, the modelling guidelines can be applied to all parametric 3D CAD modelling packages.

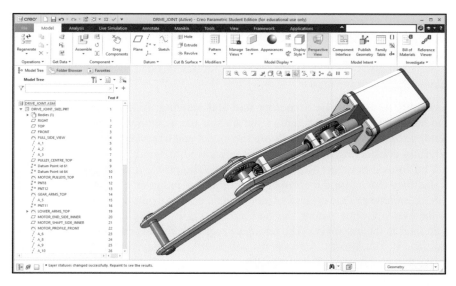

Image 2 In the model tree, seen on the left side of the window, the DRIVE_JOINT_SKEL.PRT part file has been expanded to show all its features. Model created in Creo (version 7, 2020) parametric 3D modelling system from Parametric Technology Corporation (PTC)

Below are modelling guidelines for parametric CAD assembly files:

1. File Naming—Before starting a CAD model, the first thing is to decide on a naming convention for all the files. This is important so that files can be managed, and file names are not repeated, and can be easily found.

2. Assembly file—With products that have more than one part, the CAD modelling process typically starts with an assembly. In CAD, an assembly file should generally only contain some datum features and the part files. The datum features are typically a datum coordinate system and or x, y, z datum planes that are used as references to place all the part files into the assembly, as seen in Image 3.

3. Referencing—It is important to employ a simple assembly referencing regime. This can be achieved by using the first part file in the assembly as a reference file that contains all the references to 'Drive and define the Model'. In CREO, this reference file is called a Skeleton file; it can be automatically created and is placed after the assembly file as seen in Image 2. The reference file should only contain basic parametric elements that will be used as references by all the other part files in the assembly. Basic parametric elements are sketched lines, sketched curves, point, centrelines, axis, surfaces (as opposed to faces from solid geometry) and planes that have been explicitly created by the user to be utilised as reference entities to 'Drive and define the Model'. What is meant by 'Drive and define the Model' is that by editing any Basic Parametric Element in the reference file the rest of the part files in the assembly and their feature will automatically update.

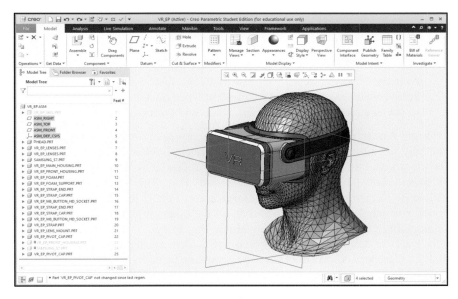

Image 3 In the model tree, seen on the left side of the window, the assembly datum features are highlighted in blue and are seen in green in the model window. Model created in Creo (version 7, 2020) parametric 3D modelling system from Parametric Technology Corporation (PTC)

4. The only features that should not be included in a driving file are solid features.
5. Solid features should only exist in the part files following the reference-file in the assembly.
6. Do not take references from solid features.
7. Reassemble (reuse) the same part. A typical case is when there are several of the same fastener in an assembly. It is much faster, less complex (less features in the assembly) and easier to simply reassemble the same part in another location in the assembly than continually rebuilding the part.

General Modelling Guidelines for Part Files: When a product consists of only one part then the following guidelines can be applied:

1. Break down the part to determine what are the basic functional geometric elements (squares, rectangles, cubes, cuboids, spheres, etc.). Decide the features (CAD tools) that will be used to model each element, how they interface with each other and their reference links.
2. Construct the references- The first features in a part file should comprise basic parametric elements that will be used as references by all the other features in the part file. Basic parametric elements are sketched lines, sketched curves, points, centrelines, axis, surfaces (as opposed to faces from solid geometry) and planes that have been explicitly created by the user to be utilised as reference entities to 'Drive and define the part'. Keep sketches of basic parametric elements as simple as possible.

3. Create the solid geometry by taking references from basic parametric elements. Try not to take references from any solid geometry.
4. If needed, it is always best to edit the basic parametric elements or insert some more basic parametric elements rather than taking references from solid geometry.
5. Create rounds, shells and draft angles last when modelling your parts.
6. Do not take references from rounds, shells and draft angles.

Example Assembly and Part File Structure: An example of a CAD parametric assembly file structure with parent–child reference links is shown in Diagram 2. The file structure starts with an 'Assembly File' followed by 'Reference Part File-A' which is the only file referenced to the 'Assembly File'. This means that all the basic parametric elements can reside in 'Reference Part File-A' to 'Drive and define the model' (the whole assembly). Therefore, all the other parts and sub-assemblies in the assembly can be referenced to 'Reference Part File-A'.

In Diagram 2, all the part files in 'Group 1' were created in the assembly with all their references, to construct the solid geometry, taken from the basic parametric elements in 'Reference Part File-A'.

All the part files in 'Group 2' were created in the 'Sub-Assembly File' with all their references, to construct the solid geometry, taken from the basic parametric elements

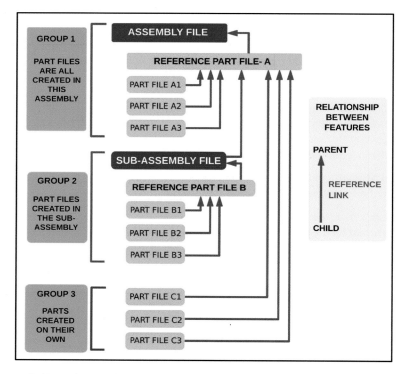

Diagram 2 Example assembly file structure with parent–child reference links

in 'Reference Part File-B'. The sub-assembly was created outside the 'Assembly File' and imported into the 'Assembly File'. The 'Sub-Assembly File' was then referenced to 'Reference Part File-A'. This means that the position of the sub-assembly is defined by 'Reference Part File-A'.

Finally, all the parts in 'Group 3' were created separately outside the 'Assembly File' and then imported into the 'Assembly File'. Each 'Group 3' file contains its own basic parametric elements to reference the solid geometry within the file. In the 'Assembly File' the position of each 'Group 3' file is defined by referencing them to 'Reference Part File-A'.

The benefit of this file structure and referencing method is that all the files in the 'Assembly File' (the assembly model) can be managed via 'Reference Part File-A'. This method greatly reduces the problem of having to make changes to every part file in the assembly to modify the model.

CAD Model Validation: Today most CAD systems have easy-to-use design validation tools. With these tools' materials can be assigned to each part complete with material properties such as density, tensile strength, compressive strength, Young's modulus, Poisson's ratio, specific heat capacity, thermal conductivity, hardness, etc. With these properties, the weight of parts and the whole model, including its structural integrity, thermal integrity, etc. can be analysed and validated. An example of a finite element analysis (FEA) is shown in Image 4.

Validation tools can also apply colours and surfaces finish, such as textures, to the model surfaces. The model can then be photo realistically rendered to validate its aesthetic quality as seen in Image 5.

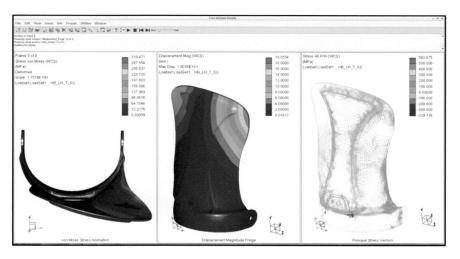

Image 4   Finite element analysis (FEA) of a Snowboard Binding Highback made from carbon fibre and Kevlar. The analysis shows the deflection and stresses resulting from a force applied to the part that simulates what would happen in the real world. Analysis was generated in Creo (version 7, 2020) parametric 3D modelling system from Parametric Technology Corporation (PTC)

Image 5 Rendering of a Hexapod Robot. Modelled in Fusion 360 parametric 3D modelling system from Autodesk. (Derivative of 'Anansi Hexapod Robot' by Bryce Cronin/CC BY 4.0. www.cronin. cloud/hexapod)

The model can also be sectioned to check how the parts fit together (see Image 6) and exploded to see all the parts (see Image 7).

12.8   CAD Parametric Modelling—Drawing Files

CAD drawing files are used to generate technical drawings for communicating manufacturing and assembly instructions. Today there is less need for technical drawings to manufacture parts. This is because CAD part files can be converted into various formats and sent directly to manufacture (Xometry, 2020). However, it's important to always contact the manufacturer to see what file format they require or what technical drawings are needed.

Generating technical drawings of CAD models is a fairly easy process but some basic knowledge is required as follows:

Technical Drawing Standards: Technical drawings need to follow strict standards to ensure instructions are unambiguous. These standards vary from country to country, so it is important to become knowledgeable with the standards in your country before preparing technical drawings to send out to manufacture. The technical drawing standards are comprehensive in describing the way that objects, assemblies, parts, features (holes, shafts, chamfers, countersinks, threads, fasteners, fillets, centrelines, assemblies, etc.), dimensions, text, line thicknesses etc. are presented on drawings. If the standards are incorrectly applied, it can lead to faulty parts being manufactured.

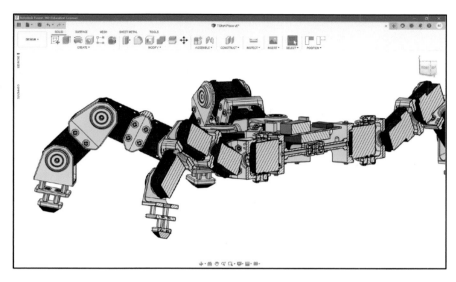

Image 6   A section view of a Hexapod Robot. Modelled in fusion 360 parametric 3D modelling system from Autodesk. (Derivative of 'Anansi Hexapod Robot' by Bryce Cronin/CC BY 4.0. www.cronin.cloud/hexapod)

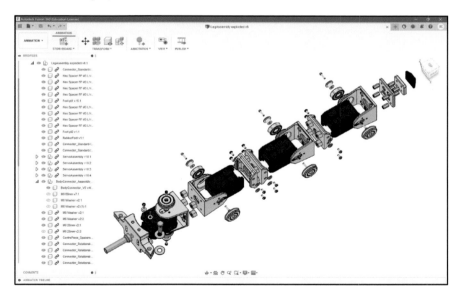

Image 7   An exploded view of a Hexapod Robot. Modelled in fusion 360 parametric 3D modelling system from Autodesk. (Derivative of 'Anansi Hexapod Robot' by Bryce Cronin/CC BY 4.0. www.cronin.cloud/hexapod)

Country	Standard
Australia	AS1100
CANADA	CAN3-B78. 1-M83 technical drawings
US	ASME Y14.5 and Y14.5 M and ISO 8015
UK	BS 8888

Table 12.2 Technical drawings standards for some countries

Although most countries have their own technical drawing standards, most are based on the International Organisation for Standards (ISO) standard ISO 128–1:2020 (ISO, 2021). In Table 12.2 are technical drawings standards for some countries.

Technical Drawing Sheet Structure: When generating technical drawings for products that consist of more than one part (in an assembly), it is important to have some structure to the set of drawing sheets. One way is to organise the sheets into three groups as follows: **Group one** contains the assembly sheet(s) that shows the product in its assembled form. This is the first drawing sheet(s) in the set. Generally, only the overall dimensions of the product are shown on the assembly drawings. This allows the manufacturer to understand what the product looks like, get a sense of its size, how all the parts fit together and how to assemble the product, as seen in Image 8.

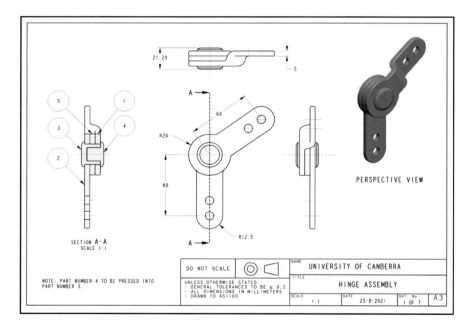

Image 8 Example technical assembly drawing of a hinge. Created in Creo (version 7, 2020) parametric 3D modelling system from Parametric Technology Corporation (PTC)

5	HINGE_SPACER	STAINLESS STEEL 304	7	I
4	HINGE_PIN_CAP	STAINLESS STEEL 304	6	I
3	HINGE_PIN	STAINLESS STEEL 304	5	I
2	HINGE_ARM_STRAIGHT	STAINLESS STEEL 304	4	I
I	HINGE_ARM_BENT	STAINLESS STEEL 304	3	I
ITEM	DESCRIPTION	MATERIAL	SHT	QTY

UNLESS OTHERWISE STATED:
- GENERAL TOLERANCES TO BE ±0.2
- ALL DIMENSIONS IN MILLIMETERS
- DRAWN TO AS1100

NAME   UNIVERSITY OF CANBERRA

TITLE   HINGE ASSEMBLY PARTS LIST

| DATE | SCALE | SHT. NO. | |
| 25-8-2021 | NA | 2 OF 7 | A4 |

Image 9 Example technical assembly parts list drawing of the hinge. Created in Creo (version 7, 2020) parametric 3D modelling system from Parametric Technology Corporation (PTC)

Group Two contains the parts list sheet(s). Here a description, material, quantity required and other information for each part is listed in a table, as seen in Image 9.

Group Three contains the drawing sheet(s) for each part. The technical drawing sheet(s) for each part need to contain all the information required to manufacture the part without having to refer to the assembly and parts list, as seen in Image 10. This enables drawing sheets for parts that require different manufacturing processes (fabrication, casting, forging, machining, etc.) to be separated and sent to different manufacturers who have the requisite manufacturing capabilities. For example, if a product contains two parts in an assembly where part one is sand cast and part two is fabricated from sheet metal. Then in the drawing set, Sheet 1 can be the assembly drawing, sheet 2 the parts list, sheets 3 to 5 for part one and sheets 6 and 7 for part two. Consequently, to make the two parts: sheets 3 to 5 can be sent to a manufacturer who specialised in sand casting while sheets 6 and 7 can be sent to a sheet metal fabricator.

Technical DrawingSheet Sizes: The sheet sizes for technical drawings are ISO-A and ISO-B. It is important to refer to the standards in your country to see which one is preferred. It is recommended that all technical drawing sheets have a border that denotes the extent of the information contained on the sheet. The typical recommended sheet sizes and borders are as shown in Tables 12.3 and 12.4.

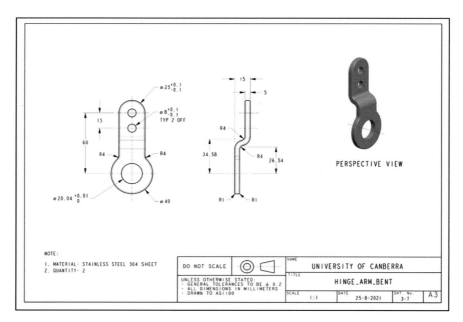

Image 10 Example technical part drawing of the Hinge_Bent_Arm part. Created in Creo (version 7, 2020) parametric 3D modelling system from Parametric Technology Corporation (PTC)

Table 12.3 ISO-A technical drawing sheet and border sizes. All dimensions in millimetres

Sheet designation	Sheet size	Border size—Top/Bottom and Sides
A0	841 X 1189	20 and 20
A1	594 X 841	20 and 20
A2	420 X 594	10 and 10
A3	297 X 420	10 and 10
A4	210 X 297	10 and 10

Table 12.4 ISO-B technical drawing sheet and border sizes. All dimensions in millimetres

Sheet designation	Sheet size	Border size—Top/Bottom and Sides
B1	707 X 1000	20 and 20
B2	500 X 707	20 and 20
B3	353 X 500	10 and 10
B4	250 X 253	10 and 10

Each drawing sheet must also have a title block that can be customised to a certain degree to suit the requirements of the author or organisation. Examples of title blocks can be seen in Images 8, 9 and 10. It is important to refer to the standards in your country to see the requirements and options for title blocks.

Technical Drawing Orthographic Projections: In technical drawings 3D objects are typically represented by 2D views known as orthographic projections. The purpose of orthographic projections is to clearly communicate the shape and dimensions of objects. This is achieved by representing the object in various orthographic views, such as front view, side views, top view, bottom view, section views, etc. and applying dimensions to the views as seen in Images 8 and 10. Orthographic views are typically represented as unshaded line drawings, however 3D rendered perspective views can also be included to help communicate the shape of the object as seen in Images 8 and 10.

There are two types of orthographic projection—first angle and third angle projection as seen in Image 11.

On a technical drawing sheet, the type of orthographic projection applied must be stated with a symbol as shown in Image 12. Furthermore, in a set of drawings all the views must be presented in either first or third angle projection. There must not be a mix of first and third angle projections in a set of drawings. The application of third angle projection in technical drawings can be seen in Images 8 and 10. Although third angle projection is the most common method used, it is important to refer to the standards in your country to see which one is preferred.

TechnicalDrawing Scales: It is recommended that technical drawing orthographic views should ideally be drawn to the following scales. The recommended ISO scales for mechanical technical drawing are shown in Table 12.5.

Reading a Technical Drawing: For products that consist of more than one part (in an assembly), the first step is to understand what the assembled product looks like, its size, how many parts there are, what material each part is made of and how all the parts fit together. This is revealed by reading the assembly sheet(s) in conjunction with the parts list. The parts list also list which sheet(s) in the set contains the details for each part.

The second step is to read the drawing sheet(s) for each part to understand the shape, dimensions and all the necessary information to manufacture the part. As seen in Image 10, there are two orthographic views of the HINGE_ARM_BENT part that describe its shape and dimensions. For more complex shapes, there may be more views including section views and detail views to fully describe the part. Furthermore, this information may be spread over more than one sheet. Also seen on Image 10 is a shaded perspective view of the part that is useful for visualising what the parts looks like. The other information on the sheet is the material that is specified as 'STAINLESS STEEL 304 SHEET', and therefore, an appropriate manufacturing process would be sheet metal cutting and bending. The remaining information is contained in the title block where the general tolerance, dimension units (millimetres) and the applied standards are stated. To fully understand how

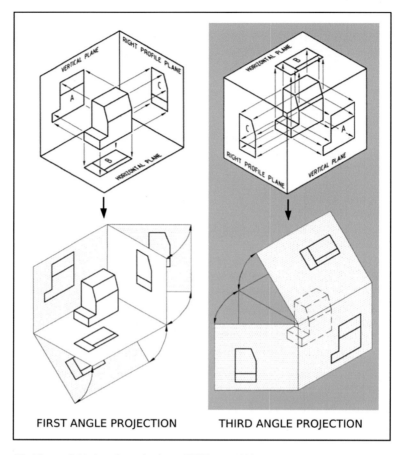

Image 11 First and third angle projections. (Williams, 1993)

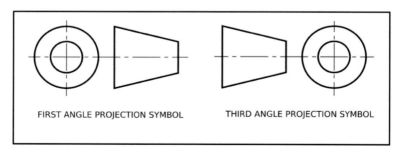

Image 12 First and third angle projection symbols. Examples showing the placement and size of these symbols can be seen in Images 8 and 10 that contain the 'third angle projection symbol'

Table 12.5 Recommended technical drawing scales (Standards Australia)

Category	Recommended scales		
Enlargement Scales	50:1	20:1	10:1
	5:1	2:1	
Full size			1:1
Reduced Scales	1:2	1:5	1:10
	1:20	1:50	1:100
	1:200	1:500	1:1000
	1:2000	1:5000	1:10,000

to read technical drawings, it's important to refer to the relevant standards in your country.

12.9   CAD File Transfer

The problem with all CAD systems is that their files are not directly transferable between systems because they use different algorithms. Therefore, CAD files must first be converted to formats such as STEP, VDA, STL, etc. before they can be opened in another system (Jezernik, 2003). Although the converted files and all their details can be viewed in other systems, their parametric associations are lost leaving limited ability for the files to be edited and the model to be modified (Tornincasa & Di Monaco, 2010, p II-12).

12.10   VR and AR for CAD

Virtual reality (VR) is a digitally simulated environment of the real world where the user is completely immersed in the experiences through a head-mounted display (Aloor et al., 2016, March; Farshid et al., 2018; Cabero-Almenara et al., 2019) or in a room where a projector projects the VR image on the walls, ceiling and floor.

Today VR is used with CAD to mainly provide close-up inspections, interact, and validation of 3D CAD models aesthetics, proportions and ergonomics. The benefit of VR is that it can deliver much better insights than are possible from a flat screen (Keane, 2019; Wong, 2019).

Augmented reality (AR) allows objects from the real world to be enhanced (seamlessly overlayed, blended, interwoven) with digitally generated objects in real time (Cabero-Almenara et al., 2019; Farshid et al., 2018).

Today AR is used with CAD for end-user feedback on design concepts, to enhance technical drawings (an AR markers is placed on 2D drawings so that when the drawing is viewed through a smartphone an overlayed 3D image of the product appears), verification of digital prototypes by overlaying them onto real objects or users (e.g. with wearable prototypes, the users can see themselves on a screen wearing the prototype in real time), manufacturing assembly processes to visualise the position and orientation of parts, monitoring quality control on production lines, ergonomic monitoring of workers posture, and employee training (Spasova & Ivanova, 2020, p 498–499).

An example of AR to verify CAD models was a project undertaken at the University of Canberra. Prescription glasses were designed on CAD and their aesthetic appeal was tested by the public using AR. From a laptop touch screen monitor, anyone could select a pair of digital prescription prototype glasses. Then by placing themselves in front of the laptop monitor they could see themselves in the monitor wearing the glasses in real time, just like looking into a mirror.

12.11 CAM and CNC

Computer-aided manufacturing (CAM) uses computer systems to automate manufacturing processes to make products with very high accuracy and precision. These manufacturing processes are performed by computer numeric control (CNC) machines (Latif et al., 2021, p 2549–2550). The manufacturing processes that these machines perform are milling, turning, cutting (laser, waterjet and plasma), CNC routing, electrical discharge machining (EDM), welding, 3D printing, etc..

The CAM process typically starts with CAD where the geometry for the part to be manufactured is generated (Elser et al., 2018, p 1514). As previously noted, the CAD geometry file is then converted to a suitable format that a CNC machine can convert into machine language. The machine language contains all the instructions needed for the CNC machine to make the part. Two common machine languages are G-code and STEP-NC (Latif et al., 2021, p 2563–2564).

The benefit of CAM is that it can speed up the manufacturing process and reduce costs. However, equipment can be costly and often skilled staff are required.

3D Printing: 3D printing is a process that enables a 3D digital model to be made into a physical object. Also known as additive manufacturing, it is a process where successive layers of a material are laid down over each other to make the object (Shahrubudin et al., 2019). Typically, 3D digital models are created in CAD and in most cases are 3D printed to verify the design aesthetics, fit, ergonomics, functionality, etc. Today, 3D printing is used in various industries to make not only prototypes but increasingly, as the technology evolves, mass manufactured end-use parts and products (Ngo et al., 2018, p 172–173).

In the design process, the benefits of 3D-printed prototypes have greatly reduced the time and cost to bring new products to market.

The benefits of 3D printing in design and manufacturing are as follows:

1. In the design process, 3D-printed prototypes have greatly reduced the time and cost to bring new products to market. It allows designers to quickly verify their design ideas at any stage of the CAD modelling process.
2. In manufacturing: 3D printing enables the production of custom-made products; transport costs are reduced as products can be manufactured closer to the end-user; provides companies more flexibility and greater quality control as they can better manage the whole process; can print a wide range of materials with new 3D printing materials regularly being released (Shahrubudin et al., 2019, p 1286–1287); there is no need for expensive tooling, such as dies needed for traditional casting processes; 3D printing can produce any shape regardless of its complexity as opposed to conventional manufacturing processes, such as pressure die casting and injection moulding (Ngo et al., 2018, p 173).

However, when 3D printing parts for mass production there are still some disadvantages compared to conventional manufacturing processes. For example, 3D-printed parts take a considerable amount of time to print, from a few minutes for small parts to several hours for large parts. In comparison, conventional processes like pressure die casting and injection moulding, small parts take seconds and large parts 1–2 min (Kridli et al., 2021, p 100–112).

A further problem with 3D-printed parts is that their mechanical properties are not as good as parts made using conventional methods such as pressure die casting and injection moulding. Therefore, 3D printing is best suited for making parts that are subjected to low structural loads (Chen et al., 2020, p 7).

3D Printing Methods: There are several 3D printing methods, below are seven of the main ones used today:

1. Fused deposition modelling (FDM) is a process where a filament of thermoplastic polymer is fed through a moving (x, y, z) heated nozzle that melts the plastic and deposits it layer-upon-layer to create a 3D object. The benefits of this process are that it is inexpensive, simple and reasonably quick compared to other 3D printing processes. However, parts have poor mechanical properties and poor surface finish (Ngo et al., 2018, p 174).
2. Selective laser sintering (SLS) involves covering a bed with successive layers of a very thin, closely packed powder. After each layer is laid down, a laser beam fuses it to the previous layer. This is repeated until the 3D part is created. The excess powder is then removed with a vacuum. SLS is used to create parts made from a variety of plastic polymers, metals and alloys.
      The advantages of SLS are that the parts have a high resolution and good surface finish. If the powder is not fused with a binder, parts can also have good mechanical properties. The disadvantages are that the process is slow and costly.
3. Selective laser melting (SLM) is the same as SLS except that it is only suitable for various metals and metal alloys. Whereas SLS only partially melts the powder, SLM fully melts and fuses the powder together. Consequently, SLM produces parts with superior mechanical properties.

4. Stereolithography (SLA) uses a UV laser beam to selectively cure a photosensitive thermoset polymer resin one layer at a time to create a 3D object. SLA was the first 3D printing technology.

    The advantage of SLA is that it produces parts of high quality, fine resolution and smooth surface finish. The disadvantages are that it is comparatively expensive and slow (Ngo et al, 2018, p 174; Ge et al, 2020, p 2–4).

5. Digital Light Processing (DLP) is basically the same as SLA except that the UV light is projected as patterns onto the surface of a photosensitive thermoset polymer resin to cure it one layer at a time to create a 3D object. The advantage of DLP is that it creates high-resolution parts at fast speeds. Consequently, it is suitable for large parts (Ge et al., 2020, p 2–4).

6. Inkjet Printing selectively deposits micro-size liquid droplets of material through a moving ink-jetting head one layer at a time until the 3D object is completed. The droplets are solidified by a UV light immediately after they are deposited. The process is low cost and parts have high precision and resolution. This process can print a large variety of materials such as polymers, ceramics, biological (materials that mimic living tissue & cellular embryonic components) (Ge et al., 2020, p 2–4; Ogunsanya et al., 2021, p 427–428; Jammalamadaka & Tappa, 2018, p 2).

    A disadvantage of inkjet printing is lack of adhesion between layers (Ngo et al, 2018, p 176).

7. Direct energy deposition (DED) or direct metal deposition (DMD) uses a moving nozzle to simultaneous deposit material (in the form of a wire or powder) and deliver a beam of energy (laser, arc or electron beam) to melt the metal one layer at a time until the object is built. If a laser or arc is used, the process is performed in an inert atmosphere while for an electron beam a vacuum is required. The process is suitable to 3D print not only metals (particularly high-performance super-alloys) but also plastics and ceramics (Ngo et al., 2018, p 174–175; Liu et al., 2019 p 1, 2; Dávila et al., 2020, p 3379–3380). The advantages and disadvantages of the three DED methods (laser, arc or electron beam) are as follows:

    Laser based- When powder is used as the deposition metal, the parts created are of a higher quality than those produced using a wire. However, using wire is more suited for higher deposition rates. The geometry of the part can affect its thermal behaviour and therefore impact the quality of the part. This can necessitate post-processing that adds time and costs (Dávila et al., 2020, p 3379).

Arc based- Advantage is the low cost of equipment compared to laser and electron beam methods; has a high deposition rate that makes it suitable for printing medium to large components; produces better surface finish and dimensional control than laser and electron beam methods. A disadvantage can be porosity between layers; however, new techniques are resolving this problem (Dávila et al., 2020, p 3379–3380).

Electron beam- Advantages are higher deposition rates compared to laser or arch methods that make it suitable for printing large components. The disadvantage is higher costs due high vacuum and need for protection from X-rays (Dávila et al., 2020, p 3380).

12.12   Workshop

For design and engineering, the workshop is where all the equipment and space to make physical prototypes by either traditional and or digital means resides. Traditional methods are typically used to verify design concepts before committing to CAD. Whereas digital methods are generally used to make physical prototypes from CAD models. However, physical prototypes of CAD models can be built employing both Traditional and digital methods. The workshop at the University of Canberra in Australia is outfitted with both Traditional manufacturing equipment and digital manufacturing equipment (as seen in Images 13 and 14). The University of Canberra staff and students use the workshop facilities for research, teaching and learning.

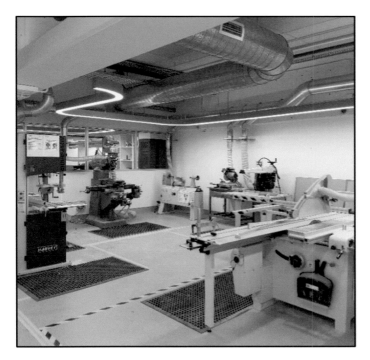

Image 13   Traditional cutting, milling and turning

Image 14 CNC 3 axis router

12.13   Case study- Hexapod Robot Project

Introduction: This case study will look at how two honour students (Christopher Lane and Bryce Cronin) from the University of Canberra (UC) applied the design process and CAD to design and manufacture a Hexapod Robot. The aim of the project was to design and manufacture a robot that could be used as a practical teaching resource for the Robotics course at the University of Canberra. The culmination of the project resulted in the design and manufacture of a Hexapod Robot as seen in Image 15. Unlike traditional hexapod designs that usually have two or three Degrees of Freedom (DoF) in each limb, the UC Hexapod Robot has four DoF in each limb.

Design Process for the Hexapod Robot

Typically, all design projects start with a problem. In this case, the problem was that there was no practical teaching resource that the UC could use for its robotics classes. Consequently, it was decided to design and manufacture a Hexapod Robot that would cater for both introductory and advances robotics classes. It was also decided to design and manufacture the Hexapod Robot in house at UC. This was because UC researchers had:

1. Expertise and experience in designing and manufacturing robots.
2. Suitable CAD software (Fusion 360) and expertise in using the software.

Image 15 UC Hexapod Robot 3D CAD model. Built in Fusion 360 CAD software. (Bryce Cronin/CC BY-NC-ND 4.0. www.cronin.cloud/hexapod)

3.   A fully equipped robotics laboratory and workshop with all the facilities to prototype (3D print), manufacture and test a robot.

The five design process steps for the design and manufacturing of the new Hexapod Robot project were as follows:

Briefing and Task Clarification (Design Process Step 1)

- Understanding the project brief. This involved discussion, formulation and agreement on the design criteria that required a robot that:

  - Could be used as a practical teaching resource for both introductory and advanced robotics classes offered at the UC.
  - Can be upgraded to cater for future developments and research.
  - All code and physical components can be easily modified by students to assist with their understanding of robotic concepts.
  - The code can facilitate simple actions such as walking and waving as a starting point for students to develop their own code.
  - Wherever possible, all the components and code to be designed and manufactured in house at UC.
  - Is safe to operate

- A timeline, budget and deliverables were prepared.

- A research report was prepared—It began with a brief affirmation of the project aim and the design criteria. It then investigated other robot designs and assessed their positive and negative attributes to meet the requirements of the brief. The report included suitable materials, manufacturing processes, off-the-shelf components, desirable features and functions, software architecture, maintenance, safety considerations and a risk mitigation strategy for the new device.

The risk mitigation identified possible risks associated with the project and how they were mitigated. Risks included issues resulting from Covid-19, software security, hardware security, injuries to all persons. The risk mitigation strategy used the risk analysis matrix shown in Table 12.6 to categorise identified risks throughout the project.

The report concluded with a list of design guidelines that would drive the following stages of the design process including a timeline, budget and deliverables.

Concepts Generation (Design Process Step 2)

Based on the research report and further brainstorming, the robot's flexibility of movement was identified as a primary design objective. This would allow the robot to perform a range of tasks such as traverse uneven terrain and preform gesture-based communication for human–robot interaction (HRI). It was also decided that a modular design would be best suited to meet all the design criteria. To achieve this, several pages of concept designs were generated as rough hand-drawn sketches as seen in Image 16. From each concept, the features that best fulfilled the design

Table 12.6 Risk analysis matrix

Risk Rating Matrix		Likelihood				
		Rare (1)	Unlikely (2)	Possible (3)	Likely (4)	Almost Certain (5)
Consequences	Severe (5)	Moderate	Moderate	High	Severe	Severe
	High (4)	Low	Moderate	Moderate	High	Severe
	Moderate (3)	Low	Low	Moderate	Moderate	High
	Low (2)	Ex Low	Low	Low	Moderate	Moderate
	Ex Low (1)	Ex Low	Ex Low	Low	Low	Moderate

- Extremely Low (2-3): Easy to fix, no danger.
- Low (4-5): OK to fix but document, no danger.
- Moderate (6-7): Moderate fix, careful not to break anything.
- High (8): Difficult fix, could be expensive to replace or jeopardise project.
- Severe (9-10): Extremely difficult to fix with likely jeopardy of project, severe harm or loss of human life.

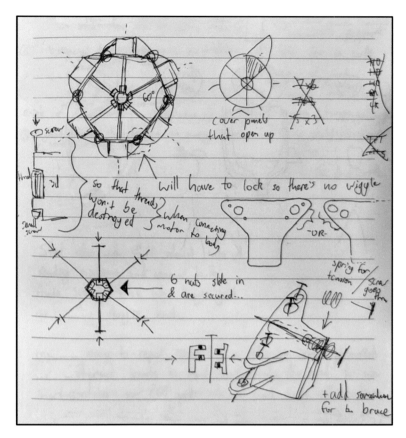

Image 16 Concept design as rough hand-drawn sketches

criteria were then combined into one initial Hexapod Robot concept. A quick CAD modelled of the initial concept was generated on Fusion 360 (Parametric) as seen in Images 18, 19. The model also included the off-the-shelf components electronic (as seen in Image 17) and all the fasteners. The electronic parts and fasteners were modelled as basic shapes to their exact dimensions to ensure their correct fitment within the model. All the parts, except for the of-the-shelf parts, were designed so that they could be 3D printed in house. The CAD model enabled the fit, functionality and manufacturability of all the parts and the assembly to be closely investigated.

Refinement CAD Model and Prototyping (Design Process Step 3)

From the initial concept CAD model, several design refinements were identified to simplify and reduce the number of parts, improve the functionality of the joints, extend the modularity of the design (so that it can be reconfigured) and make it easier to assemble the robot.

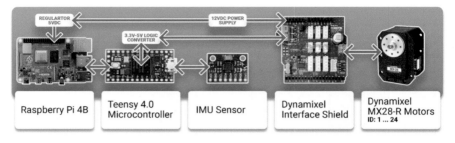

Image 17 Off-the-shelf electronic components

Image 18 UC Hexapod Robot initial concept design. (Bryce Cronin/CC BY-NC-ND 4.0. www. cronin.cloud/hexapod)

Image 19 UC Hexapod
Robot initial concept arm
designs. (Bryce Cronin / CC
BY-NC-ND 4.0. www.cro
nin.cloud/hexapod)

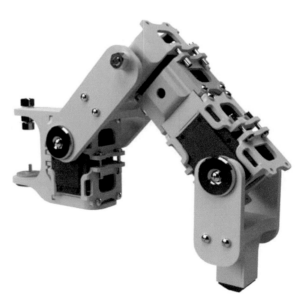

A new fully detailed 3D CAD Model of the Hexapod Robot was generated using Fusion 360 (Parametric) 3D CAD software. The new model incorporated all the design refinements identified from the initial CAD model. The new CAD model demonstrated that the new hexapod design fulfilled all the design criteria as follows:

- The design is extremely modular and intuitive that makes the robot suitable for both introductory and advanced robotics classes, future development upgrades and research.
- The modular design allows parts to be rearranged so that legs can be extend, shorten or the joints in each leg can be completely relocate.
- The radial symmetrical design allows for additional legs to be added, or for legs to be removed.
- Six legs provide plenty of movement flexibility, especially in navigation uneven terrain or continuing to operate even if some legs become disabled. This is because it can adjust its posture and centre of gravity on uneven terrain, providing it with better static stability.
- The software architecture of the hexapod is built upon the commonly used open-source software known as ROS (Robotic Operating System). ROS is a simple and flexible middleware that allows for high-speed synchronous control of the robot's motors and sensors. The advantage of building the system architecture with ROS is that it can integrate a huge collection of libraries and tools making future upgrades easier and better integrated.
- It can walk in a variety of ways (gaits) depending on requirements for speed and stability. The initial design moves in a tripod gait where it alternates three legs on the ground and the opposing three legs in the air at any one time in its movement as seen in Image 20. However, the design allows future upgrades to be made to the ROS to enable the robot to have a dynamic gait. This means that it will be able to change its footing and posture based on the terrain thereby achieving better balance and natural movement.
- Unlike traditional hexapod designs that typically have two or three DoF in each limb, this design has four DoF in each limb. This enables the UC hexapod to perform a variety of programmed actions, such seen in Images 20 and 22.

Step 1
6 legs down → Step 2
3 legs down → Step 3
6 legs down → Step 4
3 legs down

Image 20 Walking motion. (Bryce Cronin/CC BY-NC-ND 4.0. www.cronin.cloud/hexapod)

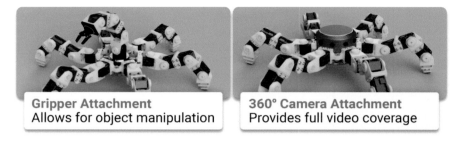

Gripper Attachment
Allows for object manipulation

360° Camera Attachment
Provides full video coverage

Image 21 Potential attachment options. (Bryce Cronin/CC BY-NC-ND 4.0. www.cronin.cloud/hexapod)

- It can accommodate a variety of additional attachments such as a gripper, 360° camera, etc. as seen in Image 21.
- It can flip itself over and continue to perform actions, such as walking while upside down.
- It can gesture and pose to human collaborators as a method of expressing intention and communication to non-technical personnel in the field such as waving, tapping, beckoning, etc.
- All the major components were designed to be easy to 3D CAD model, modify and 3D Print. The reason for 3D printing the parts was because it was more cost effective than CNC machining or traditional injection moulding (for such relatively small number of parts). It also avoided having to use expensive of-the-shelf robotic components. Furthermore, because 3D printing was available in house turnaround times could be greatly minimised making it more suited for teaching and research applications .
- The only off-the-shelf components were the electronic components, all the fasteners, bearings and rubber feet. The off-the-shelf electronic components can be seen in Image 24. The internal off-the-shelf electronic components were configured so that additional electronics could be easily added in the future that can communicate with the existing electronic components. Up to 2 kg of additional electronics and equipment could be supported by the Hexapod Robot design.

Step 1
Fully folded

Step 2
Fully unfolded

Step 3
Standing position

Image 22 Unfolding motion. (Bryce Cronin/CC BY-NC-ND 4.0. www.cronin.cloud/hexapod)

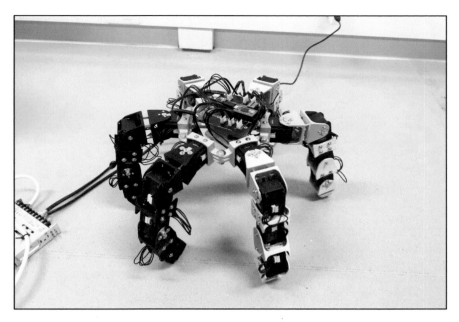

Image 23 Hexapod Robot working prototype

The new 3D CAD model of the Hexapod Robot enabled further refinements to the CAD model and verified the fitment of all the parts (tolerances).

Prototyping the Redefined Design: From the refined CAD model, one fully functional prototype of the Hexapod Robot was made as seen in Image 23. This involved 3D printing all the components designed by the two UC honours students on an in-house FDM printer. The material used was polylactic acid (PLA), a thermoplastic made from renewable resources in the form of a filament. The FDM process and PLA material were selected because of their low cost, the parts produced have adequate strength to meet the design requirements, and the process is simple. All the electronic components (see Image 24) and fasteners were purchased off-the-shelf. The prototype was fully tested to verify the assembly, fitment, articulation of the parts, the electronics and the ROS.

Testing the Redefined Design: The tests found that the Hexapod Robot met all the requirements of the design criteria. The tests only found one issue with the design, where the bolts around the joints would unscrewing themselves.

Final Design Prototyping and Documentation (Design Process Step 4)

To address the issue of the bolts around the joints unscrewing themselves, the Hexapod Robot CAD model's leg members were redesigned to include bearings at the joints. The difference in the design of the leg member can be seen in Images 25 and 26 with the latter being the redesigned version. The bearings can be seen in exploded CAD model of the leg assembly seen in Image 27.

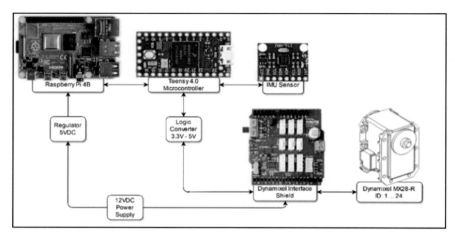

Image 24 Off-the-shelf electronic components and diagram

Image 25 Leg design with
problem of the bolts
unscrewing themselves
circled in red

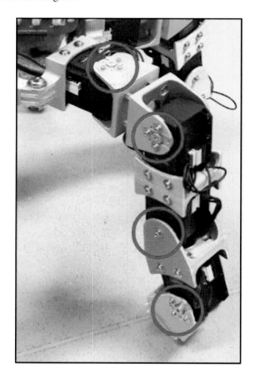

Image 26 Final leg design with unscrewing problems resolved circled in red. (Bryce Cronin/CC BY-NC-ND 4.0. www.cro nin.cloud/hexapod)

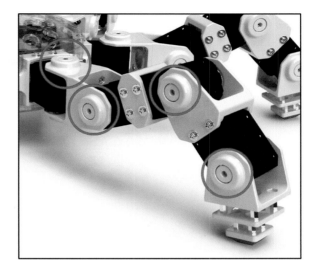

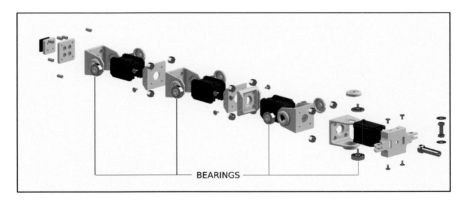

Image 27 UC Hexapod Robot Leg 3D CAD model exploded view. Built in Fusion 360 CAD. (Derivative of 'Anansi Hexapod Robot' by Bryce Cronin/CC BY 4.0. www.cronin.cloud/hexapod)

The final 3D CAD model of the Hexapod Robot is seen in Images 28, 29 and 30. In the CAD exploded views of the Hexapod Robot model, see Images 27 and 29, the modular design of the parts and the assembly can be seen.

The final 3D CAD model parts for one leg assembly and the central body can be seen in Image 31 with a description of the parts in Table 12.7. Because of the modular design of the Hexapod Robot, there are only 11 different 3D-printed parts. This included three different printed parts for the central body and eight different printed parts for all the leg assemblies as shown in Image 31.

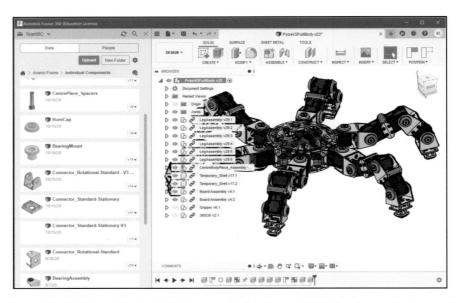

Image 28 UC Hexapod Robot 3D CAD model. Built in Fusion 360 CAD software. (Derivative of 'Anansi Hexapod Robot' by Bryce Cronin/CC BY 4.0. www.cronin.cloud/hexapod)

Image 29 UC Hexapod Robot 3D CAD model exploded view. Built in Fusion 360 CAD software. (Derivative of 'Anansi Hexapod Robot' by Bryce Cronin/CC BY 4.0. www.cronin.cloud/hexapod)

Image 30 UC Hexapod Robot rendered 3D CAD model. Built in Fusion 360 CAD software. (Bryce Cronin/CC BY-NC-ND 4.0. www.cronin.cloud/hexapod)

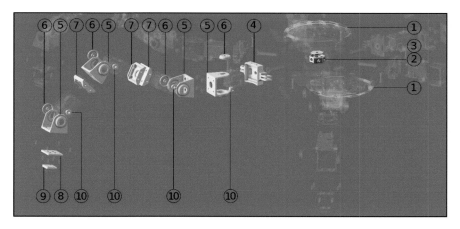

Image 31 UC Hexapod Robot final design showing the 3D-printed parts for one leg and the central body. For the description of all the parts, refer Table 12.6. (Derivative of 'Anansi Hexapod Robot' by Bryce Cronin/CC BY 4.0. www.cronin.cloud/hexapod)

Prototyping and Testing the Final Design: The new parts that make up the robot's central body and one leg assembly were individually saved to STL format and 3D printed. The printing process and material for each part are shown in Table 12.7, importantly these are same as those used in the final manufactured Hexapod Robot to test their validity. The printed parts together with their off-the-shelf electronic components, bearings and fasteners were assembled and tested. The test results found

Table 12.7 Parts list for Image 35

Item no	Description
1	Central body translucent cover—SLA printed from Accura (acrylate-based plastic)
2	Central body anchor—SLS printed from nylon 12
3	Central body anchor cover—SLS printed from nylon 12
4	Central body connector—SLS printed from nylon 12
5	Leg pivot connector—SLS printed from nylon 12
6	Leg pivot connector CAP—SLS printed from nylon 12
7	Leg mid connector—SLS printed from nylon 12
8	Leg upper foot pad—SLS printed from nylon 12
9	Leg lower foot pad—SLS printed from nylon 12
10	Leg bearing mount—SLS printed from nylon 12
11	Leg bearing spacer washer (Not shown)—SLS printed from nylon 12

that the unscrewing problems at the joints had been resolved and that the final design fulfilled all the design criteria.

Documentation: This included- 1. A Fusion 360 (parametric) 3D CAD model of the complete Hexapod Robot assembly, 2. user manual and 3. final report with an overview of the development process and recommendations for future development for the Hexapod Robot.

The Contents of the User Manual Were as Follows:

1. Cover page with an image of the Hexapod Robot
2. Images of all the parts complete with parts list. The parts list included a description of each part, materials, manufacturing process, suppliers and quantity.
3. Assembly instructions.
4. Operating instructions.
5. Images of the Hexapod Robot performing all its different functions.
6. Software.
7. Programming (coding) instructions.
8. Technical specifications.
9. Manufacturing instructions for the 3D-printed parts.
10. Maintenance instructions.

Recommendations for Future Development Included:

1. To develop a portable method of powering the robot such as batteries. Currently, the Hexapod Robot is powered by a tethered power source. Importantly, all the electronic components used in the Hexapod Robot will support batteries and require no further modifications.
2. To further develop the electronic hardware to expand the Robot's functionality.

3.   To further develop the software to refine the walking gaits and gesturing movements.

Manufacturing (Design Process Step 5)

To manufacture the Hexapod Robot, all the parts modelled on Fusion 360 (parametric) CAD software by the two UC honours students were individually saved to STL format. All except two of the parts were printed on an SLS 3D printer from Nylon 12 (as seen in Table 12.7). The reason for selecting this printing process and material is because the parts produced would have very good mechanical properties that are suitable for working applications along with high-dimensional accuracy and high-quality surface finish. The remaining two parts were printed on an SLA printer from translucent and ridged acrylate-based plastic (as seen in Table 12.7). This process and material were chosen because of the almost transparent look and very good mechanical properties making the part suitable for functional use.

All the electronic components, as seen in Image 24, and fasteners were purchased of-the-shelf.

Assembly, Testing and Delivery: A fully functional Hexapod Robot was assembled as seen in Images 32 and 33. The Hexapod Robot was fully tested and fulfilled all the design criteria.

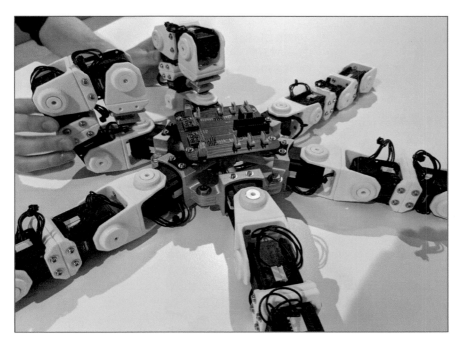

Image 32 UC Hexapod Robot final design being assembled. The description of all the parts refer Table 12.6. (Bryce Cronin/CC BY-NC-ND 4.0. www.cronin.cloud/hexapod)

Image 33 Final working UC Hexapod Robot. (Bryce Cronin/CC BY-NC-ND 4.0. www.cronin. cloud/hexapod)

A fully functioning Hexapod Robot (see image 36) complete with documentation (1. A full Fusion 360 (parametric) CAD model, 2. User manual and 3. Final report with an overview of the development process and recommendations for future development for the Hexapod Robot) was delivered on time and on budget.

12.14   Revision Questions

Question 1. When in the design process should you identify the materials and processes and why?

Question 2. When during the design process can CAD be employed?

Question 3. What do you need to do before sending files to a manufacturer?

Question 4. Where should a working prototype be tested?

Question 5. How many different 3D CAD modelling systems are there, what are they called, and what are their comparative advantages and disadvantages?

Question 6. Name the different types of 3D parametric CAD model files?

Question 7. What is a circular reference and how do you avoid them?

Question 8. What elements should be used as references and what are they?

Question 9. What features should not be used as references?

Question 10. What features should be creates last in a part file?

Question 11. What are CAD drawing files used for?

Question 12. Why are technical drawing standards important?

Question 13. Technical drawing sheets are typically organised in how many groups and what are they?

Question 14. What are orthographic projections and what is their purpose?

Question 15. What is VR and AR typically used in CAD for?

Question 16. What are some advantages and disadvantages of 3D printing?

Question 17. What is the difference between the following two 3D printing methods: SLA and SLS?

Question 18. Why is it important to include all the part in a model including off-the-shelf parts?

References

Alba, M. (Retrieved 2018, March). https://www.engineering.com/story/whats-the-difference-between-parametric-and-direct-modeling.

Aloor, J. J., Sahana, P. S., Seethal, S., Thomas, S., & Pillai, M. R. (2016, March). Design of VR headset using augmented reality. In *2016 international conference on electrical, electronics, and optimization techniques (ICEEOT)* (pp. 3540–3544). IEEE.

Ault, H. K., & Phillips, A. (2016). Direct modeling: Easy changes in CAD?

Bodein, Y., Rose, B., & Caillaud, E. (2014). Explicit reference modeling methodology in parametric CAD system. *Computers in Industry,65*(1), 136–147.

Bonollo, E. (2016). *Product design: a course in first principles.* Upfront Publishing, Calwell, A.C.T.

Cabero-Almenara, J., Barroso-Osuna, J., Llorente-Cejudo, C., & Fernández Martínez, M. D. M. (2019). Educational uses of augmented reality (AR): Experiences in educational science. *Sustainability,11*(18), 4990.

Chen, M. Y., Skewes, J., Daley, R., Woodruff, M. A., & Rukin, N. J. (2020). Three-dimensional printing versus conventional machining in the creation of a meatal urethral dilator: Development and mechanical testing. *BioMedical Engineering OnLine,19*(1), 1–11.

Cronin, B. (2020). www.cronin.cloud/hexapod.

Dam, R., & Siang, T. (2018). What is design thinking and why is it so popular. *Interaction Design Foundation.*

Dávila, J. L., Neto, P. I., Noritomi, P. Y., Coelho, R. T., & da Silva, J. V. L. (2020). Hybrid manufacturing: a review of the synergy between directed energy deposition and subtractive processes. *The International Journal of Advanced Manufacturing Technology,* 1–14.

Design Council. (2021). https://www.designcouncil.org.uk/news-opinion/what-framework-innovation-design-councils-evolved-double-diamond. Accessed 11 Dec 2021

Elser, A., Königs, M., Verl, A., & Servos, M. (2018). On achieving accuracy and efficiency in additive manufacturing: Requirements on a hybrid CAM system. *Procedia CIRP,72*, 1512–1517.

Farshid, M., Paschen, J., Eriksson, T., & Kietzmann, J. (2018). Go boldly!: Explore augmented reality (AR), virtual reality (VR), and mixed reality (MR) for business. *Business Horizons,61*(5), 657–663.

Ge, Q., Li, Z., Wang, Z., Kowsari, K., Zhang, W., He, X., Zhou, J.& Fang, N. X. (2020). Projection micro stereolithography based 3D printing and its applications. *International Journal of Extreme Manufacturing, 2*(2), 022004.

ISO 128. (Retrieved 2021). https://www.iso.org/ics/01.100.01/x/.

ISO. (Retrieved 2021). https://www.iso.org/home.html.

Jammalamadaka, U., & Tappa, K. (2018). Recent advances in biomaterials for 3D printing and tissue engineering. *Journal of Functional Biomaterials,9*(1), 22.

Jezernik. (2003). A solution to integrate computer-aided design (CAD) and virtual reality (VR) databases in design and manufacturing processes. *The International Journal of Advanced Manufacturing Technology, 22*(11–12). https://doi.org/10.1007/s00170-003-1604-3.

Keane, P. (May 2019). VR in CAD: Where are we now? https://www.engineering.com/story/vr-in-cad-where-are-we-now.

Kridli, G. T., Friedman, P. A., & Boileau, J. M. (2021). Manufacturing processes for light alloys. In *Materials, design and manufacturing for lightweight vehicles* (pp. 267–320). Woodhead Publishing.

Latif, K., Adam, A., Yusof, Y., & Kadir, A. Z. A. (2021). A review of G code, STEP, STEP-NC, and open architecture control technologies based embedded CNC systems. *The International Journal of Advanced Manufacturing Technology*, 1–18.

Liu, Z., Zhang, H. C., Peng, S., Kim, H., Du, D., & Cong, W. (2019). Analytical modeling and experimental validation of powder stream distribution during direct energy deposition. *Additive Manufacturing,30*, 100848.

Ngo, T. D., Kashani, A., Imbalzano, G., Nguyen, K. T., & Hui, D. (2018). Additive manufacturing (3D printing): A review of materials, methods, applications and challenges. *Composites Part b: Engineering,143*, 172–196.

Ogunsanya, M., Isichei, J., Parupelli, S. K., Desai, S., & Cai, Y. (2021). In-situ droplet monitoring of inkjet 3D printing process using image analysis and machine learning models. *Procedia Manufacturing,53*, 427–434.

Prud'homme van Reine, P. (2017). The culture of design thinking for innovation. *Journal of Innovation Management,5*(2), 56–80.

Shahrubudin, N., Lee, T. C., & Ramlan, R. (2019). An overview on 3D printing technology: Technological, materials, and applications. *Procedia Manufacturing,35*, 1286–1296.

Spasova, N., & Ivanova, M. (2020). Towards augmented reality technology in CAD/CAM systems and engineering education. In *The international scientific conference eLearning and software for education* (vol. 2, pp. 496–503). " Carol I" National Defence University.

Standards Australia, AS1100.101–1992 (1992).

Tornincasa, S., & Di Monaco, F. (2010, September). The future and the evolution of CAD. In *Proceedings of the 14th international research/expert conference: trends in the development of machinery and associated technology* (vol. 1, No. 1, pp. 11–18).

Williams, R. A. (1993). *Engineering drawing handbook*. Standards Australia.

Wong, K. (2019, December). Is AR/VR ready to go beyond visualization? https://www.digitalengineering247.com/article/is-ar-vr-ready-to-go-beyond-visualization/cad.

Xometry (Retrieved 2020, May 27). https://xometry.eu/en/choosing-right-file-formats-for-manufacturing/.

Eddi Pianca Eddi Pianca is an Assistant Professor and Industrial Design discipline lead at the University of Canberra where he teachers in Industrial Design, conducts research and supervised PhD students. He has interest and experience in high level CADD/CAM, user centred design, ergonomics, High Performance Composites design and advances in design technology. He maintains strong industry links by running industry-based project for the industrial design course.

Eddi holds a PhD from the University of Canberra on the design and simulation of complex full body sports activities.

Prior to the University of Canberra, he worked for the Civil Aviation Authority in Australian (Mechanical Design office manager), for Electronics Research Australia in Canberra Australia (designing computer equipment) and for General Electrics Company in Sydney Australia (Industrial Designer, lighting research and development division).

He also has an Industrial Design Degree from the University of Canberra and a Mechanical Engineering Certificate from ACT TAFE Canberra Australia.

Part III
Interaction Design

Chapter 13
Social Robots: Principles of Interaction Design and User Studies

Janie Busby Grant and Damith Herath

13.1 Learning Objectives

This chapter introduces you to the basic steps in designing and conducting social robotics research. By the end of this chapter you will:

1. Be able to describe why you are conducting your research project, including your motivation for conducting the research, who the audience is for your findings, and the key research questions you will be addressing.
2. Be able to identify the key variables you want to focus on and understand how to operationalise these variables in real research environments.
3. Be able to recognise different types of research designs, know advantages and disadvantages of each, and work out which is appropriate in which situation.
4. Be aware of the concepts of validity and reliability, and be able to both identify and address issues that can emerge.
5. Understand the key principles to consider when designing and conducting ethical research.
6. Identify key factors when analysing and interpreting data.

13.2 Introduction

This chapter considers when, why and how you can conduct user-focused research when working with robots. While exciting, high-quality human–robot interaction

J. Busby Grant (✉)
Discipline of Psychology, Faculty of Health, University of Canberra, Academic Fellow, Graduate Research, Canberra, Australia
e-mail: janie.busbygrant@canberra.edu.au

D. Herath
Collaborative Robotics Lab, University of Canberra, Canberra, Australia
e-mail: Damith.Herath@Canberra.edu.au

© The Author(s) 2022
D. Herath and D. St-Onge (eds.), *Foundations of Robotics*,
https://doi.org/10.1007/978-981-19-1983-1_13

research is carried out in laboratories and real-life settings around the world, often those working in the field of robotics feel unsure or unprepared to conduct user-focused research, as research design and analysis is typically not included in undergraduate robotics courses. Conducting well-designed research studies can allow you to identify potential issues, benefits and unexpected outcomes of real-world interactions between robots and users, and provide evidence for efficacy and impact. Being able to confidently and competently design and conduct user studies with robots will be an advantage in a myriad of workplace roles.

This chapter is designed to be an introduction and practical guide to conducting human–robot interaction research. In this chapter, we consider why you should conduct research examining human–robot interaction, how you can identify the best research design to answer your question, and select and measure appropriate variables. Throughout the chapter we provide examples of relevant real-world research projects. While we can necessarily only "scratch the surface" of each topic we discuss in this single chapter, and you are encouraged to explore deeper into the issues of relevance to you, the tools in this chapter will allow you to understand and implement well-designed research projects in human–robot interaction.

An Industry Perspective

Martin Leroux, Field Application Engineer

Kinova inc.

I was doing my bachelor's degree in engineering physics, but I didn't quite like it because I found it too abstract. At the time, I happened to find an internship for a robotics lab which took me in not so much for my then non-existent background in robotics, but rather for the advanced math skills I developed in physics. That internship was the part that really clicked for me; I finally was able to explain to my family what it is that I do in terms they could understand. I came to the field for that feeling of satisfaction, but I ended up staying because it is so wide and I get to learn new things constantly.

When I first started at Kinova, my job was to evaluate alternative control input methods to help our assistive users manipulate our robots. Once, I started working on eye-tracking and a colleague kept insisting that I let one of our users try it as soon as possible, which I found was too early, especially since I also have eyes. When I asked why he was so insistent, he told me that a while ago, their team spent multiple weeks tuning a program to help assistive users drink from a bottle - adjusting the rotation of the arm, the lift speed, and so on and so forth. Then, when they were all done and went to show it to a user, his reaction was: "Oh, I'd never bother with that. I just use a straw." Weeks of work went down the drain. Now, we always involve our end users right from the get-go.

When I first entered the field, human-robot interfaces were already very varied and fairly functional. However, their purpose was to specifically control the robot in terms that only made sense for robots. You would move individual joints or sometimes switch between translation and orientation for end-effector cartesian control. Nowadays, although the hardware hasn't evolved much, the interface itself often leverages artificial intelligence to make the entire system much smarter. Instead of asking users to think like roboticists, they can keep thinking like human beings with task-oriented commands. Where people used to need to think "I want to get my gripper there", now they can think "I want to grab this" and the robot can deduce some or all of the motion that is expected of it.

13.3   Cobots, Social Robots and Human–Robot Interaction

Considered the father of Robotics, Engineer and Entrepreneur, Joseph Engelberger, after commercialising the first industrial robot arm, stipulated in his book "Robotics in Service" that the main thrust of the industry should be towards the development of what are called service robots. He argued that technological developments in robot perception and artificial intelligence should enable the replacement of human work that is labour intensive and unpalatable with robotic devices. In fact, a considerable portion of such mundane human work is now automated using robotic technologies. In the process, robots have come to be established in increasingly human spaces. This necessitates the designers of these robotic devices to consider such elements as perceived safety, human factors and ergonomics on top of the engineering capabilities of the robots. The modern cousins of the original Unimate (like the Kinova Gen3 robots) embody such nascent technologies, providing the ability to directly interact and work alongside humans in a safe and intelligent manner. A new category of industrial robots is emerging called *collaborative robots* or cobots. These technologies are considered to belong to the fourth industrial revolution (industry 4.0) which represents the evolution of traditional industrial manufacturing technologies and practices combining with emerging smart technologies such as the Internet of Things (IoT), cloud computing and artificial intelligence (AI).

On the other hand, the origins of the word "robotics" and as we see in popular culture, robots are meant to be a reflection of the humanity. As we discussed earlier, humans have entertained the idea of human-like machines for millennia. With recent advances in related domains in computing and hardware, there is growing interest in the engineering community to explore the technical development of socially intelligent robots. We have already seen several commercial *artificially socially intelligent* robots appearing in the market with various success. This new genre of machines called "*social robots*" are meant to interact with humans at a higher cognitive and emotional level, as compared to a typical industrial robot in a factory.

In either scenario, robots are increasingly required to interact with humans, and importantly with end users who are not technically trained to operate robots (such as is currently the case with industrial era robots). As we have already seen in Chap. 3, empathetic thinking is required when designing such robots with lay end-users in mind. And at the other end of the pipeline, before deploying these robots, it is required that we not only validate the robot's engineering functions but also its ability to interact with humans as intended. The latter requires an understanding of human psychology and associated disciplinary expertise. The study of human–robot interaction is thus an emergent field that not only encompasses many fields of engineering, but yields a broader disciplinary net towards psychology, sociology, design and the humanities. Abbreviated, HRI, the new field is gaining considerable attention from the robotics community with several key journals and conferences dedicated to the subject already, including the ACM/IEEE International Conference on Human–Robot Interaction and the International Conference on Social Robotics and their respective journals, ACM Transactions on Human–Robot Interaction and the International Journal of Social Robotics.

13.4   Why Conduct Research?

Any research project should begin with the "why?". Why do you want or need to conduct the research, what questions do you want to answer, what will the findings allow you to know or do, and who will be interested in the answers you generate?

13.4.1   *Motivation for the Research*

Your reasons for conducting the research may be diverse and could include:

- You want to know how people interact with the robot in order to improve it.
- You are not sure how the robot will perform in real-world settings.
- You need evidence to give to investors that your product will be successful.
- You are conducting a project in a university or research-focused setting.
- Your boss/supervisor/funding partner said so!

All of these and many more are legitimate reasons for conducting research, but how you design your project will depend on your "why", which determines your research question and your audience. For instance, you might have questions about: efficacy (how well the robot performs in controlled conditions), effectiveness (how successful the robot is in the real world), safety (both technical and perceived by the user), or perceptions of the people interacting with them (such as ratings of likeability or animacy).

Research Examples: Why and Audience

Sylax: A university-based research robotics project.

- Why: The researchers want to understand what factors play a role in converting an industrial robot into an interactive user-friendly robot.
- Audience: The researchers themselves, the wider academic field.

Tommy: A robot product designed by a start-up for commercialisation.

- Why: The researchers want to design a conversational agent for use by the general population.
- Audience: Initially the researchers, eventually the general population.

Coramand: A tech company robotics product for large-scale industry-based roll-out.

- Why: The researchers want to assess and improve perceived and operational safety of their collaborative robot.
- Audience: Initially the development company itself, then roll-out to companies that use automation at scale (individuals in those companies: executive, skilled technologists, investors).

13.4.2 Target Audience

Who is the intended target of the outcomes of the research? Remember that there may be several uses for, and audiences of your output from a particular piece of research.

Typically, the research you conduct will inherently inform your own knowledge, so the first audience is generally **yourself**. Carefully consider what you need to know in order to move the project forward—what information will directly inform the next stage in your project? All too often researchers get carried away with their own cleverness and design complex experiments, only to find that they have tweaked their design to the point that their answers don't quite tell them what they need to know anymore! Always come back to your key research question and what answering it will allow you to conclude.

If you are a student or university-based researcher your target audience will also be **others in your academic field**. You should be familiar with the key literature, and in a fast-moving field like robotics, this also means attending the main conferences, reading new abstracts for related projects, taking part in competitions, new business ventures and start-ups. There may be typical research paradigms in your area that you are expected to use, or controversies or debates that you need to be aware of, or terminology or technological basics you should adopt to better communicate with those in your field. This applies both in terms of the design of the research itself and how you report and distribute your findings.

You may also have an audience in the form of **investors**, whether this be venture capital firms, government funding bodies or individuals. If this funding is currently supporting your research, be clear on what outcomes and reporting you have already committed to, and in what timeframes, as you may well need to set up research to directly address those requirements. If you are designing research to obtain future funding, look at the previous projects that the organisation or individual has financed, and consider what kind of evidence they provided. For instance, were they interested in projects with lab-based proof-of-concept completed, or data on projected uptake from the general public, or testing conducted in unpredictable real-world environments? Knowing the answers to these questions will help frame your own research.

Enterprise may also be an audience for your work, for example companies that use large-scale automation in industry settings. This relationship may take the form of an existing contract or potential sales. You need to be very clear on what factors will be most important to the key individuals in that organisation so that you can focus your research to demonstrate capability in those areas. You also need to identify what forms of evidence will be most convincing to that audience—are they looking for large lab safety trials, or real-world human–robot error rates, or expert review?

Consumers and/or the general public may also be the target audience for your research, particularly if you are demonstrating efficacy in a real-world environment or interest in solving a particular problem. In this case you need to think about what the target consumer would find persuasive and meaningful, and incorporate those elements into your study. For example, if safety is a concern among the general population, then part of your research should investigate the safety of the robot in general use settings so you can make an evidence-based statement about safety at the conclusion of your project.

13.4.3   Research Questions

The Scientific Method is the process used by researchers to create an accurate representation of the world. By working collaboratively, building on and sharing evidence and theories, we can assemble an understanding of how things work. The Scientific Method involves iteratively generating theories and hypotheses, gathering data and analysing that data to draw conclusions, which then feed back into our theories about the world, which then continues the process, refining our knowledge as it continues (Fig. 13.1).

In HRI research typically you begin a research project with theory about how you believe a particular interaction between a robot and a participant will play out. This may be on the basis of previous research and theory in the field, or based on observations you have personally made. A theory is a set of explanatory ideas that integrate a variety of evidence. Theories pull together facts into a general principle or set of principles—they not only help us to explain *what* is happening, but also *why* it is happening. Theories also allow us to make predictions about what will happen in a new situation. As a multidisciplinary research area, theories in HRI could be informed by many and at times conflicting ideologies. Some examples are theory of mind from psychology, perceptual control theory from cybernetics or more speculative ideas such as the Uncanny Valley effect in aesthetics.

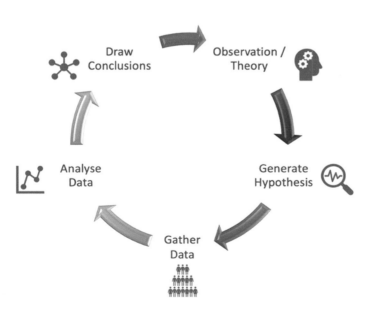

Fig. 13.1 A visual depiction of the scientific method used to conduct research

Within a particular research study, you will generate a specific hypothesis, which is typically based on a theory. A hypothesis is a specific statement about the relationship between variables in your study—it is what you expect to happen. Crucially, hypotheses are testable—that is, the findings of your study either support the hypothesis or contradict it. If a hypothesis (or a theory) is not supported, this is called falsification. The hypothesis is how you narrow down your broader theory to assess one particular effect or relationship in your research study.

Research Examples: Theories and Hypotheses

Sylax: A university-based research robotics project.

– Broad research question: What is the effect of factors like behaviour and appearance of the robot on peoples' perception of agency?
– Theory: Theory of mind (Leslie, 1987)
– Hypothesis (for one particular study): People will perceive the robot to have more agency when it shows purposeful movements rather than when it displays random movements.

Tommy: A robot product designed by a start-up for commercialisation.

– Broad research question: How can we maximise consumer satisfaction with our companion robot product?
– Theory: Attachment theory (Bretherton, 1985).
– Hypothesis (for one particular study): When given a robot to use in the home, people who have greater previous exposure to robots will report higher attachment to the product.

Coramand: A tech company robotics product for large-scale industry-based roll-out.

– Broad research question: How can perceived safety and operational safety of the robot be maximised?
– Theory: Behavioural Decision Theory (Slovic, et al., 1984)
– Hypothesis (for one particular study): People's perceived safety of the robot will be greater when the robot displays fluid movements rather than more robotic movements.

What You Should Know Going into a Research Project

I am conducting this research study because:

_____

The audience/s for my research are:

_____

> The key theory or theories relevant to my study is:
> _____
>
> My main hypothesis for the study is:
> _____

In summary, before you begin your study consider your motivation for conducting the study, your target audience/s and the theories that are relevant to your study. Use these to determine your hypothesis. Remember it needs to be *specific* to your study and *testable*. For example, predicting that robots with faces will be more acceptable to the general public than robots without faces is a theory—whereas making the statement that in your study the robot with a face will have higher likeability ratings than the same robot with the face obscured is an hypothesis. The hypothesis refers directly to what you are *manipulating* and *measuring* in your study, and we'll explore this in more detail below.

13.5   Deciding on Your Research Variables

13.5.1   *Variables*

When conducting research, one of the major tasks is to decide what you are going to manipulate and what you are going to measure. This obviously depends on your research question. For example, if you want to investigate perceived safety in the home setting, this will be measured very differently compared to a research question is focused on operational safety in an industrial setting. Assessing the likeability of a robot by young children will use a different approach from measuring the sense of agency attributed to a robot by an adult.

A variable is a characteristic which can be measured or changed. Age, object preference, experimental condition, reaction time and performance are all variables. Some of these variables are inherent to an individual and cannot be assigned (e.g. age), others change depending on the task (e.g. performance) and others can be manipulated by a researcher (e.g. exposure to different conditions). Deciding what variables you want to measure/manipulate is one of the key issues in research design and will dictate what conclusions you are able to draw.

A helpful distinction is between Independent Variables (IVs) and Dependent Variables (DVs).

- The IV is the variable you believe has an effect on the other variable/s. In some studies, the IV is manipulated (changed) by the researcher. In robotics research, this is often some aspect of the robotic system, for example your IV might be whether the robot has an anthropomorphic face or not.

- Within an IV you often have conditions (also called *levels* or *groups*)—you "do different things" to the different groups. There can be any number of conditions. Sometimes there are clear experimental and control conditions, such that the experimental condition is the group receiving the treatment, and the control condition is the comparison group (or "usual" group). For instance, if your IV is the type of robotic face, you may have two conditions—one in which the participants view the default robot face (control condition) and one in which they view a new version of the face (experimental condition).
- The DV is the variable which is measured (observed) by the researcher. Often there are many DVs in a single study—for example you may want to measure participants' ratings of likeability, animacy and safety.
- *Note*: In some research designs, such as descriptive research, all the variables are simply measured and the researcher doesn't believe one has an effect on another. In this case, all can be considered DVs.

Research Examples: IVs and DVs

Sylax: A university-based research robotics project.

- IV: Behaviour

  - Condition 1: Random behaviour
  - Condition 2: Purposeful movement—algorithmic behaviour

- DV: Perceived agency measured by the number of interactions initiated by the participant

Tommy: A robot product designed by a start-up for commercialisation.

- IV: Reported previous exposure to robots based on a questionnaire response scale.
- DV: Attachment measured using a questionnaire response scale.

Coramand: A tech company robotics product for large-scale industry-based roll-out.

- IV: Motion type

  - Condition 1: Fluid motion
  - Condition 2: Robotic motion

- DV: Perceived safety measured as how far away people stand from the robot (in m).

13.5.2 Operationalisation

So you know what you want to investigate and why, you've identified your key theory and hypothesis, you've decided on your IV and your DV—surely you're ready to go, right? Well, not quite yet! The next key step is to operationalise your variables—which involves defining what you are manipulating and measuring, and how. For the purposes of the research project you are conducting, you are deciding very specifically how your variables will be changed or measured. This might involve a survey question or set of questions combined into a single value, or a measure of reaction time, or performance on a task.

Let's say your DV is safety—how safe people feel interacting with a particular robotic construct. There are many different ways you could operationalise this variable, for instance you could use a self-reported survey question (How safe did you feel during this interaction? 5 Very Safe to 1 Not Safe at All), or you could assess how close participants stand to the robot (with people standing closer indicating a higher level of safety), or you could measure participants' heart rate or cortisol level to indicate their level of stress during the interaction. All of these different ways of operationalising safety, and many others(!), could be appropriate depending on your "why", your research question and your audience. However, operationalisation is often driven by logistical and contextual issues as well as deeper theoretical approaches. For instance, what equipment do you have? What expertise do you and the other researchers have? What time do you have to collect the data? Are your potential participants able and willing to be measured in this way?

How you operationalise your variable will have fundamental implications for the conclusions that can be drawn from your study. Also keep in mind that how you operationalise the variable directly impacts what statistical tests you can run during analysis (see later discussion), for instance whether you use a choice task, a yes/no scale, a categorical scale or a continuous number, may require different statistical analyses.

13.5.3 Relevance-Sensitivity Trade-Off

What we're looking for in a variable is a measurement that is sensitive enough to show a difference that you're interested in (i.e. don't measure something that is unlikely to change). But you also need to be careful not to pick a variable that is so specific that it isn't relevant beyond the study to other contexts (i.e. don't measure something that isn't meaningful in the wider world). This is called the relevance-sensitivity trade-off. This can be a real issue in lab-based experiments, where the focus is on finding a way of measuring the variable that is easy in that environment, rather than operationalising the variable in a way that is more relevant to the real-world environment (Fig. 13.2).

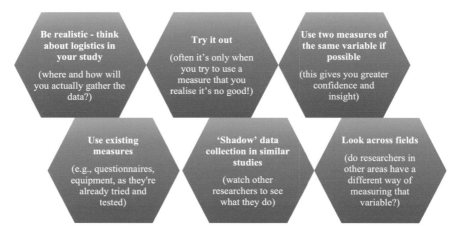

Fig. 13.2 Approaches that can help you decide how to operationalise variables

Examples of Relevance-sensitivity trade-off

Take the example of *Tommy*, the companion robot aimed at the general population. In a carefully controlled experiment you could ask participants how much they like the robot on a scale from 1 to 100. Let's say you added new capabilities and likeability scores went up from an average of 50 to an average of 60. That sounds big! But does this actually mean more people will now purchase the product? Not necessarily. The attachment scale is *sensitive* enough to pick up changes based on the improvements you made, but is that change in score *relevant* if what you really want to know if whether people will buy it or not?

13.5.4   Research Designs

There are countless ways research can be conducted, and different ways these approaches can be grouped together. One common categorisation of research designs is into descriptive, correlational, experimental research, and reviews and meta-analyses (Fig. 13.3).

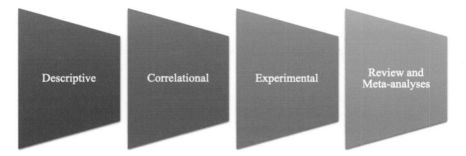

Fig. 13.3 One categorisation of the different types of research designs

13.5.5   Descriptive Research

When you conduct descriptive research you don't manipulate any variables—instead you take advantage of the natural flow of behaviour to measure what you are interested in. Some of these approaches are highly flexible, so if unexpected events happen in your research, or you develop new ideas, you can alter how you collect your data to take advantage of this. Some examples of descriptive research are observation (where you record the behaviours of your participants without interfering—e.g. examining video footage of home use of a robot), archival research (using existing data— e.g. studying web browser search history), and program evaluation (e.g. analysing outcomes of a robotic system embedded into a workplace). Focus groups are also a way of conducting descriptive research in which you sit down with small groups of participants and ask them to give their opinions and describe their behaviours. Asking descriptive information from large groups of participants is often done using surveys. In the case of the Coramand robot example used above, the company itself may want to conduct descriptive research in the form of focus groups, to get an understanding of how its employees feel about incorporating robots on the factory floor, before they commit to implementing them.

One common descriptive research method used in robotics is case studies. Case studies focus on the behaviour of a single individual or single context. They are useful in that they provide very rich information about a particular experience, so all the nuances of that person's situation, behaviour and cognition can be explored. They are also sometimes the only option, where the situation or experience or context is so unusual that other research cannot be used, such as discussed in Design Chap. 3. However, it is difficult to generalise from case studies (will this same pattern of behaviour or outcomes be seen in other contexts or individuals?) and you cannot be sure what caused any changes in the individual's behaviour—it may not have been the variable you are focused on, but could have been something else in their environment or unique to them.

13.5.6   Correlational Research

Correlational research is research which asks whether there is a relationship between two or more variables, but where the variables are not under the researcher's direct control (for logistical or ethical reasons). Often survey research falls into this category—although it can be purely descriptive (see above), sometimes surveys are used to look at whether two variables "go together", such as assessing whether age is related to perception of how dangerous robots are. Surveys are useful when you want to look at naturally occurring patterns in the world, and are relatively easy to conduct. They also allow you to measure a lot of variables at the same time. The Tommy example discussed above, in which the researchers wanted to know the relationship between previous exposure to robots and attachment to the robot product, is an example of a survey design; the researchers would conduct a correlational analysis to look at whether those two variables "go together". However, correlational research like this can only tell you about the correlation between two variables, not whether one variable causes changes in another variable. For that, you need to conduct an experiment!

13.5.7   Experimental Research

Experimental research studies the effect of an independent variable on a dependent variable. So the Sylax case discussed above, which looked at the effect of behaviour type (the IV: random or purposeful) on perceived agency (the DV) would be an experiment. Often when you manipulate an IV you measure a whole range of DVs (so for example likeability, animacy, safety). Experiments are normally conducted in highly controlled lab-based settings, but can also be conducted in natural environments—these are known as field experiments.

In an experiment, you typically want to find out if a particular manipulation (e.g. adding a face to a robot) has an effect. Let's say you want to compare a condition where you do that manipulation with one where you don't. There are several ways of doing this:

- A control condition is a group where there is no treatment or manipulation of the IV.
- A placebo condition is a group which receives what looks like the treatment/manipulation but *it is not*. It is a "look-a-like" treatment without the active component/ingredient.

13.5.8 Between-Subjects and Within-Subjects Designs

Consider the Sylax research discussed above, in which a researcher has an IV which is the type of behaviour, and measures perceived agency. In this study, the researcher has two conditions—one in which the participants are exposed to random behaviour (control condition) and one in which they are exposed to purposeful behaviour (experimental condition). The researcher then has a choice:

- They could recruit 40 participants and assign 20 to the control condition (random behaviour) and 20 to the experimental condition (purposeful behaviour), and compare the scores of the two groups.
- They could recruit 20 participants and have those participants complete both the control and experimental conditions at different times, and compare the scores of the participants on the two conditions.

In one of these cases, different people experience each of the different conditions; in the other, the same people complete both conditions. Both of these options are legitimate under particular circumstances, but they each have benefits and possible drawbacks you should be aware of when designing your study.

In a within-subjects design participants are assigned to all of the conditions of the IV, so the experimental manipulation takes place within subjects. If you had three conditions in your IV, for example, the participants complete all three conditions (they could do these sequentially in the same testing session, or take part in your study on three different days). You then compare the scores of the *same participants* across the three different conditions.

In a between-subjects design participants are only assigned to one of the conditions of the IV, so the experimental manipulation takes place between subjects. If you had three conditions in your IV, then you would have three groups of participants, and each group would complete a different condition (no individual would participate in all three conditions). You would then compare the scores of the three *different groups* of people (Fig. 13.4).

Both within- and between-subjects designs can be appropriate depending on the situation. Often, the nature of the study will determine this for you. For example, if you

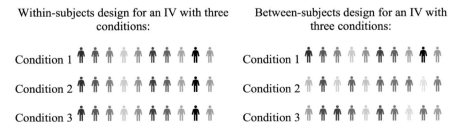

Fig. 13.4 Illustration of the differences between within-subjects and between-subjects research designs, with different individuals represented as different coloured symbols

	Pros	Cons
Within-subjects designs	• Fewer participants required. • Greater statistical power. • Less random variance.	• Exposure to one condition can affect responses to another. • Can be more difficult to set up.
Between-subjects designs	• No learning or interference across conditions. • Can be easier logistically.	• More participants required. • Lower statistical power. • More random variance.

Fig. 13.5 Pros and cons of within-subjects and between-subjects research designs

are testing whether there is a difference in learning between two interfaces, you may be unable to use a within-subjects design as learning on one will transfer to learning on another—performance on the second interface they see may be better than on the first, regardless of the interface itself. So, if exposure to one condition can potentially affect responses in other conditions, it may not be appropriate to use a within-subjects design. Similarly, if participants are only available for a single testing session, and taking part in the conditions involves a high intensity task, participants may be too fatigued to complete more than one condition. If you are using a within-subjects design one way of addressing potential ordering issues is to use counterbalancing, which involves presenting the conditions in alternating (or random) order to the participants, so that half of participants experience condition 1 first, and the other half condition 2 (and etc. if any additional conditions).

The organisation of elements such as these, and potential time constraints on participant involvement, means within-subjects designs can be more logistically difficult than between-subjects designs. However, they are statistically more "powerful" (see later discussion of power) in that this design reduces the random variance in the data collected, and means you are likely to find a significant effect if one is there than using between-subjects designs. Within-subjects designs can also be more efficient and potentially cheaper to run, as fewer individual participants are required compared with between-subjects designs (Fig. 13.5).

13.5.9   Random Assignment

If you are conducting a study in which participants are allocated to one condition of an IV only (i.e. a between-subjects design), you need to decide how to assign participants to a particular condition. That is, when a participant comes through the door, which condition are they exposed to? You could decide based on a whole range of factors—for example alternating order of arrival, surname, day of the week. Often in robotics research this is determined by technical factors—for example if the condition takes a long time to set up, the first six weeks of data collection for a study could involve participants being assigned to Condition A, and the second six weeks to Condition B.

However, if at all possible, use random assignment to allocate participants to conditions. That is, ensure participants are randomly allocated to either condition (or all conditions, if there are more than two). This can be done by a random number generator or similar (or you can even go old-school and pick a number out of a hat!). Random assignment is important because it *rules out the possibility of systematic differences between the groups*. For example, let's say the first 10 participants who take part in your study you put in a "faceless robot" condition, and the second 10 participants who take part in your study you put in a "face robot" condition. You ask both groups of participants how safe they feel on a 7-point scale from "not safe at all" to "completely safe". The issue with this is that there may well be pre-existing differences between the groups—participants who sign up earlier for an experiment may be (for example) more enthusiastic, conscientious, have more positive views towards robots, more likely to be female. *So you may end up with significant differences between the groups based on existing differences, not on the effect of the IV you are actually testing.* **If you use random assignment to groups you mitigate systematic bias between the groups, and this means you can more confidently say that the IV causes the changes in your DV.** So if you really need to demonstrate a casual effect you need to use random assignment. If you can't use random assignment, try to "match" the participants in each condition as much as you can—so try to have a similar mix of gender, age, ethnicity etc.—but without random assignment you can't say for sure that your IV causes the change in the DV.

Random and Non-random Assignment

Coramand: Researchers are examining the effect of motion type (IV: two conditions—fluid and robotic) on perceived safety. Because of technical constraints, they expose 10 participants (people from around the office) to the fluid motion condition first, then one week later expose a different 10 participants to the robotic motion condition. They compare perceived safety as reported by participants in each group. Each participant was therefore **not randomly assigned** to the different conditions. The researchers must acknowledge that existing differences between the two groups of participants could have played a role in any differences between the groups of responses (e.g. people who said yes first to taking part could be more interested in the project and have lower safety expectations than people who said yes later).

Tommy: Researchers want to know if the robot with enhanced capability (IV: two conditions—original and enhanced) produces more attachment in participants (DV). When every participant is delivered a robot they are **randomly assigned** via an algorithm to receive either an original capability product or an enhanced capability product. When they compare the attachment ratings of the two groups of participants, they can conclude that any differences they find are not due to existing differences between the groups.

13.5.10 Reviews and Meta-Analyses

Some research doesn't collect any new data, but instead "pulls together" all the published research on a particular topic into a single article, and summarises it—these are called reviews. Reviews are considered secondary sources, as no new raw data is collected. Review articles are incredibly useful, as they collect together the key research that has investigated a particular question. Some reviews are called systematic reviews, if they use systematic methods to search the literature—they will list what search terms they used and what databases they searched, and have pre-specified eligibility criteria about which studies to include in the review. Systematic reviews give you a rigorous assessment of findings on a participant topic, so they are the best evidence available to answer a particular research question. If you are working to sell a robot product to an organisation, for example, you could conduct a systematic review of available safety studies, to establish and communicate clearly to the potential customer the current evidence on safety in that setting. Or you could conduct a systematic review before you start a research project, as if there is enough evidence already existing, you may not need to conduct the study! However, keep in mind systematic reviews can be very narrow in focus, so they may not address the particular question you're interested in, and they sometimes don't give you the "big picture" of what's going on in that field.

Meta-analyses are typically systematic reviews in which the author also statistically analyses the data they've found in the studies they review. In this way they can generate new data that numerically summarises the findings. Meta-analyses provide an objective assessment of evidence in a field, however if the original selection of the studies is biased, this means the outcomes of the meta-analysis can be influenced.

13.5.11 Which Research Design Is Best?

A theory can be explored using a wide range of designs—one design isn't better than another, they just provide different ways of investigating the theory. Different designs will also give you different types of information and allow you to draw different conclusions! So make sure you have a clear idea of your motivation, audience and research question before you design your study.

How Different Research Designs can be used to Investigate a Research Question

Let's take the example of Sylax, a university-based research robotics project investigating the effect of factors like behaviour and appearance of the robot on peoples' perception of agency. You could use a broad range of research designs to investigate this question, depending on the particular factors you were interested in.

Descriptive Design: If you were just starting out looking at this research question and wanted to better understand human–robot interactions, you could bring in a single participant and ask them to interact with the robot for an extended period, and video record that process. You could then sit with the participant and watch the video together, examining and discussing all interactions, gaining an understanding of their thoughts and emotions during the interaction, focusing on the aspects of the interaction you're particularly interested in (appearance and agency).

Survey Design: You could recruit a large number of people to interact with the robot, then after the interaction give them a survey about what they noticed about the robot's appearance and how high they rated the perceived agency.

Experimental Design: You could run an experiment where one group of participants interact with a robot with no face, and the other group of participants interact with a robot with a humanoid face, and compare the groups' perceived ratings of agency.

Review Design: You could look at previous research which has explored this topic before. If you wanted to focus on the effect of facial appearance on perceived agency, for example, you could identify all studies published in the last 30 years which involved presenting different types of faces to participants and measuring agency, and summarise their findings to come to a conclusion about the evidence that faces affect agency.

13.6   Sampling, Reliability and Validity

13.6.1   Sampling

For purposes of generalisation, you should do your best to ensure that the people in your study—your sample—is a representative sample of the population you want to apply your findings to—the population. That is, you're getting your data from a sample that has *the same characteristics as the population* (e.g. same gender break-down, same age distribution, same ethnic distribution). If the sample isn't representative in this way, it can cause serious errors when you try to apply your findings to the population. For instance, if you use university students as your sample when testing a particular user interface, you might find that when you roll out your product to the general population, users who are older or younger than your sample may engage very differently with the interface.

There are two broad types of sampling approaches, known as probability samplingand non-probability sampling. Probability sampling is when you select from your intended population so that any member of the population has a specifiable probability of being sampled—for example, you could get a list of the entire population, and select every 10th person on the list to contact. When you use non-probability sampling there is not an identifiable probability of each member of the population being included in the sample. One common example of non-probability sampling is convenience sampling, where you just select your sample from whoever is available around you! Purposive sampling is another type of non-probability sampling, where you deliberately recruit people who meet a certain requirement—such as interviewing elderly people if that is who the robot is aimed at, or recruiting factory workers if the robot is an industrial product.

Using non-probability sampling like convenience or purposive sampling can be fine, as long as the sample you end up with is representative of the population on the particular variables you are interested in. Generally, the bigger the sample is, the better it will reflect the population and so you'll be able to better generalise to the population. But if there is systematic bias in your sampling you'll just make incorrect inferences more confidently… For example, many robots designed in universities or start-ups are only tested using convenience sampling with students or other people involved in the business. This means often the participants are only people who are young, and already interested in and knowledgeable about robots. Findings from studies using samples like these won't necessarily apply to the general public! Also keep in mind that the size of the sample will be reduced by non-response—so people drop out of the study, or forget to enter data.

13.6.2   *Reliability*

Reliability is our confidence that a given finding can be reproduced again and again—that it isn't a chance finding. For example, if you find that people respond more positively to a robot with child-like features than a robot with adult features, that finding is reliable if other researchers consistently find the same pattern. You can think of reliability as similar to *consistency*. However, just because an effect or test is *reliable* doesn't mean it is *valid*.

13.6.3   Validity

Validity is our confidence that a given finding shows what we think it shows. There are four key types of validity: construct validity, external validity, internal validity and ecological validity.

Construct validity asks whether we measured what we were trying to measure. This is harder than it seems! Often people do not interpret the task or question the way you intend, or other factors affect how they respond. For example, often when asked about the usability of a robot, people's responses will actually reflect their judgements about safety instead.

External validity asks how well we can generalise what we have found in our study to other times, populations and places. Let's say you conduct a survey examining attitudes to games involving a robot, using undergraduate students from Canada as participants. Will the findings be the same if I used a sample from a nursing home? Undergraduate students from China? In 10 years?

Internal Validity asks did the outcome reflect the IV we manipulated. Did the variable we are interested in cause the result? Let's say you run a study in which you want to look at the effect of working as part of a team on performance. So you have some participants complete a difficult task on their own and others complete it with other people. Can I conclude that lower performance in the teamwork group is because teamwork per se lowers performance? Not necessarily—the effect could be due to embarrassment, personal space factors, cognitive load, for example, rather than teamwork itself.

Ecological validity asks how well the findings of the study apply to real-world settings—how well does this finding actually work in the real world? For instance, if you test how people use a new type of technology in the laboratory, will they actually use it that way at home? On the bus? At work? *Note: This is different from external validity, which is about generalising to other populations/places.*

13.6.4   Things that Can Go Wrong with Validity

As we talked about above, sampling bias can affect external validity. If only certain types of people respond to a questionnaire or take part in a study, this limits who the findings apply to. It is notoriously difficult to recruit middle-aged people into studies, for example, as they are busy with young children and full-time jobs. If this is who is going to purchase your product you need to make sure your sample includes that group. Mortality—when people drop out of a study—can also affect external validity. For example, if taking part in the study requires an hour per day, those who have less time to take part will drop out of the study, meaning the results will only apply to people like those who remained in the study. If you are running an experiment and more people in one condition drop out than the other, this can also affect internal validity. This is particularly an issue if you are conducting longitudinal studies (research that is conducted over a long period), where you can get *differential drop out*, with participants more likely to drop out of the study if they are in one condition rather than another. Reactivity is when something about the study itself means only particular people respond or influences how people respond. So, if you advertise your study as "Come and play with robots!" you will only get people taking part who already are positively disposed to interacting with a robot, missing a large section of society.

You also need to keep in mind that social desirability can affect how people respond in a study, as people tend to behave based what they think will look good to others. For instance, people tend to over-report their vegetable intake and under-report how many hours of TV they watch! People also change their responses based on cues in research which suggest how they *should* respond—known as demand characteristics. For example, if the researcher repeatedly asks a participant how much they liked the way the robot's eyes moved to follow them, the participant will tend to provide more (and more positive!) information on this element, even if other aspects of the robot were more interesting to them. Sometimes they may even unconsciously change their responses to match what they think the researcher wants. You also need to be aware that often people change their behaviour simply because they are being watched! This is called the observer effect (e.g. if you knew you were being monitored for how much sugar you're consuming, are you likely to eat less sugar?).

Testing effects are when a previous testing situation affects the subsequent testing situation. A gap between pre- and post-test can cause practice effects (i.e. people tend to get better on the same task when they complete it for a second time) and fatigue effects (e.g. are people going to be concentrating all the way through a two-hour testing session?). Maturation effects are when changes occur just because we are measuring things over time. For example, if you are measuring a how a child with a long-term illness interacts with a robot over time, the simple effects of the child ageing are likely to influence the outcomes. Changes in society can also occur during a study, and this can result in history effects (e.g. 9–11, COVID-19, social views on technology). You need to take care that changes in the data due to these factors are not misinterpreted.

Confounds are another threat to internal validity, and reflect changes in your DV that are due to another variable, NOT your IV. Consider if you introduce a learning robot into a school to improve mathematical skills. You compare maths skills in the classrooms without the robot to those with the robot. However, when adding the robot into the classroom this involves change, excitement, new staff etc. Any improvements in "maths ability" could be due to any of those factors, rather than the robot itself.

13.6.5 Ways to Address Problems with Validity

The above section, with all the many potential problems, might make it seem like it's impossible to design a "perfect" study! While an individual study can never avoid absolutely every potential source of bias, there are simple steps you can take that will address many of these issues.

Unobtrusive measures can be used to address reactivity, demand characteristics and the observer effect—these are measures of behaviour that are not obvious to the person being observed. Examples are using one-way mirrors, measuring factors people are usually unaware of (such as how far they stand from a robot interface, or how often they touch it) or using other methods altogether, such as archival records, which are data which was collected for a different purpose. You can help reduce demand characteristics and reactivity by hiding the real purpose of an experiment. You can use deception, where you lie to participants about the purpose of an experiment, or you can use concealment to avoid telling them the whole truth (but be careful of ethics! See below). Using blinding is also really useful in addressing a range of potential biases. Blinding is when key people involved in the study don't know information which could affect their responses. In single-blind studies, the participants don't know which treatment group (level of the IV) they are in. In double-blind studies, both the participants and the researchers don't know which treatment group participants are in.

Focus on Living Labs

- Often when you try to increase internal validity (e.g. control the study tightly) you end up decreasing external validity, so there can be a trade-off between the two.
- A recurring critique of HRI has been the lack of consideration given to the ecological validity of experiments, resulting in poorly designed robots and interactions for the intended task.
- Attempts to address this should consider use of ecologically valid approaches (real-world conditions), such as field experiments or use of "living labs".
- In-the-wild is another term used to describe such experiments. One useful approach is to compromise by situating your experiments in more accommodating venues such as technology museums and gallery spaces, where you find populations open to such experimentation but with reasonable diversity, providing a context that more closely resembles a real-world environment.

13.7 Ethics

13.7.1 Ethics and Ethics Review Boards

All throughout this chapter, we've been talking about designing and evaluating studies—recruitment, sampling, randomisation, etc.; however, it is also important to consider ethical principles in research design. Although most lab-based robotics studies are relatively benign, it is still crucial to be able to identify and address ethical issues and demonstrate to a research ethics committee or institutional review board that your study is appropriate to be conducted. This is in addition to being aware of the broader implications of ethical robot design considerations discussed in Chap. 16 and safety of robot deployment discussed in Chap. 14.

> **Common Ethical Issues in Robotics User Studies**
> So what are some of the typical ethical issues you might face when conducting human–robot interaction research?
>
> - Some of the most important risks are around data: Who has access to the data? Where is it stored? If you are using video recordings this is particularly important, as it is difficult to ensure confidentiality when people are identifiable from their images.
> - There can also be physical danger from proximity to robots that needs to be carefully considered, and this risk communicated to participants when they give consent to participate.
> - Remember participants won't know most of the terminology you are used to using, so write all participant-facing information in easy-to-understand language.
> - If you are asking people you know to participate (friends, relatives, other students) make sure they actually want to take part and don't feel coerced! Avoid asking people you have an unequal power relationship with, such as students you are teaching or supervising.

There are regulations that govern what research can be conducted, typically at the institutional, national and international level, so you need to check based on where you are located. Most broadly the World Medical Association Declaration of Helsinki (*Ethical Principles for Medical Research involving Human Subjects*) is applicable to any research you do in which you recruit participants.

The ethics review process is a formal procedure in which you write a statement including details about the research project and addressing any ethical concerns, which is then submitted to a research ethics board for approval. The board will approve, reject or ask for changes to the study or more information to be provided before approval. Many researchers view applying for ethical approval as purely a logistical a stumbling block, but ethics boards will sometimes identify very real ethical concerns that the researcher has not considered. Ethics boards are typically composed of experienced researchers, legal experts and laypeople. Each of these groups can give insight from those perspectives that you may not have thought about, that necessitate changes to your project.

13.7.2 Ethical Principles in Research

There are many different ways of considering ethical principles, all of which are based on the fundamental idea that you show respect to the people who take part in your study. This means that you are considerate of their experiences and take all the steps you can to be sure that they are consenting freely to participate, and are protected from harm. Some of the key principles to keep in mind are to use informed consent, minimise risk, ensure confidentiality and provide debriefing.

13.7.2.1 Informed Consent

People should participate in your study based on free, informed consent. This means that you provide them with all relevant information about the study (including any potential risks), that they understand this information, and that they are not pressured into participating in the study. You need to ensure that your participants are able to consent. For children or people with cognitive impairments (e.g. dementia), this requires additional checks—it typically means informed consent from both the guardian and the person themselves. You also need to provide information in a way that ensures people understand what you are saying (i.e. free of technical terms or jargon). Participants should also be free to discontinue the study at any time. This means that they can withdraw from the study, without any penalty, and do not have to provide a reason for doing so.

13.7.2.2 Minimise Risk

When designing your research (and writing your ethics application), you will need to carefully clarify the benefitsand risks of your research. In terms of the benefits, be explicit—what will this particular study tell us that we don't already know? Who will benefit from what you learn from the study? Will there be any broader social value? Will the participants get any benefit out of it? Remember that this all depends on your study being well designed in the first place, so that it gives you accurate data about what you're trying to assess.

You then need to weigh these benefits against the risks. One common risk is stress. Try to minimise unintended or unnecessary stress, and remember that what might not be stressful to you, might be for participants! So consider all the ways in which you can reduce stress for the participants. If you are using any deception—giving participants information that is false—this is a potential source of risk as it violates informed consent and can cause harm. You should only use deception if required, and if you do, you need to undertake debriefing (see below) to disclose that deception and the reason for using it to the participant. Obviously if you are doing anything which is invasive, this risky! Invasive research is any research which changes the participants, such as administering drugs, inserting a recording device into a person's body, or exposing them to a situation where they could potentially be hurt or physically impacted. You need to ensure that what you are doing is absolutely *necessary* (the study won't achieve its aims without it), that you have *minimised* any risk during the study, and *removed* any long-term negative effects.

13.7.2.3 Confidentiality

Participants often provide sensitive information during a study and it is your responsibility to keep this information confidential. Remember also that information you may personally not consider sensitive (e.g. weight, performance on a task) may be considered sensitive to others. There are various ways you can ensure confidentiality. The easiest is to ensure participants are anonymous—that the data they provide is not identifiable. Using a participant ID number rather than names is good practice, for example. If you're conducting case study research, you may choose to refer to that person by their initials or a pseudonym rather than their name. If you are using audio or video recording, you should specifically ask for the participants' permission for this on the consent form. Data storage and protection also needs to be considered—where is the data stored? No one should have access to the data that isn't part of the project. Ensure storage devices are appropriately and securely protected.

13.7.2.4 Debriefing

Debriefing is explaining to the participants who took part in your study exactly what the study was about and what occurred during the study. This will counteract any deception that took place during the study (i.e. you tell them the truth about what happened in the study and why) and hence minimise potential harm. You should also encourage them to ask questions about the study and you should answer them fully.

The Wizard-Of-Oz Paradigm

When conducting user studies, at times researchers need the participating robots to exhibit capabilities beyond their technical abilities (either because it is not possible at the current state of the technology or due to non-availability of resources). In such situations, a commonly used technique in HRI is to augment the missing skills through the integration of a human "wizard". Let's use Tommy (see above) as an example. If we were to examine the impact on participant–robot attachment by comparing a version of Tommy which converses verbally compared with a version which uses only non-verbal cues, it might be difficult to implement a fluid Natural Language Processing system that could mimic human competency appropriately. In such a situation, a confederate (another researcher) could be placed behind a curtain to converse with the research participant through the robot, giving the illusion that the participant is conversing with the robot. The concept comes from the classic fantasy novel "The Wonderful Wizard of Oz" by L. Frank Baum. While helping researchers to circumvent technical difficulties in conducting user studies, it should be cautioned that the practice has ethical and social implications. Ethical, in that you are potentially deceiving a research participant into believing the interaction is purely with a robot. Socially, when such research is presented in the wider media, there is a risk of misrepresenting the current state of the technology, leading to false understandings about the capabilities of robots. This has implications for research funding and the formation of exaggerated expectations or fears towards robots in society.

13.7.3   Data, Analysis and Interpretation

The earlier sections of this chapter have introduced you to the key factors to consider when designing a research study. While this text does not attempt to teach you data analysis (that's several other texts just on its own!), this section walks you through some of the fundamentals of data analysis, so that you can work out what analyses you need to conduct and can use other resources to follow up how to do those analyses with whatever data analysis program you are using.

In this section, we are assuming that you have conducted your study and collected your data. You are probably looking at a data file listing a big bunch of numbers and wondering what to do now! This chapter will help you understand what you need to know to take the next steps.

13.7.3.1   Research Data

One of the first things you need to know is what type of data you have, as this will allow you to work out what analyses you can do with it.

Although there are many definitions, qualitative data is generally considered to be data that describes or characterises what it is measuring—it is typically descriptive information about attributes in the form of words, that you can't easily summarise in numbers. You often collect qualitative data if you are conducting case studies or observational research, or using other open-ended ways of gathering data in real-world settings. There are many different ways of presenting and analysing qualitative data, including approaches like grounded theory, thematic analysis and discourse analysis. In contrast, quantitative data is data that can be represented as numbers. But although this sounds simple, not all quantitative data is the same! Overall, knowing what *type* of data you have is important and depends on how you chose to measure your variables (see *operationalisation* above).

If you're going to conduct statistical analyses on your data, you need to work out which of the following four types it is.

- Nominal variables are variables which measure what category people fall into. Examples are gender (female, male, non-binary etc.) and the condition someone is in an experiment (control, experimental, etc.).
- Ordinal variables are categorical variables that are sequenced in a certain way, such as grades in school (A, B, C etc.) or outcomes in a running race (1st, 2nd, 3rd). In other words, the categories "go" in a certain order. However, these ordered categories do not have consistent intervals between each category.
- Interval variables are variables in which responses are quantitatively related to each other, with equal intervals between them but no true zero. For example, IQ is an interval scale as the "0" is not a true absence, but just the lowest score on that measure.
- Ratio variables are variables in which the numbers are quantitatively related to each other and have a true zero. This includes variables such as weight and height.

Once you are clear what type of variables you have, you can work out what descriptive and inferential statistics you can conduct on that data.

Examples of Common Variable Type in Robotics

Many of the variables below could be several different variable types –

it depends exactly how you've chosen to measure (operationalise) them.

Reaction time (milliseconds between robot movement and participant reaction) Ratio

Perceived agency (5 levels strongly agree to strongly disagree) Ordinal.

Reported safety (combined score across 10-item questionnaire) Interval.

Distance (centimetres between robot and participant) ratio.

Experimental condition (robot without a face, robot with a humanoid face) Nominal

Previous exposure to robots (none, occasionally, frequently) Ordinal

Interactions (number of times the participant initiated conversation): Ratio

13.7.3.2   Descriptive Statistics

Descriptive statistics are numerical statements that summarise the data you've collected from your sample. You will need to report the descriptive statistics of your data when you communicate your findings to your audience. Think about it—if you collect data from 40 people, you can't just give all those "raw" numbers in your presentation or your report! You need to summarise them in some way that tells your audience what your data "looks like" in a simple overview.

How you describe your data depends on what kind of data you have. If you have categorical variables (nominal or ordinal), you typically report the number of people in each category, and/or the percentage of people in each category. For example, if you asked participants whether they trusted robots, and they were given the option of yes or no, this would be a categorical variable. You would then report as your descriptive summary the number of people (the n) who said yes and the number of people who said no. You could also report the percentage of participants who fell into that category. For example, "Of the participants, 10 people (25%) reported trusting robots, and the remaining 30 (75%) reported did not."). Remember also to report how many people didn't answer that question, if that occurred.

If you have numeric variables (interval or ratio), you report the "middle" of the values for that variable, and how spread out they are around that middle point, as this is more meaningful than n or percentages with these kinds of variables. The "middle" of the data set is usually the mean, the median or the mode. The mean (M) is calculated by adding all values together, and dividing by the number of values. The median is the central value when all values are ordered from smallest to largest, and the mode is the most common single value for that variable. Of these the mean is most frequently reported. The most common ways to report how spread out the data are the range and the standard deviation. The range is the difference between the smallest and largest value for that variable. The standard deviation (SD) is a measure of how much the values vary around the mean—a larger standard deviation means the values are more spread out, a smaller standard deviation means they are less spread out. All of these descriptive statistics can be easily calculated using available statistical software. For example, if you measured how close people stood to a robot (in m), you could summarise that data as "The participants stood on average 2.17 m ($SD = 0.73$) from the robot."

13.7.3.3   Inferential Statistics

While it's useful to report what the results of the study are for the participants in your sample (descriptive statistics), usually you want to make a statement that goes beyond the people in your sample, and talk about what this means for the entire population you want to apply your findings to. Inferential statistics are numerical statements that draw conclusions about the broader population based on your sample data. While teaching you inferential statistics and the associated statistical theory is (well!) beyond the scope of this chapter, the following should give you a quick overview so that you know what kinds of questions to ask, and how to seek help, when you begin to conduct analyses.

The first thing you should do is recognise your limits! Appropriately conducting statistical tests requires a good understanding of the theory those analyses are based on, what those analyses represent and what they can tell you (and what they can't!). You should first find guidance in terms of a statistical advisor, a senior supervisor, or take part in an introductory statistics course (there are many available online) to upskill you in these factors. If you don't understand what you are doing you are likely to conduct inappropriate analyses and/or draw inaccurate conclusions.

Once you have a basic knowledge of inferential statistics in general, you then need to decide how to apply that knowledge to your particular study. The place to start is your hypotheses. You should have pinned down particular hypotheses at the beginning of the study—what exactly did you predict? These will form the basis for your analyses. You normally have several different hypotheses in the one study, and you will need to go through this process for each one to decide which analysis is appropriate for each hypothesis. For example, if you are manipulating what motion type the robot exhibits and measuring perceived safety, you might also in the same study be measuring acceptability and how far people stand from the robot. Or you might also be manipulating the colour of the robot in the same study. You would have separate hypotheses for each of these effects and so these would be different analyses.

Steps when deciding on and conducting analyses.

- What was your initial hypothesis?
- How did you operationalise these variables? Identify the IV(s) and DV(s) involved in this particular hypothesis (how exactly did you measure or manipulate them? Do have conditions, if so how many, and are they between-subjects or within-subjects?)
- What types of variables are those IVs and DVs? (i.e. nominal, ordinal, interval or ratio?)
- Work out which analysis is relevant for you to run, given your IVs and DVs (you can use a decision tree like the one provided below).
- Check the assumptions of that particular analysis (e.g. some tests require normally distributed data) and change analyses if necessary (e.g. to a test that doesn't require that assumption).
- Conduct analysis and interpret output.
- Write up the output, conveying all necessary information to your audience (Fig. 13.6).

Research Examples: Selecting and Conducting Analyses

Let's take the example of Sylax, the university-based robotics project looking at the effect of factors like behaviour and appearance of the robot on perception of

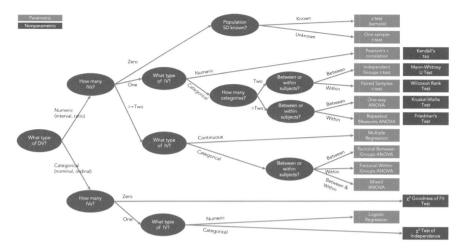

Fig. 13.6 Decision tree representing common statistical analyses

agency. The researchers have lots of questions about these different variables, but in one study they focus particularly on comparing random versus purposeful behaviour.

- Their *hypothesis* for this particular study is that people will perceive the robot to have more agency when it shows purposeful movements rather than when it displays random movements.
- They conduct a study in which they manipulate behaviour and measure perceived agency. Their *IV is behaviour*, which has *two conditions*: random and purposeful. They use the same participants in each condition, so this IV is manipulated *within-subjects*. They assess their *DV perceived agency* by measuring the number of interactions initiated by the participant.
- The IV of behaviour is *nominal* (as there are two conditions/groups). The DV of perceived agency is *ratio* (as it is the number of interactions).
- Given that the researchers have a ratio DV, have one IV which is nominal, with two conditions, manipulated within-subjects, the researchers decide they should run a *paired-samples t-test*.
- They check the *assumptions* for a paired-samples *t*-test (e.g. that their data is normally distributed) and find that it meets those requirements.
- They conduct a paired-samples *t*-test using their chosen statistical analysis program (e.g. The Jamovi Project, 2021) and conclude that although there is a difference between the means of the number of interactions between the two conditions, the analysis shows that this is not *statistically* significant.
- They write up their findings including all the relevant information and values. They conclude that their hypothesis that people's perception of

agency will differ depending on the robot's movement type is not supported, and in fact movement type does not affect perception of agency.

13.7.3.4   Presenting Your Findings

When presenting your findings, you need to tell the person reading it (or watching, if it is a presentation) all of the relevant details so that they understand what analysis you conducted, provide them with the key values so they can see for themselves what you found, and clearly communicate what this means.

Some key information you should present:

- The name of the test you conducted.
- The variables involved (and name the conditions you are contrasting if relevant).
- The outcomes of any assumption testing.
- The key descriptive statistics (e.g. the means and standard deviations of each group).
- The key values output from your analysis (typically the test value such as r, t or F, the number of people or degrees of freedom, the p value).
- Additional information such as confidence intervals and effect sizes.
- Whether or not the outcome was statistically significant.
- Additional information to aid interpretation (e.g. the effect size, confidence intervals).
- Appropriate graphs or figures to illustrate your findings.

Examples of Typical Presentations of Results of Statistical Analyses

A paired samples t-test was used to compare the perceived agency (number of interactions initiated) between 20 participants exposed to a randomly moving robot ($M = 4.75$, $SD = 1.59$) and an algorithmically-driven robot ($M = 5.55$, $SD = 2.01$). Assumption checks confirmed the data was normally distributed. There was a mean difference of 0.80, 95% CI [-1.96, 0.36] between the number of interactions generated in the two conditions, but this difference was not significant ($t(19) = 1.44$, $p = 0.166$, $d = 0.32$).

A Pearson's r correlation analysis was conducted to examine the relationship between self-reported previous exposure to robots and attachment to a robot product. The assumption of normality was supported. There was a moderate, significant, positive correlation between previous exposure and attachment scores ($r(15) = 0.54$, $p = 0.026$), such that participants with more contact with robots before the study tended to report higher attachment to the robot.

13.7.4 Common Mistakes and Pitfalls

This section introduces you to some of the common errors that both novice and experienced researchers show from time to time. By knowing what they are hopefully you can avoid them!

One key way researchers run into trouble is not planning the analyses. You should have formulated your key hypotheses before conducting your study. What are the main effects you are looking for? What are the key variables? You may have four or five key hypotheses that you want to test in a single study. Your analyses should then be pretty straightforward—you are running the analyses that test those hypotheses! This will also protect you from what is called "data fishing" or "p-hacking", which is when researchers run many different analyses on a single data set, but only report the significant ones. This is highly problematic statistically as it means many of those findings may actually then be false positives. It is also problematic in that it reflects a poor understanding of non-significant results—just because a finding isn't significant doesn't mean it's not interesting or useful! *Not* finding a significant difference between two conditions, for example, may tell you that you don't need to incorporate that additional capability to improve safety, or that people can't differentiate between different robotic faces.

When you first get your data from your study it is very exciting! You are keen to see what you've found and it's all too easy to rush into conducting analyses without understanding the data. If you do this, you end up with whole pile of outcomes that are a big mess! And are largely uninterpretable! Take your time to get to know the data set—what type is each variable? Should any variables be recoded to make them more useful? (e.g. turning age from a number into a category) Do you have any missing data and is this problematic? Take a look at your data using graphs and descriptive statistics. Does it "look" ok? Are there any weird values that shouldn't be there or outliers that might suggest equipment failure or data entry error? Are your numeric variables normally distributed or will you violate some assumptions? What is your plan for this?

After you have conducted your analyses there are a few common issues that can emerge. One error you see frequently is people assuming correlation equals causation. For example, just because you find that people who have previous exposure to robots also are more attached to their companion robot, this doesn't mean one causes the other—it doesn't mean that increasing people's exposure will cause their attachment to increase. Maybe it is a positive perception of robots that causes both ratings to rise? Also be careful in interpretating all your findings more generally. If your sample size is too small, you will have low power, which means that you don't (statistically) have the ability to find some effects even if they are really there. There are programs available (e.g. *G*Power*) which will enable you to calculate how many people you need in a study to have a particular level of power. This is often a problem in robotics user research where we tend to have small sample sizes. Running a power analysis before you conduct your study, to work out how many participants you need, is an important step. Finally, even if you find a significant effect, make sure you consider effect size as well (most statistical programs will calculate this for you as well, for each analysis). Some effects can be significant, but not meaningful! For example, if you find a significant difference between the perceived safety of two robots, if the difference itself is only .5 of a point on a 1 to 20 scale, it is unlikely to be a *meaningful* difference at the end of the day.

The key thing to remember is that you have to be confident in what you're presenting. Are you sure that the data reflects your conclusions? Are there any issues the reader should know about in order to interpret your findings appropriately? Remember others will use your work and build on it, so you want it to be an accurate representation of the world!

13.8 Chapter Summary

This chapter introduced you the fundamentals in conducting research in human–robot interaction. We have highlighted the importance of carefully designing research projects so that you get the most accurate and useful information out of the research you conduct. This chapter should provide you with the necessary knowledge and confidence to design your own studies in this field.

13.9 Revision Questions

Q1: You are conducting a research study focusing on the effect of robotic faces (child-like or adult-like) on perceptions of animacy. You predict that you'll find that child-like faces have higher animacy than adult-like faces. This is an example of a:

(a)   Theory
(b)   Hypothesis
(c)   Control condition
(d)   Relevance-sensitivity trade-off.

Q2: You want to work out which of three voice options for a companion robot elicits the most positive response from the general public. You give a group of people the same robot with one of the three difference voices and after a week ask them how positively they view the robot on a scale from 1 to 7. What is the IV and what is the DV in this study?

(a)   IV positive rating; DV voice type
(b)   IV robot type; DV safety rating
(c)   IV voice type; DV positive rating
(d)   IV before and after rating; DV voice type.

Q3: You figure out the best type of articulation to use on your robotic design by reading all the previous studies conducted looking at articulation and summarising them. The research design you are using is:

(a)   Descriptive
(b)   Correlational
(c)   Experimental
(d)   Review.

Q4: To establish what factors are important in designing an industrial robot for a particular company you send out a survey to all the company employees. You particularly want to know about the relationship between how long people have worked there and how important they think particular design features are. The research design you are using is:

(a)   Descriptive
(b)   Correlational
(c)   Experimental
(d)   Review.

Q5: You are testing how people react to the new robotic interface you have designed. You ask some friends if they can drop by the lab to help you test it out. You are using what kind of sampling?

(a)   Convenience
(b)   Purposive
(c)   Probability
(d)   Random.

Q6: You are testing how people react to the new robotic interface you have designed. You ask them to do a task with the help of the old interface, then do the same task with the new interface. They report they found it easier to complete the task with the new interface. What is one explanation for this difference?

(a)   Observer effects
(b)   Mortality
(c)   Practice effects
(d)   History effects.

Q7: You place two video cameras in the corner of your laboratory to record the interactions between your participants and the robot. What ethical issues do you need to consider when using these?

(a)   Informed consent provided by participants to be videoed
(b)   Secure storage of the video files
(c)   Protecting confidentiality of the participants in the videos
(d)   All of the above.

Q8: You are conducting a pilot test installing a robot in a manufacturing setting. You measure how many times people physically touch the robot during the course of a day. This is what type of variable?

(a)   Ordinal
(b)   Nominal
(c)   Interval
(d)   Ratio.

Q9: You conduct a research project and gather some data. You present your data as the percentage of participants who responded "Strongly agree" to each question. You are using:

(a)   Inferential statistics
(b)   Descriptive statistics
(c)   The mean and standard deviation
(d)   Significance testing.

Q10: A survey of the general population finds that people who are afraid of robots are more likely to say they don't need robotic help around the house. The researchers conclude that if they make people less afraid of robots they will then want more robots to help in the household. What error are they making?

(a)   They are p-hacking
(b)   They have low power
(c)   They are assuming correlation equals causation
(d)   They have low effect size.

References

Bretherton, I. (1985). Attachment theory: Retrospect and prospect. *Monographs of the Society for Research in Child Development, 50*(1/2), 3–35. https://doi.org/10.2307/3333824.
Leslie, A. M. (1987). Pretense and representation: The origins of "theory of mind". *Psychological Review, 94*(4), 412.
Slovic, P., Fischhoff, B., & Lichtenstein, S. (1984). Behavioral decision theory perspectives on risk and safety. *Acta Psychologica, 56*(1–3), 183–203.
The Jamovi Project. (2021). *jamovi* (Version 1.6) [Computer Software]. Retrieved from https://www.jamovi.org.

Janie Busby Grant is a research psychologist based at the University of Canberra. She is a cognitive psychologist whose research focuses on future-oriented thought, mental health and human-robot interaction. She has experience designing and conducting collaborative cross-disciplinary research projects and engaging with stakeholders with complex relationships and data sensitivities. Janie has taught research methods at different levels across three universities and is currently an Academic Fellow mentoring Ph.D. students across all Faculties in her current university. She specialises in the use of new technologies in both research and teaching contexts.

Damith Herath is an Associate Professor in Robotics and Art at the University of Canberra. Damith is a multi-award winning entrepreneur and a roboticist with extensive experience leading multidisciplinary research teams on complex robotic integration, industrial and research projects for over two decades. He founded Australia's first collaborative robotics startup in 2011 and was named one of the most innovative young tech companies in Australia in 2014. Teams he led in 2015 and 2016 consecutively became finalists and, in 2016, a top-ten category winner in the coveted Amazon Robotics Challenge—an industry-focused competition amongst the robotics research elite. In addition, Damith has chaired several international workshops on Robots and Art and is the lead editor of the book *Robots and Art: Exploring an Unlikely Symbiosis*—the first significant work to feature leading roboticists and artists together in the field of Robotic Art.

Chapter 14
Safety First: On the Safe Deployment of Robotic Systems

Bruno Belzile and David St-Onge

14.1 Learning Objectives

The objective at the end of this chapter is to be able to:

- recognize the different standard organization and their publications;
- conduct a risk-assessment procedure on a robotic system and propose risk mitigation measures;
- know the difference between an industrial robot and a cobot as well as their respective potential hazards;
- differentiate the types of collaborative operation methods;
- conduct a risk assessment on a mobile robotic system.

14.2 Introduction

The deployment of robotic systems always brings several challenges. Among them, safety is of uttermost importance, as these robots share their environment with humans at a certain degree. In this chapter, you will get an overlook of some standards relevant to robotic systems, pertaining mostly to their scope and the organizations issuing them. These standards and others documents such as technical specifications are relevant to conduct the risk assessment of a new system and mitigation of the identified hazards, two critical steps in the deployment of robot cells, mobile manipulators, etc. While we will first focus on conventional industrial robots, we will then move to collaborative robots (cobots), with which human operators' safety is even more critical considering the intrinsic close proximity, as well as mobile robots. It is important to understand that the information presented in this chapter is only a

B. Belzile (✉) · D. St-Onge
Department of Mechanical Engineering, ÉTS Montréal, Montreal, Canada
e-mail: bruno.belzile.1@ens.etsmtl.ca

D. St-Onge
e-mail: david.st-onge@etsmtl.ca

© The Author(s) 2022
D. Herath and D. St-Onge (eds.), *Foundations of Robotics*,
https://doi.org/10.1007/978-981-19-1983-1_14

brief introduction to the process leading to the safe deployment of a robotic system, whether it is a conventional industrial robot, a cobot or a mobile robot. You will need to refer to existing standards, technical specifications, guidelines and other documents that are yet to be released, as it is a field constantly adapting to new technologies. Moreover, a safe deployment goes beyond any written document, as a thorough analysis is critical, which includes elements that may not be considered by any standard.

An Industry Perspective

Camille Forget
Quality assurance manager, Suppliers

Kinova inc.

I have a bachelor's degree in automated production. I worked a few years in the metal industry and went back around robotics when I joined Kinova's quality assurance team four years ago. I always found robots fascinating, which made my job even more interesting to do.

Since we design and manufacture both medical and industrial robots, a big challenge that we have at Kinova in terms of quality is to optimize the quality management system to meet the requirements of medical standards while still allowing rapid and constant development required by the industry. We need to go back to the essence of the requirements and make sure to fulfill them while also keeping the system efficient and flexible.

Absolutely, the robot safety field is in constant evolution as the technology progresses to help create safer and better human-robot collaborative applications. Standards are evolving to help structure the industry and push (things) forward. New technologies are being developed and refined to help integrate the robots more safely and efficiently.

14.2.1   Terms and Definitions

First, some terminology must be defined. Table 14.1 includes terms commonly found in the field of robotic system safety and they will be used throughout this chapter. The definitions provided here are based on several standards, including new ISO 8373 ISO (2021c) and ISO/DIS 10218-1.2 ISO (2022).

14.2.2   Challenges with the Safe Deployment of Robotic Systems

Several challenges arise in regard to the deployment of robotic systems, particularly in environments where they were not commonly found in the past. Moreover, existing and well-known standards are not necessarily adapted for every new situ-

Table 14.1   Definitions

Term	Definition
Robot	Programmable, with more than one DoF, autonomous and versatile, designed for locomotion, manipulation or positioning
Automated guided vehicle (AGV)	Mobile platform that follows fixed routes designed to transport loads
Autonomous mobile robots (AMR)	Mobile robot that travels under its own control using sensors to avoid obstacles
Industrial robot (IR)	At least one robotic arm and its controller used in industrial automation applications in an industrial environment
Industrial robotic system (IRS)	An industrial robot, its end-effector (tool), and the task program
Industrial robotic system application (IRSA)	Industrial robotic system completed with workpieces, machinery and equipment
Industrial mobile robot (IMR)	Autonoumous mobile robot used in an industrial environment
Standard	Established norm including definitions, technical guidelines on how to approach particular problems, questions and systems
Technical specification (TS)	Document whose publication aims to make public work which is still under development and which is not mature enough to be the subject of a standard; The TS is intended to give rise to the future publication of a standard when the research work is sufficiently completed

ation. For instance, before late 2020, if you were working with industrial mobile robots (IMR), you would not have found exactly what you were looking for, as IMRs did not fell within the scope of a particular standard. Then, the ANSI/RIA R15.08 ANSI/RIA (2020) was released in December 2020. Both autonomous mobile robots (AMR) and automated guided vehicles (AGV) with a manipulator used in an industrial environment fall under the definition of IMRs, thus are covered by this standard. However, AGVs and AMRs have different scopes: the former follow fixed routes, while the latter use sensors to avoid and go around obstacles by autonomously computing its own trajectory. The ANSI/RIA R15.08 is based on relevant guidance from ANSI/RIA R15.06 ANSI/RIA (2012) and ANSI/ITSDF B56.5 ANSI/ITSDF (2019), which focus on industrial robot safety and guided industrial vehicles, respectively.

Moreover, even if a technical specification, precursor of a full standard, is already established, as it is the case for collaborative robots, the increase in interactions between robots and operators in close proximity also makes risk assessment more complicated, as topics such as the onset of pain must be considered, which was not the case for conventional industrial robots, with which any contact with the robot is prohibited by design.

14.3   Standards

Standards are established norms agreed by experts and published by an organization. They cover a large spectrum of topics, ranging from environmental management to IT security, including sustainability as well and, of course, safety in robotics. While reading a standard, you will find similarities regarding the content, such as a section defining clearly the scope of the document with respect to related standards, as well as the definition of the critical terms used.

14.3.1   Organizations

Organizations issuing standards, technical specifications, norms and codes relevant to safety of robotic systems can be classified into three categories, i.e., 1. international; 2. national; 3. local. Organizations acting at the international level are obviously those usually most well known. It should be noted that standards published by national organizations can be relevant internationally. For instance, the ANSI, an American organization, is well known and is viewed as a reference well beyond the borders of the USA. It is even more true in emerging topics where national institutions have not considered yet. However, you should know that these organizations do not have the power to make their standards compulsory. It is at the local level, through codes and regulations, that legal obligations may appear. Nevertheless, the latter may refer explicitly to standards/technical specifications, as well as generally requiring you to follow well-known good practices, which include them implicitly.

You can also look at the different organizations and levels from another angle. At the local level, there are common rules and good practices from the field. These rules can give rise to codes and regulations. If these codes and regulations are found to be common to a domain, a branch, a type of machine at the national level, this can give rise to standards established by national organizations, such as CSA Group. At another level, if the rules and codes around a subject are of interest beyond the borders of a country, they may fall under an international standardization process and eventually give rise to a technical specification and an international standard.

You should remember that standardization work at both national and international level is the result of the work of experts in the corresponding field from different backgrounds. For example, for the international standardization work of the ISO/TC/299/WG3 group in charge of standards associated with industrial robots, the group consists of

- manufacturers of industrial robots;
- industrial robot integrators;
- companies using industrial robots;
- academics/researchers;
- government agencies;
- prevention organizations.

The international standard thus obtained is the result of a consensus between different experts from different countries. It represents the best practices that can be applied in that field.

International

You have probably heard already about ISO. The ***International Organisation for Standardisation* (ISO)** is an federation dating back to 1946 with a membership of more 150 national standard organizations. It has published tens of thousands of standards in numerous fields and has several hundreds of technical committees working on revising and publishing new standards, notably the ISO/TC/299 on robotics mentioned above. This technical committee also relies on the input from more grounded organizations, such as the ***Robotic Industries Association* (RIA)**,[1] which publishes guidelines designed for robotics applications based on existing standards.

National

There are more than a hundred national standards organizations around the world; it would be pointless to just list them all here. It should be noted, however, that they issue their own norms and also participate in the elaboration of international norms, released by ISO for example. Among them, we can mention the ***American National Standards Institute* (ANSI)** which "is a private, non-profit organization that administers and coordinates the US voluntary standards and conformity assessment system."[2] There is also the ***CSA Group***, an organization accredited by the *Standards Council of Canada*, a governmental corporation promoting voluntary standardization

---

[1] Now part of the Association for Advancing Automation (A3).

[2] www.ansi.org/about/introduction.

in Canada. Standards from CSA Group also have an international reach, being used notably in China, but the main objective of this organization is to adapt international standards to the Canadian reality.

14.3.2   Classification and Relevant Technical Specifications/Standards

You can classify ISO's safety-of-machinery-related standards into three categories, types A, B and C (see Villani et al., 2018 for more information). In Fig. 14.1, we have added standards from other organizations and technical specifications as well (which are precursors to standards, as mentioned above). As you can see in Fig. 14.1, type A standards focus on basic safety. For instance, ISO 12100 ISO (2010) "specifies basic terminology, principles and a methodology for achieving safety in the design of machinery." It also proposes principles of risk assessment and risk reduction, which we will see later in this chapter and will help you design safe robotic systems. Type A standards apply to all types of machines. Type B standards give you more techno-logical specifications for the design of machinery, and therefore have a more limited scope. For example, you can refer to ISO 13850 ISO (2015) if you need to design emergency stops (*e-stops*). In other words, type B standards give recommendations and safety requirements that can be applied to different types of machines. They relate to safety aspects or a type of protection device that can be used for a series of machines. For example, emergency stops are not specific to a given machine nor are movable guards.

While basic (type A) and generic safety (type B) standards are not specifically written for mobile and collaborative robots, they should still be taken into account in all deployment scenarios of robotic systems. Finally, type C standards focus specif-ically on machine safety and are of particular relevance for robotics. They can be categorized depending if it is an industrial robot, a mobile robot, a collaborative robot, a service robot or a personal-care robot. Many of the current-standards target manufacturing robots, e.g., fixed (and recently some industrial mobile robots), collab-orative devices, automated guided vehicles (AGV), automated agricultural machines, etc. The scope of a standard or a technical specification is one of the first element defined in the corresponding document. A chart, taken from ANSI/RIA R15.08 and displayed in Fig. 14.2, details the different scopes of IMRs (ANSI/RIA R15.08), AGVs without an attached manipulator (ANSI/ITSDF B56.5) and IRSs (ANSI/RIA R15.06), based on their characteristics.

The ANSI B11 series of standards and technical reports are particularly inter-esting, as they focus on machinery safety. Similarly to ISO publications, they are classified in types A, B and C. A full list can be found online.[3]

You will find a summary of the some relevant standards to robotic systems' safety in Table 14.2. For example, ISO/TS 15066 is of particular interest, because it focuses

---

[3] https://www.b11standards.org/current-standards.

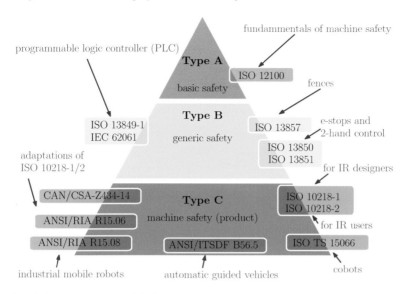

Fig. 14.1 Safety standard pyramid (ISO equivalent, adapted from Villani et al., 2018)

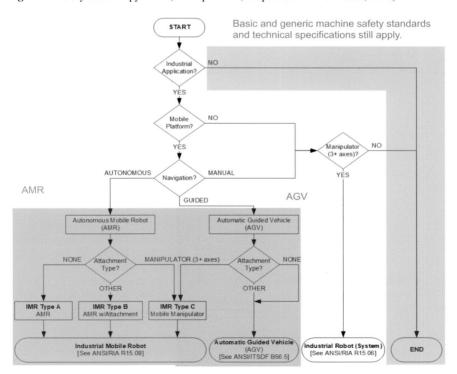

Fig. 14.2 Scope of application of several notable standards (adapted from ANSI/RIA R15.08)

Table 14.2 Some relevant standards and technical specifications for robots used in industrial environments

Document	Title	Field of application
ISO 10218 (2011)	Robots and robotic devices—Safety requirements for industrial robots—Part 1: Robots and Part 2: Robot systems and integration	This standard specifies requirements and recommendations for intrinsic prevention, protective measures and information for the use of industrial robots. It describes the basic hazards associated with robots and provides the basic requirements for reducing or eliminating the risks associated with these hazards. Part 1 is intended for designers and manufacturers, while Part 2 is intended for integrators and users
ISO/TS 15066 (2016)	Robots and robotic devices—Collaborative robots	This technical specification specifies the safety requirements for collaborative industrial robot systems and the working environment and complements the requirements and guidance on collaborative industrial robot operation given in ISO 10218-1 and ISO 10218-2
ANSI/RIA R15.08 (2020)	Industrial mobile robots safety	This standard defines the safety requirements for manufacturers of industrial mobile robots Part 1; Part 2 describes the requirements for integrators working on the design, installation and integration of a safe mobile robot system in a user's facilities; and Part 3 defines the safety requirements for the end user of industrial mobile robots
ANSI/ITSDF B56.5 (2019)	Safety standard for guided industrial vehicles	This standard defines the safety requirements relating to the elements of design, operation and maintenance, industrial vehicles with automatic guidance without mechanical restraint and unmanned and the system of which the vehicles are part
CSA/Z434-14 (2019) (Canada)	Industrial robots and robot systems	This standard replicates ISO 10218 with some specifications for Canada
ISO 8373 (2021c)	Robotics—Vocabulary	This standard defines terms used in relation to robotics
ISO 19649 (2017)	Mobile robots—Vocabulary	This standard defines terms used in relation to mobile robotics

on collaborative robots, which are addressed later in this chapter. You should not see this list as exhaustive, because only a selection of standards/technical specifications are included and new ones are currently being written/revised.

14.4 Industrial Risk Assessment and Mitigation

In this section, we will focus on isolated industrial robotic system first and how to conduct a risk assessment. The global procedure is depicted in Fig. 14.3. The following section covers the elements shown in this figure. Particularities pertaining to collaborative and mobiles robots will be addressed in the subsequent sections.

14.4.1 Risk Assessment

Risk assessment is a critical and essential process before deploying a new robotic system. Your **first step** will be to **identify the limits** of the robotic system application in terms of use, space and time throughout its life cycle. This step amounts to defining the expected use of the machine and the environment in which it is to perform these functions. It is therefore essential to have proceeded, before the risk identification stage, to a **functional design process** of the robotic system. For instance, this stage includes defining the place where the robot will be installed, the surrounding objects, the parts handled, the number of operators and their training, the tasks that the robot will perform, etc.

Then, you must identify every potential sources of harm, known as hazards. It is necessary to conduct a first analysis to estimate, i.e., quantify, the risk posed by each hazardous situation. It is an iterative process, therefore the analysis is conducted again after implementing the risk-reduction measures (mitigation) to validate the desire outcome has been reached. The approach is unique to every industrial robotic system application, which means you should avoid a "one-size-fits-all" solution, as it may be too restrictive for the application, ultimately leading to frequent bypass of some safeguards to accomplish a task. The risk analysis is, in fact, specific to a particular machine **and** installation. Therefore, it is necessary to carry out a new risk assessment if the environment, tasks or operators change (as part of a machine move, for example).

Potential Causes of Hazards
A non-exhaustive list of potential hazard causes involving isolated industrial robotic system (as opposed to collaborative and mobile robots, which will be considered later in this chapter) is detailed in Table 14.3. This table not only considers injuries to the human body, but also material damages. You should note, however, that international standards only refer to the former when "harm" is mentioned. For your information, you will find a list of significant hazards in the Annex A of ISO 10218-2.

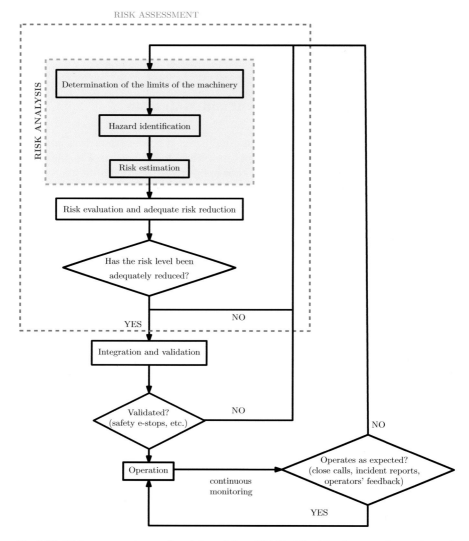

Fig. 14.3 Risk assessment procedure (adapted from ISO 12100) with subsequent integration, validation and monitoring

Initial Analysis

The initial analysis is an hypothetical exercise done by the integrators where potential hazards are identified. No risk mitigation measures should be considered while conducting the initial risk analysis: you will therefore have to consider unauthorized access to the robot workspace and unqualified operator, risks that are easily prevented. We will cover the mitigation in the next section (step). Moreover, the operator should **always** be considered unqualified and the workspace not protected at this stage of

Table 14.3 Non-exhaustive list of potential causes of hazards involving industrial robotic systems

Categories	Examples
Workers related	Unqualified operator incapable of controlling the robot
	Human error
	Unauthorized access to the robot workspace
	Not following the manufacturer's instructions
Control	Interference
	Software error
	Bad programming
Power	Contact between various electrical cords in the robot system
	Leaks in an hydraulic system
Mechanical	Deficient/broken part
Environment	Sources of electromagnetic interferences

the risk assessment. All of this is done to avoid overlooking any potential hazard. You can see this as a worst case scenario. For each risk, you need to estimate two elements or parameters according to ISO 12100:2010:

1. the severity of harm and
2. the probability of that harm.

   The latter normally comprises three subparameters:

1. the exposure of the person(s) to the hazard;
2. the occurrence of a hazardous event;
3. the possibilities to avoid or limit the harm.

   Various risk estimation tools exist in order to rank the severity and probability of the harm. However, not all of those tools cover the same number of parameters. For example, the RIA TR R15.306 proposes the chart illustrated in Fig. 14.4 which comprises three parameters, namely severity, exposure and avoidance:

- Severity of injury;

  – serious (death, chronicle disease, amputation, etc.)
  – moderate (broken bone, short hospitalization, etc.)
  – minor (bruises, etc.)

- Exposure to the hazard;

  – high (more than 1 time par day)
  – low (less than 1 time par day)
  – prevented (not used in the initial analysis, since we ignore risk mitigation measures at this stage)

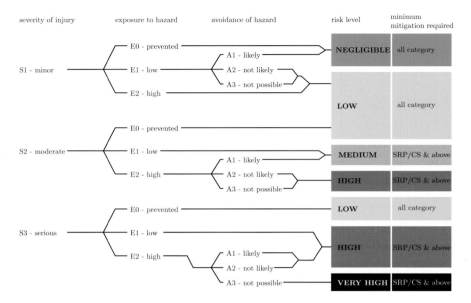

Fig. 14.4 Chart to evaluate the risk level for each dangerous phenomenon (adapted from RIA TR R15.306 RIA, 2016, which is a supplement of ANSI/RIA R15.06-2012)

- Avoidance of the hazard;

  - impossible (insufficient space, caged operator)
  - improbable (insufficient space, but under robot speed limitation, obstructed exit)
  - probable (sufficient space, under robot speed limitation, early warning).

After completing the potential hazards identification and the three criteria quantification, the next step of the initial analysis is to determine the risk level. You can do this with the chart mentioned above (Fig. 14.4).

14.4.2   Risk Mitigation

Preventive and corrective measures must be put in place and the risk index will then be reevaluated accordingly. These measures can be classified into eight categories according to RIA TR R15.306-2016 RIA (2016) (Task-based Risk Assessment Methodology), in this precise order:

1. elimination;
2. substitution;
3. limit interaction;
4. safeguarding and safety-related part of a control system (SRP/CS);
5. complementary protective measures;

6. warnings and awareness means;
7. administrative controls;
8. personal protection equipment (PPE).

We will see examples for the categories listed above. Some are displayed in Fig. 14.5. Only a brief summary of some mitigation measures will be given here and you must refer to the relevant standards/technical specifications/guidelines for more information.

Elimination: Modifying the industrial robot system's design (hardware, software, process, layout, etc.). It should be emphasized that this involves the modification of the design of the machine to **eliminate the risk inherently**. For example, eliminating obstacles that may be the cause of the risk of jamming and thereby eliminating the risk of jamming in an intrinsic way (without resorting to guards and barriers for instance).

Substitution: You can mitigate the risk by a substitution, namely by changing materials handled, and replacing the robot by another less powerful, slower or with a smaller workspace.

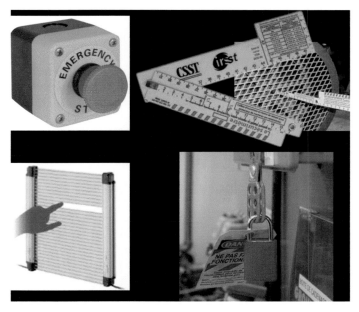

Fig. 14.5 Mitigation measures that can be applied to industrial robot cell (upper left, emergency stop; upper right, safety distance related to projectors; lower left, safety light curtain; lower right, lockout procedure)

Interaction limitation: Limit physical interactions between the operator and the industrial robot within the latter's workspace.

Safeguarding and safety-related part of a control system (SRP/CS): To reduce the risk of someone coming in close proximity of an industrial robot, the most simple and common measure is to enclose the robot with a **perimeter fence**, serving as a barrier between the robot and anyone it could harm. According to ISO 13857 ISO (2019), a barrier is necessary if the mechanism potentially dangerous, not necessarily a robot, is lower than 2.5 m above ground. In this case, the same standard recommends rigid panels with certain parameters depending on the dimensions of the system and its workspace. For instance, the minimal height of the panels is 1.8 m, regardless of the system. These fences can include openings necessary for the robot's operation as well as a door for maintenance. This door must be equipped with a locking mechanism or/and sensors to detect intrusion in the workspace. **Machine lockout and sensors** are also a possibility to avoid a robot operating at full capacities while an operator is within close proximity or inside the machine/workspace.

Complementary protective measures: Common complementary measures include elements to achieve e-stop functions, measures for safe access to the robot, handrail, mechanical blocks and additional padding.

Warnings and awareness means: Rotating beacons, alarms, warning panels are among the measures used to increase the operator's awareness of the potential dangerous phenomenon.

Administrative controls: Organizational-type measures are also essential to reduce risks involving robotic systems. Indeed, some risks cannot be completely eliminated, thus a proper training of the staff is critical to increase awareness. The information shared includes the nature of the risks, existing protection methods, proper safe ways to approach the robot, etc. Other measures include compliance with the manufacturer's instructions, regular inspections and preventive maintenance of the robot, rewarding workers for safe behavior, etc.

Personal protection equipment: Common examples include glasses, helmet, boots, etc.

While considering potential risk mitigation measures, you should consider the above eight categories **in the order they are presented**. Therefore, you should favor elimination and substitution rather than administrative controls and PPE. For hazards initially evaluated as medium and above on the risk level scale, mitigation measures must include those within the first four categories, as the four others are not considered enough to reduce the risk, as displayed in Fig. 14.4. You should note, however, that you can still apply mitigation measures, such as elimination and substitution, to potential hazards initially evaluated with a low risk level, even though it is not required by the standard.

14.4.3   Integration, Validation and Monitoring

After conducting the initial analysis and applying risk mitigation measures, the final step is to analyze again the potential hazards by quantifying the risk parameters mentioned above, but this time taking into account the risk-reduction measures applied. New risk levels will be obtained, allowing the integration of the robotic system deployment.

After the integration comes the validation. This can be done with several methods, for example, (as suggested in ANSI/RIA R15.08 for IMRs, but valid for any robotic system):

- visual inspection;
- practical tests;
- measurement;
- observation during operation;
- review of application-specific schematics, circuit diagrams and design material;
- review of task-based risk assessment;
- review of specifications and information for use.

A validation step can be, for example, measuring the real contact forces to see if the force and torque limits programmed in the power-and-force (PFL) limitation safety function make it possible to reduce the contact forces below the thresholds prescribed by ISO TS 15066 ISO (2016). Another example is to conduct safety stop tests to ensure that the response time of the stop safety function and that the robot stop time have both been taken into account during the calculation for the positioning of virtual barriers (presence detected by proximity sensors close to the robot workspace).

The continuous monitoring of the system by the users is then needed to reevaluate the risk with new information gathered from experience feedback (incidents, close calls, etc.).

14.5   Cobots

While the term *cobots*, created from collaborative robots, can be dated back to 1996 (Colgate & Peshkin, 1997), the idea of robots collaborating with humans in close proximity has been around for much longer, as can be seen in various works of science fiction. However, in reality, robots have been far more often operating in human-free environment for safety reasons. Nowadays, with technological advancements in robotics, cobots are becoming more prevalent, notably in industrial settings. There are many advantages with cobots, including reducing the space used by the robot (no physical isolation) and partially automating tasks which still require a human participation. Because of their close proximity to human workers while performing various tasks, a safe deployment is even more critical. The ISO/TS 15066, previously mentioned in this chapter, focuses on collaborative robots.

14.5.1 Human-Robot Collaboration

The literature provides various categories of human-robot collaborative tasks. In this chapter, we will consider the three following categories:

- direct collaboration—the operator and the robot work **simultaneously** on a task;
- indirect collaboration—the operator and the robot work **alternately** on a task;
- shared workspace—the operator and the robot work on **distinct** tasks for which they may need to share the same workspace.

14.5.2 Types of Collaborative Operation Methods

The classification of collaborative tasks with regards to the safety requirements can be divided into three types, detailed below, according to ISO/TS 15066.

1. Hand guiding;
   The operator manually send commands to the cobot: before the operator enters the collaborative workspace, the robot system achieves a safety-rated monitored stop (drive power remains on); operator grasps hand-operated device (includes an enabling device), activating motion/operation. Non-collaborative operation resumes when the operator leaves the collaborative workspace.
   Applications: robotic lift assist, highly variable applications, limited or small-batch production.
2. Speed and separation monitoring;
   Operator and robotic system may move concurrently in the collaborative workspace: a minimum separation distance between the operator and the cobot must be maintained **at all times** for safety. Protective devices are required to decrease the minimum separation distance. Speed is lowered (safety-rated) to keep minimum separation distance. If separation distance falls below the established threshold, a protective stop is required.
   Applications: simultaneous tasks, direct operator interface.
3. Power-and-force limiting;
   In this mode, physical contact between the cobot/workpiece and the operator is possible, either intentionally or unintentionally: the cobot must be specifically designed for this mode to take into account potential contacts and the corresponding forces must be limited. The contact (quasi-static/pressure or transient/dynamic) must be detected by sensors and the cobot must react when it occurs.
   Applications: small or highly variable applications, conditions requiring frequent operator presence.

You can find a fourth type in the literature, called safety-rated monitored stop. However, in the new version of ISO 10218-1, which will be published in 2022, it will no longer be considered a type of collaborative operation. It is defined as a direct interaction between the cobot and the operator under specific circumstances, which include a safety-rated stop condition. Before the operator enters the "collaborative" workspace, the drive power remains on, motion resumes after the operator leaves the workspace (cobot motion resumes without additional action). Protective stop is triggered if a stop condition (to configure) is violated. If the operator is outside the workspace but inside the monitored space, there is no need to stop the robot. The robot can continue to operate as long as a space monitoring safety feature is in place that prevents the robot from exiting its workspace. The potential applications include direct part loading or unloading to the end-effector (tool of the robotic arm), work-in-process inspections, when the robot or the operator moves (not both) in the same workspace, etc. However, keep in mind that it will no longer be considered a collaborative operation according to the new ISO terminology.

14.5.3   Hazards Inherent to Cobots

Beyond the risks and potential dangerous phenomena detailed earlier in this chapter, some are more specific to cobots. Obviously, the close proximity to humans is a common source for many of them, but some are linked to the task itself. A non-exhaustive list is detailed below, as well as corresponding mitigation measures:

- physical risks: collisions, crushing, jamming, repetitive impacts, tool used by the robot
  risk mitigation $\longrightarrow$ lightweight robot, rounded surfaces, safe speed limitation, safe force and power limitation, training
- psycho-social risks: isolation, pace difficult to follow by the operator, work transformation
  risk mitigation $\longrightarrow$ improving the working conditions of workers
- risks of musculoskeletal disorders: high repetitivity, excessive efforts, high precision required, inadequate posture that may be required for extended periods of time
  risk mitigation $\longrightarrow$ arranging workstations to respect the comfort zones, using appropriate handling techniques, optimizing lighting, choosing the right tools.

14.5.4   Risk Assessment and Mitigation Measures for Collaborative Applications

The risk assessment with collaborative robots is similar to the process presented earlier in this chapter. It differs by the different measures to be applied and added conditions that must be assessed, as detailed in ISO/TS 15066. Indeed, you will remember that the categories preferred for mitigation measures to obtain an inherently safe design where elimination, substitution and limiting interaction. In the case of cobots, this will translate into reduced energy, robots' surfaces made of compliant materials, modified tasks, etc. Therefore, a contact between the robot and the operator is still possible, as we mentioned above in power-and-force limiting mode, and you will have to make sure that it will not result in an injury. This is done by:

- identifying conditions for such contact to occur;
- evaluating risk potential for such contacts;
- designing robot system and collaborative workspace so contact is infrequent and avoidable;
- considering operator body regions, origin of contact event, probability or frequency, type (quasi-static or transient), forces, speeds, etc.

You must prevent contact over the shoulders, and **shall** avoid any of the robot motion above this level. Considering it may not always be realistic, experts on the standardization committee working on ISO 10218 update are proposing to replace the verb "shall" with "should," still strongly encouraging to keep the robot's movements below head level.

For other contacts, ISO/TS 15066 contains specifications on the onset of pain, as shown in Fig. 14.6, as well as transient contact speed limits. An example of risk mitigation in a power-and-force limiting operation is illustrated in Fig. 14.7. Here, we first (1) eliminated pinch and crush points, then (2) we reduced robot system inertia or mass and (3) we reduced robot system velocity. Finally, to reduce the risk of potential injuries, (4) we modified the robot posture such that contact surface area is increased and (5) moved away from sensitive upper body parts.

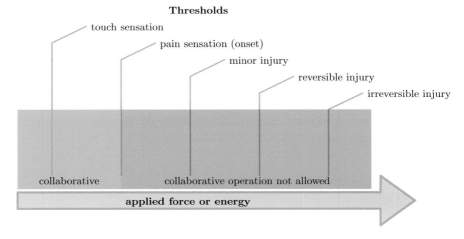

Fig. 14.6 Study of the pain onset regarding collaborative operation (adapted from ISO/TS 15066)

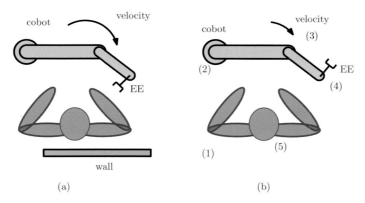

Fig. 14.7 Risk mitigation in a power-and-force limiting operation (adapted from ISO/TS 15066): **a** before risk mitigation, **b** after risk mitigation

14.6 Mobile Robots

Mobile robots used in an industrial setting, which include AGVs and AMRs, usually operate alongside operators and other workers in a shared environment. These robots fall under the scope of the ANSI/RIA R15.08. The latter refers to several other standards for particular items relevant to mobile robots. For example, regarding wireless communication, ANSI/RIA R15.08 refers to NFPA 79, IEC 60204-1 and IEC 62745 for specifications that are recommended. This is a critical component, as a mobile robot that does not react fast enough to an order sent by an operator could have catastrophic consequences, such as a collision.

By definition, mobile robots are not located within a caged environment with barriers. Therefore, with mobile robots such as AGVs, collisions must be avoided by safety functions. The robot must be equipped with safety sensors which are able to detect obstacles, including an operator. The robot must have rules to adapt its speed (safety speed monitored by a safety function) and changes its course according to the detected obstacle. If the distance between the mobile robot and the operator drops below a threshold value, a safety stop must be triggered.

14.6.1 Hazards Inherent to Mobile Robots

The risks posed by hazards identified in the previous sections are multiplied by the presence of the other workers that are not operating the robot while still sharing the same environment. Indeed, we moved from caged industrial robots which should not share their workspace with anyone except on rare occasions to collaborative robots which can share theirs with a limited number of operators to mobile robots with a quasi unlimited workspace. Therefore, making sure everyone is aware of the presence of mobile robots in certain areas is critical, and the capability of IMRs to detect potential dangers (to themselves and others) surrounding them is even more important.

14.6.2 UAV Operations

Unmanned aerial vehicles (UAVs) are a specific type of mobile robots that falls in a different category with respect to the danger they may present. From teleoperation to fully autonomous flights, the risk of not detecting an obstacle (either by the operator or by the onboard sensors) can have fatal consequences. Several commercial UAVs weights more than 1kg and can harm a person under its fall. Because of the level of danger, and inspired from the aviation field, UAVs safety standards are rather in the form of regulations. However, as UAVs become more commonly used, policy makers are struggling to keep up. A lot of countries have develop regulations (FAA, 2021; Transport Canada, 2019), but without much international cohesion. Regulations can vary, including the maximum height to which aircrafts are able to fly to, the areas they are permitted, the distance they can go to buildings and whether or not identity tags are necessary. In the vast majority of cases, UAVs should only be flown while they are still visible to the pilot. Most countries have structured their regulations with regards to the position of the individual, for instance:

- UAV pilot—follow training sessions, read the device manual, check the weather and flight zone, etc.;
- UAV owner—register the UAV and ensure its maintenance;
- UAV manufacturer—provide the documentation to prove its safety;
- piloting school—provide the documentation to demonstrate the quality of their curriculum.

14.6.3 Battery Hazards

Batteries are by their nature one of the most frequent hazards to generate incidents. This is the reason for the severe regulations on their transport, notably on airplanes.[4] Moreover, with the rapid increase in the number of hybrid and electrical vehicles on the road, as well as the number of mobile robots in operation, risks related to batteries are becoming more frequent. Considering the ever increasing power density of these batteries and their decreasing cost, they should not be taken lightly, quite the contrary. Regarding the standards related to batteries, you can consider, for example, the ISO 26262 on road vehicle for some guidelines, especially regarding the literature focusing on the compliance of batteries (and related systems) to this standard (Tiker,

---

[4] For instance, in Canada, see https://www.catsa-acsta.gc.ca/en/guidelines-batteries.

2017). You can look as well into standards such as the IEC 62133. Globally, as proposed by Ashtiani (2008), we can categorize hazards related to batteries in four categories:

- electrical (short-circuit, overcharge, soft short);
- thermal (fire, elevated temperature);
- mechanical (crush, perforation, drop);
- system (contactor fail to close, loss of high voltage continuity, chassis fault).

Furthermore, hazard identification and, more generally, the risk assessment of a robotic system with batteries, must consider their full, but limited, life cycle. Indeed, a system initially safe may become dangerous without any human intervention only because the battery has reached its end-of-life. Moreover, even if the robot is not in operation, not powered up or the battery not even installed, there are potential hazards. The risk assessment and mitigation does not end with the robot in operation, as its spare parts such as batteries must be safely kept in storage. You must therefore also consider the charging process of the battery in your risk assessment as well. Manufacturers' manual are obviously a good starting point regarding potential hazards and mitigation measures. Otherwise, there is an extensive literature on this topic to help you. For example, Ouyang et al. (2019) listed several countermeasures related to thermal hazards.

14.6.4 Risk Assessment and Mitigation Measures for Mobile Robots

As means of mitigating risks with mobile robots, you will find and implement first safety devices (sensors) which detect operators or any other human being in close proximity of the robot. Safe speed limitation functions and safety stop functions capable of stopping the robot before the collision will be used. Otherwise, here you can find a list of other potential risk mitigation measures intended for mobile robots:

- keeping the batteries at a good level of charge or else completely change the battery;
- ensuring a safe form factor of the robot;
- stable ground surface for the robot;
- automatic brake when the robot loses control;
- reducing administrative staff on the site;
- supervision of the workers' movement by a site supervisor;
- pedestrian traffic plan indicating traffic lanes, road markings;
- reversing alarms;
- sensitive bumpers for presence detection;
- permanent lighting on-site allowing easier and safer motion of the robot;
- road signs.

The validation step at the end of the risk assessment and mitigation process for mobile robots is particularly important, as they operate in unstructured environments. Therefore, the ANSI/RIA R15.08 recommends using test pieces representing adult humans instead of human subjects when conducting tests. The standard guidelines also mention that the test pieces "shall be tested in a number of orientations reflecting persons who are standing, sitting, kneeling, or lying prone." Finally, other obstacles and hazards you may need to test include

- overhanging objects;
- negative obstacles (e.g., floor grates, potholes, or steps);
- transparent objects (e.g., glass doors or acrylic walls);
- chain-link fences;
- narrow support columns (e.g., shelf or table legs, sign posts, or ladders);
- reflective and retroreflective surfaces.

14.7 Chapter Summary

In this chapter, we looked into the deployment of a new robotic system from the safety point-of-view. We went over some definitions, the major standards organizations, and some of their relevant standards. We then introduced the concept of risk assessment and mitigation, notably based on the standards discussed previously. We finally tackled briefly elements specific to collaborative robots, cobots and mobile robots regarding safety, focusing mostly on their differences with industrial robotic systems and particular risk mitigation measures.

14.8 Revision Questions

Question #1
True or false: personal protection equipment can be used alone to mitigate any level or risk.

Question #2
True or false: with cobots, safety stop functions cause the robot to stop once the operator enters the hazardous area and an automatic restart resumes its operation when they exit that area.

Question #3
With conventional industrial robots (i.e., fixed base), the perimeter fence can be made oh rigid panels having a minimal height of:

1. 1.5 m;

2. 1.8 m;
3. 2.0 m;
4. 1.0 m.

Question #4

In a warehouse, a rover is equipped with sensors to compute its path and a 6-DOF serial manipulator to perform pick-and-place tasks simultaneously with an operator. This robotic system falls within which category (more than one answer possible):

1. IMR;
2. AVG;
3. AMR;
4. cobot.

14.9   Further Reading

As mentioned at numerous places in this chapter, before deploying a new robotic system, you should always read the appropriate standards and technical specifications beforehand. While not directly addressed in this chapter, safety with personal-care robots has also been studied in the literature (Salvini et al., 2021) and falls under ISO 13482:2014 (2014). We can also mention ISO 18646:2021 (performance criteria and related test methods) and ISO 22166-1:2021 (modularity) on service robots (ISO, 2021a, 2021b).

References

ANSI/ITSDF. (2019). *ANSI/ITSDF B56.5-2019—Safety standard for guided industrial vehicles.*
ANSI/RIA. (2012). *ANSI/RIA R15.06-2012—Industrial robots and robot systems—Safety requirements.*
ANSI/RIA. (2020). *ANSI/RIA R15.08-1-2020—Industrial mobile robots—Safety requirements—Part 1: Requirements for the industrial mobile robot.*
Ashtiani, C. (2008). Analysis of battery safety and hazards' risk mitigation. *ECS Transactions, 11*(19), 1–11. https://doi.org/10.1149/1.2897967
Colgate, J. E., & Peshkin, M. A. (1997). *Cobots.*
CSA. (2019). *CAN/CSA-Z434-14 | Industrial robots and robot systems.*
FAA. (2021). https://www.faa.gov/uas/
ISO. (2010). *ISO 12100:2010—Safety of machinery—General principles for design—Risk assessment and risk reduction.*
ISO. (2011). *ISO 10218-1:2011—Robots and robotic devices—Safety requirements for industrial robots—Part 1: Robots.*
ISO. (2014). *ISO 13482:2014—Robots and robotic devices—Safety requirements for personal care robots.*
ISO. (2015). *ISO 13850:2015—Safety of machinery—Emergency stop function—Principles for design.*

ISO. (2016). *ISO - ISO/TS 15066:2016—Robots and robotic devices—Collaborative robots.*

ISO. (2017). *ISO 19649:2017—Mobile robots—Vocabulary.*

ISO. (2019). *ISO 13857:2019—Safety of machinery—Safety distances to prevent hazard zones being reached by upper and lower limbs.*

ISO. (2021a). *ISO 18646:2021—Robotics—Performance criteria and related test methods for service robots.*

ISO. (2021b). *ISO 22166-1:2021—Robotics—Modularity for service robots.*

ISO. (2021c). *ISO 8373:2021—Robotics—Vocabulary.*

ISO. (2022). *ISO/DIS 10218-1.2—Robotics—Safety requirements—Part 1: Industrial robots.*

Ouyang, D., Chen, M., Huang, Q., Weng, J., Wang, Z., & Wang, J. (2019). A review on the thermal hazards of the lithium-ion battery and the corresponding countermeasures. *Applied Sciences, 9*(12). https://www.mdpi.com/2076-3417/9/12/2483

RIA. (2016). *RIA TR R15.306-2016—Task-based risk assessment methodology.*

Salvini, P., Paez-Granados, D., & Billard, A. (2021). On the safety of mobile robots serving in public spaces: Identifying gaps in EN ISO 13482: 2014 and calling for a new standard. *ACM Transactions on Human Robot Interaction, 10*(3), 1–19.

Tikar, S. S. (2017). Compliance of ISO 26262 safety standard for lithium ion battery and its battery management system in hybrid electric vehicle. In *2017 IEEE Transportation Electrification Conference (ITEC-India)* (pp. 1–5). https://doi.org/10.1109/ITEC-India.2017.8333870

Transport Canada. (2019). *Standard 921—Small remotely piloted aircraft in visual line-of-sight (VLOS)—Canadian Aviation Regulations (CARs).*

Villani, V., Pini, F., Leali, F., & Secchi, C. (2018). Survey on human-robot collaboration in industrial settings: Safety, intuitive interfaces and applications. *Mechatronics, 55*, 248–266. https://doi.org/10.1016/j.mechatronics.2018.02.009

Bruno Belzile is a postdoctoral fellow at the INIT Robots Lab. of ÉTS Montréal in Canada. He holds a B.Eng. degree and Ph.D. in mechanical engineering from Polytechnique Montréal. His thesis focused on underactuated robotic grippers and proprioceptive tactile sensing. He then worked at the Center for Intelligent Machines at McGill University, where his main areas of research were kinematics, dynamics and control of parallel robots. At ÉTS Montréal, he aims at creating spherical mobile robots for planetary exploration, from the conceptual design to the prototype.

David St-Onge (Ph.D., Mech. Eng.) is an Associate Professor in the Mechanical Engineering Department at the École de technologie supérieure and director of the INIT Robots Lab (initrobots.ca). David's research focuses on human-swarm collaboration more specifically with respect to operators' cognitive load and motion-based interactions. He has over 10 years' experience in the field of interactive media (structure, automatization and sensing) as workshop production director and as R&D engineer. He is an active member of national clusters centered on human-robot interaction (REPARTI) and art-science collaborations (Hexagram). He participates in national training programs for highly qualified personnel for drone services (UTILI), as well as for the deployment of industrial cobots (CoRoM). He led the team effort to present the first large-scale symbiotic integration of robotic art at the IEEE International Conference on Robotics and Automation (ICRA 2019).

Chapter 15
Managing the World Complexity: From Linear Regression to Deep Learning

Yann Bouteiller

15.1 Objectives of the Chapter

At the end of this chapter, you will:

- understand the fundamentals of modern ML, and in particular deep learning,
- become familiar with linear regressions, MLPs, CNNs, and RNNs,
- be aware of the supervised techniques that are most relevant for robotics,
- understand the fundamentals of deep reinforcement learning,
- become familiar with Gym environments and DQN,
- be aware of the deep RL algorithms that are most relevant for robotics.

15.2 Introduction

Classical robot algorithms for perception and control are often based on simple, linear models of the world. These approaches are very effective for simple tasks where the system reasonably satisfies the corresponding assumptions in its domain of operation. However, they become inoperative in many high-level reasoning tasks where the complexity of the real world is relevant and needs to be captured. A typical example is the task of driving autonomously from camera pixels, which requires a deep, conceptual understanding of the environment. How can an autonomous car detect other agents such as vehicles and pedestrians, often partially when not entirely occluded? How to predict their individual behaviors and react accordingly? How to reliably detect traffic signalization in all possible variations of the environment, including

Y. Bouteiller (✉)
Department of Computer and Software Engineering, Polytechnique Montréal, Montreal, Canada
e-mail: yann.bouteiller@polymtl.ca

© The Author(s) 2022   441
D. Herath and D. St-Onge (eds.), *Foundations of Robotics*,
https://doi.org/10.1007/978-981-19-1983-1_15

light and weather? Over the past decade, the state-of-the-art solutions to these problems have emerged from statistical approximation techniques, nowadays referred to as *machine learning* (ML). Instead of relying on engineered representations of the world, ML approaches build their own representations automatically from large amounts of data, collected either directly from the real world, or from a simulator. The process of building these representations is called *learning* (or, equivalently, *training*). In modern ML, learnt representations can be so abstract that they are often interpreted as being similar to a human-like, conceptual understanding of the world. For instance, ML algorithms are able to learn high-level concepts such as *pedestrian* and *car* by analyzing a large number of images featuring road scenes and can then be used to complete tasks in which these concepts are relevant.

An Industry Perspective

Jonathan Lussier
Director, Intellectual Property and Innovation

Kinova inc.

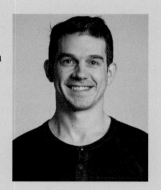

I graduated in mechanical engineering and started my career in the aerospace industry, mainly in system engineering over 10 years both in simulation and in the product. Meanwhile, in my free time, I was building robots on the side (in my basement) as I saw it as the wave of the future coming. So I started with online resources, 3D printing, and sheet metal work in order to start building some small and subsequently some larger robots arms. It was at this moment that I found out about Kinova (a company close to where I live) and the great work they were doing in the assistive field and decided to apply.

My first task when I started at Kinova was to build the proof of concept for what is now our Gen3 lite robot. Over a span of eight months, I read as much as I could and benefitted from the extremely high level of expertise from Kinova engineers to ramp up and design and build it. Afterward, I transitioned into a role ensuring the Gen3 robot was launched on time, which was very challenging but a great experience especially from the collaborations between different groups within the company, which is so great about robotics—the integration of mechanical and electrical hardware, quality assurance, and of course all the different software disciplines.

The field of machine learning in robotics is changing extremely quickly. One aspect I love is that, contrary to some other fields or even other aspects of robotics where research and development are heavily either industry-led or academic-led, we are seeing many practical applications based on the integration of machine learning and robotics launched commercially, often by people still in academia. These advances, built off the back of thousands of researchers doing more segmented AI (natural language processing, image recognition or classification, etc.), can be combined into the robotic system in an integrated way. As mentioned above, there are so many different disciplines that need to be combined when launching a robotic or automation product that there are opportunities for hardware, traditional software (e.g., machine vision), and AI-led approaches at all the different levels—it is very exciting!

Start small and simple! When working on robotics, getting a complete system up and running can be a challenge in itself. Limiting the number of hardware and software components is key and that applies to AI as well. Take advantage of existing libraries (e.g., Gym, Stablebaselines, etc.) which often combine different options for simulators, datasets, and algorithms wrapped in an easy to use and (most importantly) well-documented interface.

15.3 Definitions

ML is a vast field consisting of many techniques developed for various purposes. We can roughly separate these techniques under two categories: *supervised* and *unsupervised*.[1]

Supervised learning consists of using a set of *labeled* data points to train a *model*. In other words, given a dataset $\mathcal{D} = \{x_i, y_i\}$ of data points x_i (e.g., camera images …) and corresponding labels y_i (e.g., type of the closest agent present in the image, position estimate of this agent …), the goal is to find a model f mapping data points to labels such that $f(x) \approx y$ for all data point x and corresponding label y … including those not present in the dataset! This last property is a central objective of ML, called *generalization*: A good model is not a model that fits the dataset, but a model that fits the real phenomenon (of which the dataset is only a comparatively tiny sample). You will often hear that a good model is one that "generalizes well". Depending on the nature of the labels, the task is called *classification* or *regression*. A *classifier* produces categorical outputs (e.g., is it a car or a cat?). This is typically done

---

[1] This is a very rough categorization. In particular, semi-supervised learning and reinforcement learning are other important ML categories, which borrow aspects from both supervised and unsupervised learning.

by outputting a vector of probabilities where each dimension represents a category of interest. For instance, a vector of dimension 3 could represent three classes of interest such as "pedestrian", "car", and "none". The output $f(x) = [0.1, 0.8, 0.1]^\top$ could then mean that the closest agent in the x input image is most likely a car. A *regressor* instead produces a real-valued output (e.g., the three-dimensional relative position of the closest agent). In this case, the model directly outputs the value of interest. For instance, the output $f(x) = [3.5, -1.0, 0.0]^\top$ could mean that the closest agent is 3.5 m ahead and 1.0 m on the right. In classifiers and regressors alike, the model usually consists of a set of tunable parameters θ. Thus, finding a good model essentially consists of finding good values for these parameters. Among the most relevant types of parametric models, we can cite decision trees/random forests, which are simple ML algorithms with good properties in terms of interpretability, linear regressions, and neural networks. In this chapter, we will denote parametric models as f_θ, and we will focus our attention on neural networks, which are omnipresent in modern ML. Note that there also exist ML algorithms using nonparametric models, where the model is typically the dataset itself. For instance, the K-nearest neighbors (KNN) algorithm compares new data points to the whole dataset, so as to infer their corresponding labels from the closest labeled data available.

The locution *unsupervised learning* refers to all ML techniques that instead use an unlabeled dataset $\mathcal{D} = \{x_i\}$. Famous examples of unsupervised methods are generative adversarial networks (GANs), intensively used in image generation/transformation, and trained with unlabeled pictures. While GANs are definitely useful for robotics, they are used in very advanced situations that we will only briefly cite in this chapter.

An alternative to the aforementioned categories, called *reinforcement learning* (RL), will be covered with greater attention in the second part of this chapter. RL algorithms learn a controller from their own experience, in a near-unsupervised fashion. However, this is done by leveraging an external reward signal that remotely resembles a label, and thus RL stands somewhere in between supervised and unsupervised approaches.

In robotics, supervised methods are typically useful for perception. In particular, we use neural networks for image analysis, spatial perception, speech recognition, signal processing. ... Supervised learning is also possible for control, in particular through behavioral cloning, which consists of imitating the *policy*[2] of an expert. However, behavioral cloning is inherently limited by the expert level. Thus, policy optimization strategies based on trial-and-error, such as RL and genetic algorithms, are often preferred for learning a controller.

---

[2] A policy is a set of relations that maps observations to actions.

15.4 From Linear Regression to Deep Learning

15.4.1 Loss Optimization

The goal of supervised learning is to find a model f such that, for all input x and desired output y, $f(x) \approx y$. In ML, the quality of the model f is typically evaluated in terms of a *loss function*. A loss function takes a model and a dataset as input and outputs a real value that represents how bad the model is performing on the dataset. In this chapter, we will denote loss functions as $L(f, \mathcal{D})$ in the general case and $L(\theta, \mathcal{D})$ for parametric models. The smaller the loss is, the better the model is considered. In other words, the goal of ML is almost always an optimization problem consisting of minimizing a loss function. Many different loss functions exist, each with their own properties in terms of what they consider being a good model and how easy they are to optimize. The most common losses are the *mean squared error* (MSE) loss and the *cross-entropy* (CE) loss.

Mean Squared Error Loss
The MSE loss is typically used for regression. It is defined as follows:

$$L_{\text{MSE}}(f, \mathcal{D}) = \frac{1}{n} \sum_{i=1}^{n} (|f(x_i) - y_i|^2)$$

where n is the size of the dataset $\mathcal{D}$ (i.e., the number of (x_i, y_i) pairs in $\mathcal{D}$).

An important property of the MSE loss is that it strongly penalizes models that have a large prediction error $|f(x_i) - y_i|$ for some data point x_i. In other words, the MSE loss prefers models that do not ignore any data point. Although this is desirable in general, this also has the drawback of being strongly impacted by outliers.[3]

Cross-Entropy Loss
The CE loss is typically used for classification. It is defined as follows:

$$L_{\text{CE}}(f, \mathcal{D}) = \frac{1}{n} \sum_{i=1}^{n} -\ln(f_{y_i}(x_i))$$

where n is the size of the dataset and where $f_{y_i}(x_i)$ is the probability that $f(x_i)$ outputs for the class y_i.

A reason why the CE loss is widely used is that its gradient[4] is easy to compute (we will see why this is important later in this chapter). Since $f_{y_i}(x_i)$ is a probability,[5] its

---

[3] In ML, outliers are data points whose labels are far from what is expected.

[4] A gradient is the Jacobian of a single multivariate function, i.e., with one row. Note that, in deep learning, we often transpose the gradient to work with column vectors only. For instance, $\nabla_{[a,b]^\top}(a + b^2) = [1, 2b]^\top$.

[5] In practice, this is not really a probability, but the output of a softmax function, which also sums to 1.

value lies between 0 and 1 (0 being excluded in practice). The closer this probability is to 1, the smaller the loss is, while a probability close to 0 is strongly penalized. Indeed we want $f_{y_i}(x_i)$ to be 1, since the label of x_i is y_i in our dataset.

15.4.2 Linear Regression

One of the oldest and most fundamental supervised ML techniques is the *linear regression*. Linear regression was introduced by Legendre and Gauss who used it to predict astronomical trajectories in the early 1800s (Stigler, 1981), way before the term "machine learning" was introduced and popularized. Performing a linear regression consists of fitting a parametric linear model to a labeled dataset $\mathcal{D} = \{\mathbf{x_i}, y_i\}$ where the labels $y_i \in \mathbb{R}$ are single real values. As seen in Chap. 6, a linear model is of the form $f_\theta(\mathbf{x_i}) = \mathbf{w}^\top \mathbf{x_i} + b$, where $\mathbf{w}$ is a vector of *weights* and b is a single *bias*. The set of tunable parameters is $\theta = \{\mathbf{w}, b\}$.

Interestingly, it is possible to find the optimal solution to the linear regression problem using matrix calculus. For this matter, a useful trick is to write θ as the concatenation of $\mathbf{w}$ and b, and to append a 1 to the $\mathbf{x_i}$ vector:

$$\theta = \begin{bmatrix} w_1 \\ \vdots \\ w_m \\ b \end{bmatrix} \quad \text{and} \quad \overline{\mathbf{x}}_{\mathbf{i}} = \begin{bmatrix} x_{i,1} \\ \vdots \\ x_{i,m} \\ 1 \end{bmatrix}$$

This allows us to write the linear model as a simple vector multiplication:

$$f_\theta(\mathbf{x_i}) = \theta^\top \overline{\mathbf{x}}_{\mathbf{i}}$$

Now we can minimize the MSE loss of our parametric f_θ model. For our dataset $\mathcal{D}$ of n $(\mathbf{x_i}, y_i)$ pairs, we define $\overline{\mathbf{X}} \in \mathbb{R}^{n \times (m+1)}$ as the matrix formed by the n "augmented" data points and $\mathbf{Y} \in \mathbb{R}^n$ as the vector formed by the n labels:

$$\overline{\mathbf{X}} = \begin{bmatrix} \overline{\mathbf{x_1}}^\top \\ \vdots \\ \overline{\mathbf{x_n}}^\top \end{bmatrix} \quad \text{and} \quad \mathbf{Y} = \begin{bmatrix} y_1 \\ \vdots \\ y_n \end{bmatrix}$$

This enables us to write the MSE loss in matrix form:

$$L_{\text{MSE}}(\theta, \mathcal{D}) = \frac{1}{n}(\overline{\mathbf{X}}\theta - \mathbf{Y})^\top (\overline{\mathbf{X}}\theta - \mathbf{Y})$$

To minimize this loss, we take its gradient with respect to our vector of tunable parameters θ, which can be shown to be (the proof is out of this chapter scope):

$$\nabla_\theta L_{\text{MSE}}(\theta, \mathcal{D}) = \frac{2}{n}(\overline{\mathbf{X}}^\top \overline{\mathbf{X}}\theta - \overline{\mathbf{X}}^\top \mathbf{Y})$$

We then set this derivative to 0 to find the minimum (the convexity of the loss with respect to θ is easy to prove), which yields our optimal vector of parameters[6]:

$$\theta^* = (\overline{\mathbf{X}}^\top \overline{\mathbf{X}})^{-1}\overline{\mathbf{X}}^\top \mathbf{Y}$$

on the condition that $\overline{\mathbf{X}}^\top \overline{\mathbf{X}}$ is invertible. Note that this is the left MPGI of $\overline{\mathbf{X}}$ multiplied by $\mathbf{Y}$, as we have essentially minimized the Euclidean norm of $\overline{\mathbf{X}}\theta - \mathbf{Y}$ (c.f. Chap. 6).

15.4.3 Training Generalizable Models

Overfitting

As seen in the previous section, the linear regression problem has an optimal solution that can be written analytically. This solution is not really straightforward, though, since computing the inverse of $\overline{\mathbf{X}}^\top \overline{\mathbf{X}}$ can be challenging, especially when the $\mathbf{x_i}$ are high dimensional. Moreover, linear regressions are a very simple and special case. In advanced ML approaches, such analytical solution is virtually never available. Indeed, to find the optimal vector of parameters θ^*, we have computed the gradient of the loss function with respect to θ and found where this gradient was equal to zero. The reason why this worked is that, in a linear regression, the MSE loss is convex with respect to θ. This property is generally not satisfied in complex models such as neural networks, and thus it is not possible to apply the same strategy. But more importantly, this is not even a suitable thing to do!

Remember that we are *not* looking for the model that best fits our dataset (which is exactly what we have computed in the previous section), but the model that best fits the real world. In fact, the "optimal" set of parameters that we have computed is the worst possible example of *overfitting* that one can commit with a linear regression: We have selected our set of parameters θ^* not because it is best at describing the real world, but because it is best at describing the dataset.

This is usually not a big deal when using linear regressions: If the phenomenon of interest is indeed linear, any linear approximation using a reasonable number of data points is likely to be a good approximation. However, practical problems are scarcely ever linear. In fact, high-level reasoning tasks—such as driving from pixels—are highly nonlinear and require nonlinear models like neural networks (try to imagine what would happen if you performed a linear regression on a dataset of

---

[6] The scikit-learn Python library can be used to compute this set automatically: scikit-learn.org.

Fig. 15.1 Overfitting. Given data points sampled from a nonlinear phenomenon, a model that perfectly describes all the data points is likely to generalize poorly

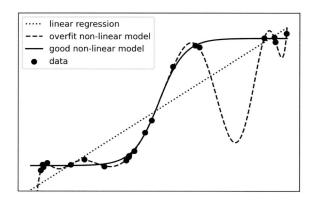

camera images x_i to compute outputs $f_\theta(x_i)$ that represent, say, the distance to the nearest car ...).

Typically, nonlinear models do not have strong *inductive biases*[7] like linearity and have a much bigger *capacity*.[8] They are able to represent crazily complex shapes, which can fit the dataset exactly and yet generalize horribly. For instance, in Fig. 15.1, we are trying to model a nonlinear phenomenon from which a dataset has been sampled (each black circle represents a data point and its label: the x_i are the values on the x axis and the y_i are the values on the y axis). Linear regression (dotted) performs poorly on this simple nonlinear problem. Using a complex nonlinear model instead and minimizing the MSE loss all the way down to zero produce a strongly overfit model (dashed). The model represented with a full line has a slightly bigger MSE loss when evaluated on our dataset, but it is likely to generalize much better to unseen data.

Minimizing a loss in ML is not a typical optimization problem where one seeks to actually find the minimum of the loss. Instead, the loss minimization procedure is merely a tool to find a good set of parameters for our model. But how exactly do we find this set of parameters, and how do we know that it is a good one?

Stochastic Gradient Descent

Although the loss function for nonlinear models is typically not strictly convex, it can usually be considered approximately pseudoconvex[9] in practice. Complex nonlinear models such as neural networks have many tunable parameters (i.e., a very high-dimensional θ), and we are unlikely to find a θ vector that cannot be improved in any of its dimensions. Thus, despite being unable to find the analytical solution to the loss minimization problem, we can select a random parameter vector θ_0 and iteratively optimize our loss from there by following the negative gradient. For a given value of θ_t, we compute the local negative gradient of the loss:

---

[7] An inductive bias is an assumption about the structure of the world that we force into our model.

[8] The capacity of a model indicates the degree of complexity that it is able to represent.

[9] A pseudoconvex function increases forever in the direction of any of its local gradients.

Fig. 15.2 Gradient descent. The GD algorithm optimizes the loss by iteratively descending its slope in the directions of its local gradient with respect to $\boldsymbol{\theta}$ (arrows)

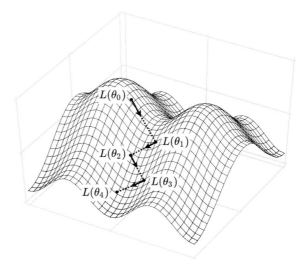

$$\boldsymbol{\nabla}_t = -\nabla_\theta L(\boldsymbol{\theta}_t, \mathcal{D}),$$

and we update our parameter vector in the direction of this local negative gradient:

$$\boldsymbol{\theta}_{t+1} = \boldsymbol{\theta}_t + \alpha \boldsymbol{\nabla}_t,$$

where the *learning rate* α is an hyperparameter.[10] This procedure is illustrated in Fig. 15.2.

Computing the true local gradient of the loss at each gradient descent iteration is very computationally intensive. The gradient needs to be averaged over the entire dataset at each iteration, which is not suitable. A better way of performing gradient descent is *stochastic gradient descent* (SGD), an important key of modern ML success. In its "pure" (vanilla) version, SGD is the same algorithm as gradient descent, except instead of averaging the local gradient over the whole dataset, the gradient is taken with respect to a single (x_i, y_i) pair sampled from the dataset. This technique produces a very rough estimate of the gradient, at a much smaller computational cost. But despite this estimate being rough, a small optimization step can still be taken in its direction. This operation can be performed rapidly and repeated over many times. Moreover, the stochastic nature of the gradient estimate enables SGD to escape from *local extrema* and *saddle points*[11] easily where vanilla gradient descent would fail. These properties make SGD much more efficient than gradient descent in practice.

However, this version of SGD is still computationally inefficient. In fact, computing an average gradient over several samples is a parallelizable task, and thus it

---

[10] Hyperparameters are parameters not learnt by the optimization algorithm (often just set manually).

[11] Point where the gradient is close to zero on all dimensions ($\nabla_\theta L(\theta_t) \approx \mathbf{0}$), but that is not a local extremum.

is not really a good idea to use something as extreme as one single (x_i, y_i) pair for our local gradient estimate. Modern GPUs enable using several of them at no additional cost in terms of computation. This is why, in practice, we never use one single sample from the dataset, but a certain number of them. The number of samples per gradient estimate is an hyperparameter, called the *batch size*. Batch sizes between 16 and 4096 are common choices. As a rule of thumb, small batches yield rough gradient estimates and work best with smaller learning rates, whereas larger batches yield better[12] estimates and can afford larger learning rates (He et al., 2019). The resulting algorithm, called *minibatch gradient descent* (or also SGD), is the basis of most state-of-the-art loss optimizers, such as *Adam* (Kingma & Ba, 2014) and *RMSProp*.

Algorithm 1 Minibatch gradient descent (SGD)

Require: $\mathcal{D}, f_\theta, L, \alpha, n$ ▷ dataset, model, loss function, learning rate, batch size
Ensure: $\theta \approx \theta^*$ ▷ near-optimal parameters for the model
   $\theta \leftarrow$ random values   ▷ initialize parameters
   repeat
      batch $\leftarrow$ n (x, y) pairs sampled from $\mathcal{D}$ ▷ sample minibatch from dataset
      $\nabla \leftarrow -\nabla_\theta L(\theta, \text{batch})$ ▷ estimate gradient on minibatch
      $\theta \leftarrow \theta + \alpha\nabla$ ▷ update parameters by descending gradient
   until convergence of $L(\theta, \mathcal{D})$ ▷ once in a while, evaluate actual loss on dataset

---

Training, Validating, and Testing

As long as the local gradient can be computed for any value θ of the parameter vector, SGD enables minimizing the loss of approximately pseudoconvex nonlinear models. However, when optimized by SGD, the loss on our dataset will still eventually converge too close to its true minimum. In other words, if not stopped early enough, SGD will overfit!

Fortunately, there is a way of stopping convergence right before this happens. This technique, called *early stopping*, also enables evaluating the true performance of the model on unseen data. The main idea is that we do not train our algorithm on the entire available dataset. Instead, we shuffle the dataset $\mathcal{D}$ and split the result into three disjoint subsets:

- a *training set* $\mathcal{D}_{\text{train}}$,
- a *validation set* $\mathcal{D}_{\text{validation}}$,
- a *test set* $\mathcal{D}_{\text{test}}$.

We then perform SGD by estimating gradients only on the training set, with a small change: instead of stopping the algorithm when we think the loss has converged on the training set, we stop the algorithm when the loss stops improving on the validation set.

---

[12] In the sense of being closer to the true gradient and thus less stochastic, which is only partly suitable!

Fig. 15.3 Early stopping. The validation set enables finding when to stop training before overfitting starts harming the generalization properties of the model

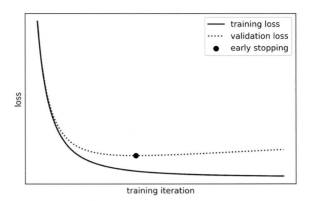

This is because, when the model starts overfitting, its performance on the validation set (on which it is not trained) starts decreasing (i.e., the validation loss starts increasing). Figure 15.3 displays a typical example of this phenomenon over training.

Note that the loss function is not necessarily what people use to determine when early stopping should happen. It is possible to use metrics we are more directly interested in. For instance, in classification tasks, we often use the accuracy or the $F1$-score.

Finally, you may wonder why we have split our dataset into three parts rather than two. Indeed, we have not used $\mathcal{D}_{\text{test}}$ at all. And there is a good reason for this: You should never use the test set before your system is final and ready for production! There are many subtle ways in which it is possible to overfit on our dataset in an ML project, and early stopping is one of them. Because we have selected our best model based on its performance on the validation set, we have slightly overfit on this subset. To really evaluate our performance, the only unbiased way is to do it on the unseen test set once the model is final. In supervised learning, this is important as a last sanity check. Of course, this can be replaced by testing the model directly in the real world when possible, in which case the real world becomes the test set. Typically, the performance on the test set is slightly worse than the performance on the validation set, which is noticeably worse than the performance on the training set.

Regularization

On top of early stopping, many existing techniques, called *regularizers*, help improve the generalization performance of a model. Some of these techniques, such as $L1$ and $L2$, add a term to the loss in order to penalize large parameter values and promote models that use as few parameters as possible (this avoids crazy models similar to Fig. 15.1). Others, such as *dropout*, introduce noisy modifications to the model during training in order to promote robustness.

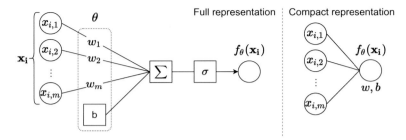

Fig. 15.4 Simple neuron

15.4.4 Deep Neural Networks

ML has attracted a lot of attention over the past few years. The main reason for this surge of interest is that modern GPUs (and, more recently, TPUs/IPUs) have provided enough computational power to train a class of complex nonlinear models invented in the 1940s–1960s (Fitch, 1944; Ivakhnenko & Lapa, 1965), whose potential had remained unknown for several decades (Krizhevsky et al., 2012). These models, called *deep neural networks* (DNNs), are an algorithmic attempt to mimic the brain. They project their input into successive, more and more abstract representations, that eventually map to the desired output. DNNs are today at the core of most ML successes. In fact, they have become so prominent that modern ML is often simply called *deep learning*.

The atomic component of a DNN is a very simple, usually nonlinear model, called *neuron*. A neuron is made of a linear model,[13] directly followed by an easily differentiable, usually nonlinear function, called *activation*. Using the same notation as for linear regressions, the operation performed by a neuron is

$$f_\theta(\mathbf{x_i}) = \sigma(\boldsymbol{\theta}^\top \overline{\mathbf{x}}_\mathbf{i})$$

where σ is the activation function. We often represent a neuron as a graph, which helps visualize the flow of operations. In particular, we will use the compact representation to understand more complex DNNs (Fig. 15.4).

The only structural difference with a linear regression model is the activation σ, which plays a central role in deep learning. Many activation functions exist in the literature, the most common being the *sigmoid* and the *rectified linear unit* (ReLU).

The sigmoid is defined as follows:

$$\text{sigmoid}(a) = \frac{1}{1 + e^{-a}}$$

---

[13] The same model as used by linear regression.

Fig. 15.5 Sigmoid activation

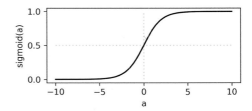

Fig. 15.6 ReLU activation

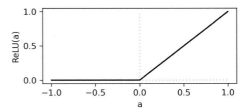

The sigmoid is generally used when one needs to squash an output between 0 and 1. However, its derivative is near zero everywhere except around the origin, which is harmful to the convergence of SGD. Plus, compared to ReLU, the sigmoid is relatively costly to compute (Fig. 15.5).

The ReLU is a simple clipping operation:

$$\text{ReLU}(a) = \max(x, 0)$$

Computing a ReLU is blazing fast and so is computing its derivative (0 for negative numbers, and 1 for strictly positive numbers, the derivative at the origin being arbitrary) (Fig. 15.6).

The point of using such a simple nonlinearity may seem unclear at first: A neuron with a ReLU activation is just a crippled linear model unable to output anything negative! But contrary to linear models, a neuron is never used alone: Its representational power comes from being coupled with other neurons to form a DNN. In its simplest form, called *multilayer perceptron* (MLP), a DNN is a stack of *layers*, each made of several parallel neurons (Fig. 15.7).

Remember that each individual neuron has a vector of tunable weights and a single tunable bias as parameters. Since a layer has several parallel neurons, this translates to each layer having a matrix of tunable weights and a vector of tunable biases. The set of tunable parameters of an MLP is thus $\theta = \{\mathbf{W_i}, \mathbf{b_i}\}_{i=1\ldots k+1}$. The operation performed by an MLP is as follows:

$$h_1(\mathbf{x_i}) = \sigma_1(\mathbf{W_1 x_i} + \mathbf{b_1})$$
$$h_2(\mathbf{h_1}) = \sigma_2(\mathbf{W_2 h_1} + \mathbf{b_2})$$
$$\cdots$$
$$f_\theta(\mathbf{x_i}) = f(\mathbf{h_k}) = \sigma_{k+1}(\mathbf{W_{k+1} h_k} + \mathbf{b_{k+1}})$$

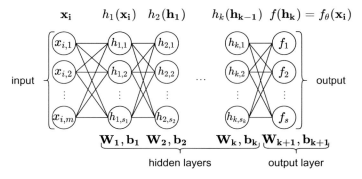

Fig. 15.7 Multilayer perceptron

Despite their fairly simple structure, DNNs perform extremely complex nonlinear projections and are typically treated as black boxes. Thus, we say that layers other than the last one are *hidden*. On the other hand, the last layer has a special role and is typically a simple linear layer with no activation (i.e., σ_{k+1} is the identity function). This layer projects the output of the last hidden layer into the output space of the DNN (for instance, into a 3D vector if we are predicting a position …).

Notice that, without nonlinear activation functions, this structure would only be a crazy way of building a linear model. This is because combining linear combinations yields other linear combinations. Yet, simple nonlinearities such as ReLUs make DNNs much more powerful. In fact, a famous result called the *universal approximation theorem* (Hornik, 1991) shows that, even with a single (large enough) hidden layer, a neural network can approximate any continuous function arbitrarily well.[14] This includes mappings from raw camera images to conceptual information about their content, or even directly to optimal control commands for our robot!

15.4.5   *Gradient Back-Propagation in Deep Neural Networks*

We know that DNNs can approximate virtually any complicated nonlinear mapping of interest, such as mappings from camera images to conceptual descriptions of their content. Moreover, we know a way of searching for this mapping: SGD with early stopping. The only ingredient we are missing for applying this strategy is an estimate of the gradient of the loss with respect to all tunable weights and biases of our DNN.

The key to the success of DNNs is an algorithm introduced in 1970 (Linnainmaa, 1970) and made practical by the use of modern GPUs/TPUs/IPUs, called *gradient back-propagation* (or *backprop* for short). Backprop is a dynamic programming algorithm that efficiently computes the gradient of the loss. To perform a backprop,

---

[14] For an animated illustration of this result: http://neuralnetworksanddeeplearning.com/chap4.html (Nielsen, 2015).

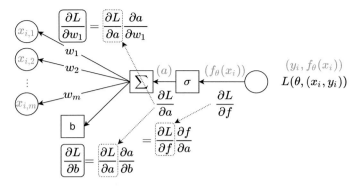

Fig. 15.8 Gradient back-propagation

one first needs to perform a *forward propagation* in the DNN, i.e., compute the $f_\theta(x_i)$ output. Then, backprop uses the *chain rule* of partial derivatives to propagate gradients backward in the graph. For simplicity, let us visualize this process on a single neuron.

In Fig. 15.8, we want to compute the gradient of the loss L with respect to the weights w_i and the bias b. Once the output f has been computed by a forward propagation, it is straightforward to compute the derivative of the loss with respect to this output, $\frac{\partial L}{\partial f}$. We can then use the result to compute $\frac{\partial L}{\partial a}$,[15] which, according to the chain rule, is equal to $\frac{\partial L}{\partial f}\frac{\partial f}{\partial a}$. Indeed, $\frac{\partial f}{\partial a}$ is just the derivative of the activation σ. The result can then be propagated further back to compute the partial derivatives we are interested in: $\frac{\partial L}{\partial w_j} = \frac{\partial L}{\partial a}\frac{\partial a}{\partial w_j}$ and $\frac{\partial L}{\partial b} = \frac{\partial L}{\partial a}\frac{\partial a}{\partial b}$. Indeed, $\frac{\partial a}{\partial w_j}$ is $x_{i,j}$ and $\frac{\partial a}{\partial b}$ is 1. To generalize this procedure to DNNs, we also use $\frac{\partial L}{\partial a}$ to compute $\frac{\partial L}{\partial x_{i,j}} = \frac{\partial L}{\partial a}\frac{\partial a}{\partial x_{i,j}}$, where $\frac{\partial a}{\partial x_{i,j}}$ is w_j, and we repeat this process in previous neurons. Note that intermediate results such as $\frac{\partial L}{\partial a}$ are computed only once and reused many times. This makes this dynamic programming procedure very efficient in DNNs.

We now master the basics of deep learning! In practice, we will not implement SGD and backprop manually, because highly optimized libraries have done all the work for us. Nowadays, the most popular such libraries are PyTorch and TensorFlow.[16]

15.4.6 Convolutional Neural Networks

We have seen how DNNs can learn extremely complex nonlinear tasks such as mapping camera pixels to relevant high-level information … in theory. In reality, using an MLP to process camera images is bound to fail.

---

[15] Here, a denotes the intermediate forward value after the sum and before the activation.

[16] pytorch.org and tensorflow.org.

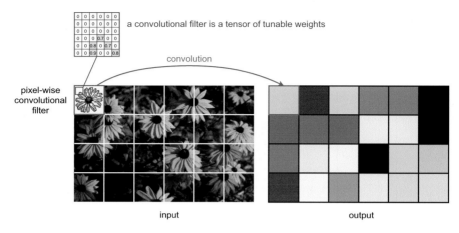

Fig. 15.9 Neural convolution

To understand why, let us consider a small 100×100 RGB image input. When flattened with all three color channels concatenated, this becomes a 30,000-dimensional vector. For an MLP to be able to extract meaning of this vector, it should have at least about as many neurons in its first layer; otherwise, a near-linear regression on pixels would happen. The number of weights in an MLP layer being the number of inputs multiplied by the number of parallel neurons, this would require around 9×10^8 tunable weights in the first layer alone. Computing their individual partial derivatives at each SGD iteration would be painfully slow.

Furthermore, deep learning is never guaranteed to converge to anything interesting. The convergence and generalization properties of SGD rely on the pseudo-convexity assumption and depend on many hyperparameters (structure of the neural network, learning rate, initial set of parameters ...). Thus, it is often a good idea to help our models learn meaningful mappings by enforcing inductive biases when possible, similar to how linearity makes linear models efficient for linear problems. In particular, camera images have a strong spatial structure that we can use to our advantage. Convolutional neural networks (CNNs) are an effective way of doing so. Instead of connecting each color channel of each pixel to each neuron of the first hidden layer, CNNs borrow a much lighter technique from traditional computer vision: image convolution.[17] This technique, illustrated in Fig. 15.9, uses *filters* (also called *kernels*) to "scan" images for specific patterns and perform local projections.

A filter is a small array of weights, plus a single optional bias,[18] all tunable. We split the image into pieces of the same size as the filter. Typically, these pieces are overlapping (e.g., shifted by only one pixel), although for the sake of clarity they do not overlap in Fig. 15.9 (they form the white "grid"). We apply the filter to

---

[17] The mathematical operation is actually a cross-correlation, but ML practitioners call it "convolution".

[18] Another (less common) version exists in which there is one bias per output pixel.

Fig. 15.10 Convolution options

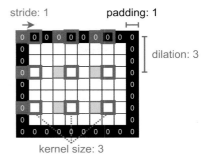

stride: 1   padding: 1

dilation: 3

kernel size: 3

each piece individually. This is done by multiplying each pixel of the piece with the corresponding weight of the filter, then summing the results into a single value, and adding the bias (NB: This operation is a linear combination).

The result is then fed to an activation function, which produces a new pixel value. Together, new pixels form an output image called *feature map*. More precisely, a 2D convolutional filter is in fact a 3D tensor[19] whose depth is the number of input channels (e.g., 3 when the input is a RGB image). It combines all input channels into a single output feature map. CNNs commonly use hundred of filters in parallel, each producing a different feature map depending on the weights and bias of the filter. These feature maps then become the input channels of the next convolutional layer. Using CNN filters greatly reduces the number of trainable parameters when compared to MLPs: Only the weights and bias of each filter are trainable, and convolutional filters are often of size 3×3 in practice. Typically, we find that filters become edge detectors or pattern detectors during training. For instance, in Fig. 15.9, the trained filter has naturally taken the shape of a flower so as to detect flower patterns.

Additionally, we commonly use the following operations in convolutional layers:

– *Zero padding*: We append zeroes to the border of the input image.
– *Stride*: We shift pixels between convoluted pieces of the input image.
– *Dilation*: We shift pixels between elements of the convolutional filter.

These options are illustrated in Fig. 15.10 (integer values describe both dimensions).

Finally, kernel-based operations other than image convolution are often used in CNNs. The most common is *max pooling*, which reduces the size of a feature map by selecting the pixel with the maximum value in the area of the kernel, as illustrated in Fig. 15.11.

CNNs are typically made of alternating convolutional and max pooling layers and are often very deep (i.e., they have many layers). They are by far the current state of the art in a wide range of computer vision tasks, some of which we will highlight later in this chapter. CNNs are a building block of many GANs used for image

---

[19] A tensor is a multidimensional array, for instance a matrix is a 2D tensor.

Fig. 15.11 Max pooling

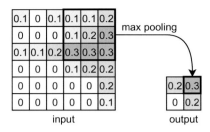

manipulation and generation, and they are not limited to 2D image processing. For instance, 1D convolutions can be used for signal processing, and 3D convolutions can be used for video processing.

15.4.7 Recurrent Neural Networks

Time series are often central in robotics: We need them to analyze the past and plan in the future. Thus far, we have seen how DNNs can analyze the present, but can they keep track of the past? Is it possible to predict and plan in the future? Can we generate coherent sequences such as paths, or even sentences? Can we process time series such as video streams, or even sound and voice?

In deep learning, the past can be analyzed by feeding the whole history of relevant observations to a DNN. For instance, a self-driving car would be unable to output a relevant command from one camera image only, as this would contain little information about the dynamics of the world (i.e., only contextual information …). On the other hand, a history of the last few camera images equally spaced in time is enough to infer simple dynamics. This concern is more generally known as the *Markov property*: The history fed as input to the model must be long enough so that any earlier observation is irrelevant to the task.

Planning in the future can be done by recursively feeding a DNN with its own last few outputs. For instance, let us imagine a model that takes a target position and a path as input. The model appends a waypoint to the path so that it gets closer to the target. The updated path can then be fed back to the model. This procedure repeated several times yields a path planning algorithm. A similar procedure can be used to generate speech or music. …

Although feeding a history of observations directly to a vanilla DNN is possible, this quickly gets inefficient and impractical. Naively processing the whole history of relevant observations at each forward propagation is computationally intensive and may perform poorly due to the lack of inductive biases. Fortunately, a better alternative exists: Recurrent neural networks (RNNs) are able to automatically detect and keep track of only the relevant information from past observations. Instead of being fed the whole history at each forward propagation, they take a single observation as input and store the relevant information directly in their hidden layers, within a persistent *hidden state*.

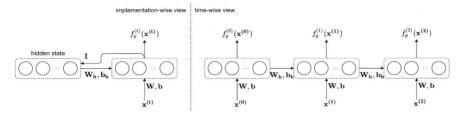

Fig. 15.12 Recurrent layer

Figure 15.12 describes a simple RNN layer. For the sake of clarity, arrows represent matrix multiplications, i.e., connections between layers, instead of individual connections between neurons ($\mathbf{I}$ is the identity matrix). Time is discretized into timesteps. The output $\mathbf{f}_\theta^{(t)}$ of the layer at timestep t is computed from both the input $\mathbf{x}^{(t)}$ and the output from the previous timestep $\mathbf{f}_\theta^{(t-1)}$. An additional set of parameters $\mathbf{W_h}$ and $\mathbf{b_h}$ handles how memorized information is combined with new information. Mathematically, the output at timestep t is as follows:

$$\mathbf{f}_\theta^{(t)} = \sigma\left(\mathbf{W}\mathbf{x}^{(t)} + \mathbf{b} + \mathbf{W_h}\mathbf{f}_\theta^{(t-1)} + \mathbf{b_h}\right)$$

To train an RNN, all observations in the relevant portion of the history are fed to the model one by one. Then, the gradient of the loss can be back-propagated through time. This operation is similar to how back-propagation is performed in MLPs, except the gradient also flows back through the horizontal arrows in the time-wise view of Fig. 15.12.

RNNs not only make the forward propagation computationally efficient (since only one observation is fed to the model at each timestep), but also constitute an inductive bias that promotes memorization of high-level concepts rather than raw inputs. Indeed, the persistent information consists of the values projected by hidden layers. In deep learning, these projections are typically seen as extracted concepts.

15.4.8 Deep Learning for Practical Applications

The field of deep learning is very competitive and evolving rapidly. This yields many high-performance models that practitioners can use directly in robot applications. Due to Python being particularly popular in the deep learning community, most readily available implementations are found in Python. Nevertheless, it is always possible, although a bit cumbersome, to extract readily trained weights and biases from Python in order to implement the model in more efficient languages for production. In fact, PyTorch and TensorFlow both provide ways of facilitating the transfer of Python models to C++. We provide a non-exhaustive list of supervised and unsupervised approaches that are relevant for robotics.

CNNs and GANs

CNNs and GANs have attracted a large portion of the ML research focus over the past few years. They are particularly often used in modern computer vision.

– ImageNet: ImageNet (Deng et al., 2009) is a benchmark on which many high-performance CNNs are compared for pure image classification. At the moment of writing this book, the best-performing such models are the EfficientNet family (Pham et al., 2021; Tan & Le, 2019).
– YOLO: YOLO (You Only Look Once) (Bochkovskiy et al., 2020; Long et al., 2020; Redmon et al., 2016) is a very popular family of CNNs combining image classification and bounding boxes. YOLO finds all instances of known categories in an image and draws a bounding box around each instance.
– Mask-R CNN: Mask-R CNN (He et al., 2017) is similar to YOLO, but even more evolved. On top of detecting all class instances with their bounding boxes in an image, Mask-R CNN draws the actual segmentation of each instance.
– PoseNet: PoseNet (Kendall et al., 2015; Moon et al., 2018) is a CNN able to extract human poses from camera images, e.g., for non-verbal communication with the robot.
– Super-resolution: Super-resolution models (Wang et al., 2020) are able to improve the resolution of input images, e.g., for low-quality cameras. They are often based on GANs.
– Image inpainting: Image inpainting models (Elharrouss et al., 2020) are able to fill gaps in images. For example, they can be used to fill gaps in depth maps generated by LIDARs, or to reconstruct partially occluded subjects. They are also often based on GANs.
– Domain adaptation: Domain adaptation models (Wang & Deng, 2018) enable transforming data from one domain (e.g., data from a simulator) into data from another domain (e.g., real-world data!). CycleGAN (Zhu et al., 2017) is a popular example.

Sequential Modeling

The RNN structure that we have described in the previous section is often informally called "vanilla RNN". In practice, much more efficient types of RNNs are available.

– LSTM: A long short-term memory (LSTM) (Hochreiter & Schmidhuber, 1997) is a special type of RNN with a more complicated, *gated* structure. In particular, it is able to selectively forget pieces of information and keep what it thinks is relevant for many timesteps.
– GRU: A gated recurrent unit (GRU) (Chung et al., 2014) is similar to an LSTM, but computationally lighter.
– Autoregressive models: Autoregressive models are not really RNNs, but evolved forms of the naive recursive procedure described in the previous section for generating sequences with vanilla DNNs. These models use their own previous outputs directly as an input sequence and implement inductive biases to process this

sequence efficiently. They can be used to generate human voice for example, as done by WaveNet (Oord et al., 2016).

- Transformers: Transformers (Vaswani et al., 2017) are not really RNNs either. They also take whole sequences as input. Nevertheless, they are the current state of the art in many sequence processing tasks. Transformers use an *attention* mechanism to efficiently process sequences by focusing only on relevant parts of the input. For instance, GPT-3 (Brown et al., 2020) is an autoregressive transformer able to generate human-like text. Another example is BERT (Devlin et al., 2018), a transformer used for language understanding. Both BERT and GPT-3 are immense models readily pretrained that one needs to fine-tune for their application.

Note that training LSTMs and GRUs is not fully parallelizable and is thus slow, but using them is fast once trained. On the other hand, training transformers is parallelizable and is thus fast, but they are slow to use once trained. This is an important concern in practice for robotic applications where the model needs to run as fast as possible once deployed.

Behavioral Cloning

Behavioral cloning is a supervised technique that can be used with any DNN to train a robot controller. First, an expert remotely controls the robot to perform a task many times. Everything is recorded into a dataset $\mathcal{D} = \{x_i, y_i\}$, where the x_i are sensor readings and the y_i are expert commands. We then use this dataset to train a DNN as seen previously. The resulting DNN maps sensor observations to expert commands. We call this DNN a *policy network* and denote its output π_θ instead of f_θ, by convention.

15.5 Policy Search for Robotic Control

15.5.1 Limitations of Supervised Learning for Control

While behavioral cloning enables learning a policy, it is inherently limited by its supervised nature. Behavioral cloning only tries to "imitate" the expert policy from a dataset of demonstrations, and thus it is unable to really match (let alone outperform) the expert level. Moreover, expert demonstrations are likely to be concentrated around a small number of interesting trajectories from which they never deviate. Robots not being perfect, they do deviate from these known trajectories and get lost in unexplored situations.

This is where the paradigm of *policy search* comes into play. Unlike supervised methods, policy search algorithms learn from their own experience. They are not limited by the expert level, and they learn in a way that is arguably closer to how natural intelligence arises. For example, evolutionary algorithms are inspired from natural genetics. They learn their own policy by continuously applying random mutations to their model, evaluating the new performance on the task after each mutation, and choosing to keep or discard the new model based on this evaluation. These algorithms

Fig. 15.13 RL transition

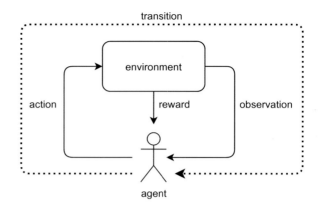

are able to easily find reasonable solutions to theoretically very complex scenarios, such as multi-robot tasks. However, because their mutations are random, they are often less efficient than the informed mutations performed by SGD. Deep reinforcement learning is a class of policy search algorithms able to find high-performance policies by leveraging SGD instead.

15.5.2 Deep Reinforcement Learning

Deep reinforcement learning is the modern conjunction of deep learning and RL (Sutton & Barto, 2018), a subfield of ML inspired from behavioral psychology, and more particularly from the concept of *reinforcement*. Reinforcement partially explains living organisms' behavior as the result of a history of positive and negative stimuli. Simply put, when facing a given situation, living organisms would try different actions and experience different outcomes. When later facing similar situations, they would become more likely to retry the actions that yielded the best outcomes and less likely to retry the actions that yielded the worst. RL emulates these outcomes by mean of a *reward* signal, which is a measure of how well the robot performs.

Interaction with the Environment
In RL, the world is framed as a special type of state machine called *Markov decision process* (MDP), or, less formally, *environment*[20] (Fig. 15.13).

Robotic environments usually discretize time into timesteps. At each timestep, the robot, called *agent*, retrieves an *observation* of the current state of the environment and uses its policy to compute an *action* from this observation. Once the action is computed, it is applied in the environment, which *transitions* to a new state. The agent observes this new state and receives an instantaneous reward. The new observation

---

[20] More precisely, real-world environments are usually partially observable Markov decision processes.

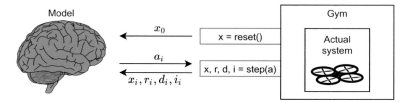

Fig. 15.14 Gym interface

can then be used to compute a new action from the agent's policy and so on. As implied by the name, the observations outputted by an MDP must have the Markov property. In other words, the observation actually fed to the policy must contain the whole history of past observations reasonably relevant to the task. Alternatively, the policy can be an RNN …

In practice, most RL environments are implemented using the Gym Python interface (Brockman et al., 2016) (Fig. 15.14).

The Gym interface is very simple and essentially consists of two methods. The *reset* method sets the environment to its initial state and returns an initial observation. The agent computes an action from this observation and feeds this action to the *step* method. The step method performs a transition of the environment and returns four values: a new observation, an instantaneous reward, a Boolean that tells whether the task is complete, and a Python dictionary that can usually be ignored (it may contain debugging information).

Importantly for robot applications, note that MDPs do not naturally take real-time considerations into account. The world is simply "stepped" from one fixed state to the next, and time is supposed to be "paused" between transitions. We can achieve a real-time behavior by means of a timer that clocks our Gym environment, but this comes with delayed dynamics that we need to handle properly.[21]

Reinforcement Learning Objective
The general philosophy of RL is to maximize the accumulated reward signal that the agent gets from the environment. More precisely, we want to find an optimal policy which, from any given observation, maximizes the sum of future rewards that the agent can *expect*. In other words, we want to find optimal parameters θ^* for the policy of the agent, such that:

$$\theta^* = \operatorname{argmax} \sum_{t=t_0}^{\infty} \mathbb{E}[r_t],$$

where $\mathbb{E}[r_t]$ is the *expectation* over the instantaneous reward r_t received from the environment at timestep t when following the policy π_{θ^*}, starting from an arbitrary initial observation.

---

[21] For a helper that handles these dynamics automatically, see rtgym (pypi.org/project/rtgym/).

Note that this sum can be infinite (e.g., if $r_t > 0$ for all t), and thus θ^* may be undefined. To alleviate this issue, we introduce a *discount factor* $0 \leq \gamma < 1$ in the rewards. This hyperparameter is very often used in RL, and understanding its role is important. Instead of trying to achieve the maximum sum of expected rewards, the agent tries to achieve the maximum sum of expected γ-*discounted* rewards:

$$\theta^* = \text{argmax} \sum_{t=t_0}^{\infty} \mathbb{E}[\gamma^t r_t].$$

Since $\gamma < 1$, this makes the optimal policy relatively *greedy*: instead of optimizing for long-term rewards, we give more importance to the rewards that are not too far in the future. The closer γ is to 0, the greedier the optimal policy becomes. We usually set γ very close to 1 (0.95 or 0.99 are common values), but the effect is still noticeable. For instance, let us imagine we design the reward to be 0 everywhere except for the single timestep when the robot completes the task, where the reward is 1. Without a discount factor (i.e., with $\gamma = 1$), any policy would be optimal as long as it completes the task someday, and it should take ten thousand years. On the other hand, when $\gamma < 1$, the only optimal policies are the ones that complete the task as fast as possible.

Several ways of maximizing this sum exist. We will focus on an algorithm published in 2015, called "Deep Q-Network" (DQN) (Mnih et al., 2015), because it introduces many of the basic concepts that more advanced deep RL techniques used nowadays.

Deep Q-Network Policy
In the DQN algorithm, we train a near-deterministic policy that maps complex observations to discrete actions.[22] For this matter, we use the concept of *Q-value*.

In RL, the *state-value* function $V_{\pi_\theta}(x)$ maps the observation x to the sum of γ-discounted rewards that the agent can expect from following its current policy π_θ after observing x. It is defined recursively as follows:

$$V_{\pi_\theta}(x) = \mathbb{E}_{a \sim \pi_\theta(\cdot|x)} \mathbb{E}_{x',r' \sim p(\cdot|x,a)}[r' + \gamma V_{\pi_\theta}(x')],$$

where p is the transition distribution of the environment, i.e., the statistical distribution of the new observation x' and the new reward r' when observing x and taking action a. The policy π_θ is also written as a distribution because it is not deterministic. The *Q-value*, or *action-value* function, is almost the same thing, except it "forces" the first action:

$$Q_{\pi_\theta}(x, a) = \mathbb{E}_{x',r' \sim p(\cdot|x,a)}[r' + \gamma V_{\pi_\theta}(x')].$$

---

[22] Determinism is not really suitable for robotics in practice: We may get stuck in unseen situations. More advanced deep RL techniques are able to train stochastic policies that output real-valued actions.

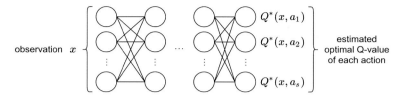

Fig. 15.15 Deep Q-network model

In plain English, the Q-value $Q_{\pi_\theta}(x, a)$ is the sum of discounted rewards that the robot can expect when it observes x, takes action a, and then follows it current policy π_θ ever after. We use the Q-value to discriminate good and bad actions from a given observation.

Let us imagine that we magically have access to the optimal Q-value function Q^*, i.e., the Q-value function under the optimal policy π_{θ^*}. When actions are discrete, the optimal policy π_{θ^*} is obviously to choose the action with the highest Q^* at each timestep, i.e.,

$$\pi_{\theta^*}(x) = \operatorname*{argmax}_a Q^*(x, a).$$

This is exactly how the DQN policy works. We train a DNN with parameters θ that maps observations to the optimal Q-value Q^* of each action (Fig. 15.15).

Once the DQN model is trained, the robot uses it with the observation received from the environment at each timestep. The optimal policy π_{θ^*} is simply to select the action with the highest estimated Q^*. Ties are broken randomly, which is why the DQN policy is not entirely deterministic.

Deep Q-Network Training
Of course, we do not magically have access to the optimal Q-value function for training our DQN model in practice. But there is a well-known method for approximating this function: *Q-learning*. The previous equations can be combined into the following form:

$$Q^*(x, a) = \mathbb{E}_{x', r' \sim p(\cdot|x, a)}[r' + \gamma \max_{a'} Q^*(x', a')].$$

This identity, called the *Bellman equation*, is very important for RL. It enables identifying the optimal Q-value function by performing *Bellman backups*. For any transition (x, a, x', r') performed in the environment, we can improve our approximator (e.g., DQN model) of the optimal Q-value function with a simple operation:

$$Q^*(x, a) \leftarrow r' + \gamma \max_{a'} Q^*(x', a')$$

Let us take a moment to unveil the full potential of this operation. The Bellman backup improves the estimate of $Q^*(x, a)$ by aggregating the actual reward r' and the estimated best Q^* under the next observation x'. We can select any transition

(x, a, x', r') to perform this backup, as long as the transition has been sampled from the environment at some point. This includes transitions collected under a policy that is not the current policy of the robot. In particular, we can use transitions collected by an expert, or simply by an older version of the current policy. In RL, this important property is called *off-policy*. An off-policy algorithm, such as DQN, is able to improve the current policy with transitions collected under another policy. This is in contrast to *on-policy* algorithms, which can only use transitions collected under the current policy. The main reason why the off-policy property is important is that it enables reusing old transitions several times at different stages of training. This is particularly important in robotic applications, where it is costly to collect environment transitions.

We want to use the Bellman backup for training our DQN model, which is a DNN. Therefore, we need to translate this backup into a loss function, so as to enable SGD. This is pretty straightforward: We can simply use the MSE loss. We select a transition (x, a, x', r'), we feed x and x' separately to our DQN model so as to retrieve $Q^*(x, a)$ and $Q^*(x', \cdot)$, and then we perform an SGD step for the $Q^*(x, a)$ output only, using the following loss:

$$L_{\text{MSE}}(Q^*, \{(x, a, x', r')\}) = |Q^*(x, a) - (r' + \gamma \max_{a'} \underline{Q}^*(x', a'))|^2$$

The reason why we underline Q^* in the right-hand part of this equation is a bit subtle. Notice that, in supervised learning, the quantity between parenthesis would correspond to the ground truth label y, which would be a constant. Here, however, this quantity depends on the parameters of the DNN, because the DNN is used to compute $Q^*(x', \cdot)$. Thus, if we are not careful, performing SGD will modify the DNN parameters such that our "ground truth" gets closer to our estimate rather than the other way round! To avoid this issue, we do not back-propagate gradients through the computation graph of $\underline{Q}^*$.[23] However, this leaves yet another issue: Updating our DNN parameters still updates the very target that we are trying to reach! Indeed, we are updating these parameters such that our estimate of $Q^*(x, a)$ gets closer to the quantity between parentheses, but this collaterally changes $\underline{Q}^*$ and thus this very quantity, making training unstable. We instead keep an old copy of our DQN model that we periodically update. This copy, called the *target network*, is used to compute $Q^*(x', \cdot)$. Since it is only updated once in a while, it becomes easier to track.

It is straightforward to generalize our loss to minibatch gradient descent. Using a minibatch $\mathcal{M} = \{(x_i, a_i, x'_i, r'_i)\}$ of n transitions, the MSE loss becomes

$$L_{\text{MSE}}(Q^*, \mathcal{M}) = \frac{1}{n} \sum_{i=1}^{n} |Q^*(x_i, a_i) - (r'_i + \gamma \max_{a'_i} \underline{Q}^*(x'_i, a'_i))|^2.$$

Since DQN is off-policy, we can sample these transitions randomly from a dataset, as we would do in supervised learning. Sampling from a fixed dataset (e.g., of expert demonstrations) is possible and known as *offline RL*. Theoretically, this could match

---

[23] In PyTorch and TensorFlow, the "no gradient" option ensures this constant-like behavior.

and even outperform the expert level, because we are not imitating the expert any-more: We are literately learning from their experience. However, this approach comes with practical difficulties, in particular because it tends to overestimate the value of unexplored situations. We often prefer letting the agent collect its own experience, by *exploring* the environment while learning. In this situation, there is virtually no theoretical limit to how good the agent can get.

In DQN, the agent explores its environment by following an ϵ-*greedy* policy: With a certain probability ϵ, the agent selects a random action, and with probability $(1 - \epsilon)$, it selects the best action as estimated by the DQN model. This scheme *exploits* the current knowledge of the environment by drawing exploration toward promising trajectories only. A higher ϵ yields a higher tendency to explore and thus slower convergence, whereas a smaller ϵ yields faster convergence at the price of being more likely to converge to local optima, i.e., to poor policies. This trade-off is known as the *exploration/exploitation dilemma*.

In practice, we store the transitions collected by the agent in a huge circular dataset called *replay buffer* (1–100 million transitions are common sizes). In parallel, we randomly sample minibatches of transitions from this buffer, and we train our DQN model by minimizing the $L_{\mathrm{MSE}}(Q^*, \mathcal{M})$ loss via SGD.

15.5.3 Improvements of Deep Q-Learning

DQN was published in 2015 and has popularized the idea of deep Q-learning, which is leveraged in many state-of-the-art deep RL algorithms nowadays. Over the years, improvements have been introduced that increased the performance of these methods. In particular, we usually dampen the updates of the target network and train two DQN models in parallel.

Slowly Moving Target
Using an old copy of our DQN model that we only periodically update makes the target easier to track. An even better strategy is to update the target slowly in a dampened fashion, by mean of an *exponentially moving average*. Instead of only periodically updating the target parameters, we update them with each SGD step, but only by pulling them weakly toward those of the current DNQ model. More precisely, the parameters of the target are updated according to:

$$\underline{\theta} \leftarrow (1 - \tau)\underline{\theta} + \tau\theta,$$

where θ are the parameters of the current Q-network, $\underline{\theta}$ are those of the target network, and τ is a small attraction coefficient (commonly 0.005). With this strategy, the $\underline{Q}^*$ target does not move erratically due to the stochasticity of SGD updates.

Double Q-Networks
Another well-known issue of the original DQN algorithm is that the Q-network tends to converge to overestimated values for some actions. This issue, known as the

overestimation bias, comes from the fact that the target network $\underline{Q}^*$ is an estimator and has a noisy error. When selecting the maximum $\underline{Q}^*$ over all actions, we also select the maximum over … the noisy error!

To tackle this issue, we train two DQN models in parallel, with different sets of initial parameters. This yields two target networks, each with different error distributions. In recent algorithms, we compute the value of each action as its minimum across both networks, which cancels the overestimation bias.

15.5.4 Deep Reinforcement Learning for Practical Applications

Despite DQN being relatively recent (2015), the research effort has been so intense in the deep RL community over the last few years that it is already vastly outperformed by more advanced techniques.

– Soft Actor Critic (SAC). Since its publication in 2018, SAC (Haarnoja et al., 2018a, 2018b) has been one of the main players in deep RL for robotics. SAC is an *Actor Critic* algorithm. This means that two neural networks are trained in an interleaved fashion: an *actor* (policy network similar to behavioral cloning) and a *critic* (Q-network similar to DQN). The policy trained by SAC is stochastic and able to output continuous actions. Moreover, its entropy[24] is maximized in parallel to the RL objective. This yields a policy that is very robust to unseen situations, which is particularly important in the real world. SAC is also an off-policy algorithm.

– Model-based reinforcement learning. More recently, another class of deep RL algorithms, called *model based*, has seen a dramatic surge of interest from the community. Model-based algorithms use a model of the world to predict the response of the environment from a given observation. In state-of-the-art approaches, this model is learnt from interacting with the environment. Once learnt, the policy can be trained without interacting with the environment anymore, and the model can be used for planning. The MuZero (Schrittwieser et al., 2020) algorithm is currently the winning player in this field.

– Non-stationary environments. Let us point out a real-world concern that we have silenced so far. We have focused on off-policy algorithms because these have tremendous advantages in situations where the collection of transitions is costly, which is typically the case in robot applications. However, "naive" off-policy algorithms only work when the environment is stationary. Indeed, these algorithms rely on a dataset of past transitions to train their current policy. But think of what happens in the real world, where other agents are continuously learning and changing their behavior. Old transitions may become obsolete, and learning from those may become counterproductive. In this situation, you might want to draw inspiration from on-policy approaches such as Proximal Policy Optimization (PPO) (Schulman et al., 2017), which only rely on the present.

---

[24] The entropy is the amount of randomness of a stochastic function.

– Sim-to-real. It is often practical to train an RL algorithm in simulation, especially for robotics. RL being fundamentally based on trial-and-error, we do not want our robot to break because of crazy random actions during training. Simulation instead provides a safe environment where arbitrarily bad actions can be tried out. Moreover, it is possible to accelerate time in simulation and collect transitions faster than real time. But this comes at a price: Simulation is never the same as reality, and a policy trained in simulation typically fails in the real world. This concern, called the *sim-to-real* gap, is one of the main challenges that has long kept deep RL from being really useful in practical robotics. However, it has recently been alleviated impressively by techniques such as Rapid Motor Adaptation (RMA) (Kumar et al., 2021), which uses a combination of deep RL and supervised learning for this matter.

15.6 Wrapping It Up: How to Deeply Understand the World

Deep RL is an elegant illustration of how deep learning enables a robot to produce its own understanding of the world. When learning a task such as driving autonomously from camera images through deep RL, the robot is never explicitly told what a car, a pedestrian, or a road is. Instead, it learns these concepts on the fly, only from maintaining its own belief about which decisions are good or bad for each possible observation. This belief is formed by the principle of gradient back-propagation in a DNN. In the case of DQN, this DNN is the DQN model, which can be seen as a black box, mapping observations to the corresponding Q-value of each available action. For instance, an image in which there is a pedestrian would likely be mapped to low Q-values for actions that lead to run over the pedestrian and to higher Q-values for actions that do not. This implies that DQN builds its own way of detecting pedestrians (with convolutional kernels that detect pedestrian features for instance) and grasps a certain understanding of what a pedestrian is, how it moves, how it interacts with the world, etc. Now, there is no magic at play here: This understanding is entirely statistical, and it is defined as the complex projections performed by the DNN in its successive layers of artificial neurons. How this fundamentally differs from a human understanding of the world, however, is a real question.

15.7 Summary

In this chapter, an introduction to the fundamentals of modern machine learning has been provided, in its aspects most relevant to robotics. We have started from simple linear regressions and have built our way up to highly expressive and nonlinear deep neural networks. This allows us to approximate complex mappings, such as optimal controls from sensor readings. Furthermore, we have introduced the basics of deep reinforcement learning, a popular approach that enables robots to look for such controls autonomously. In the course of this chapter, we have seen how a dataset can be used for training a model and how it is possible to ensure that this model generalizes well. Finally, we have introduced state-of-the-art supervised models that the reader can use out of the box for robotic perception and state-of-the-art reinforcement learning algorithms that are becoming increasingly relevant for robot control. Keep in mind however that all these techniques are uncertain in nature, due to the use of statistical approximators (e.g., neural networks). Safety is therefore always an important concern when applying modern ML techniques in the real world.

15.8   Quiz

Please find the quiz for this chapter in the Jupyter notebooks available online.[25]

15.9   Further Reading

To learn more about deep learning, we recommend "Deep Learning" (Goodfellow et al., 2016) by Ian Goodfellow, Yoshua Bengio, and Aaron Courville. To learn more about reinforcement learning, we recommend "Reinforcement Learning: An introduction" (Sutton & Barto, 2018) by Richard Sutton and Andrew Barto. Both references are available online for free.

References

Bochkovskiy, A., Wang, C. Y., & Liao, H. Y. M. (2020). Yolov4: Optimal speed and accuracy of object detection. arXiv preprint arXiv:200410934

Brockman, G., Cheung, V., Pettersson, L., Schneider, J., Schulman, J., Tang, J., & Zaremba, W. (2016). Openai gym. arXiv preprint arXiv:160601540

Brown, T. B., Mann, B., Ryder, N., Subbiah, M., Kaplan, J., Dhariwal, P., Neelakantan, A., Shyam, P., Sastry, G., Askell, A., Agarwal, S., Herbert-Voss, A., Krueger, G., Henighan, T., Child, R., Ramesh, A., Ziegler, D., Wu, J, Winter, C., ... Amodei, D. (2020). Language models are few-shot learners. arXiv preprint arXiv:200514165

Chung, J., Gulcehre, C., Cho, K., & Bengio, Y. (2014). Empirical evaluation of gated recurrent neural networks on sequence modeling. arXiv preprint arXiv:14123555

Deng, J., Dong, W., Socher, R., Li, L. J., Li, K., & Fei-Fei, L. (2009). Imagenet: A large-scale hierarchical image database. In *2009 IEEE Conference on Computer Vision and Pattern Recognition* (pp. 248–255). IEEE.

Devlin, J., Chang, M. W., Lee, K., & Toutanova, K. (2018). Bert: Pre-training of deep bidirectional transformers for language understanding. arXiv preprint arXiv:181004805

---

[25] https://github.com/Foundations-of-Robotics/ML-Quiz.

Elharrouss, O., Almaadeed, N., Al-Maadeed, S., & Akbari, Y. (2020). Image inpainting: A review. *Neural Processing Letters, 51*(2), 2007–2028.

Fitch, F. B. (1944). McCulloch Warren S. and Pitts Walter. A logical calculus of the ideas immanent in nervous activity. Bulletin of Mathematical Biophysics, vol. 5, pp. 115–133. *Journal of Symbolic Logic, 9*(2).

Goodfellow, I., Bengio, Y., & Courville, A. (2016). *Deep learning.* MIT Press.

Haarnoja, T., Zhou, A., Abbeel, P., & Levine, S. (2018a). Soft actor-critic: Off-policy maximum entropy deep reinforcement learning with a stochastic actor. In *International Conference on Machine Learning, PMLR* (pp. 1861–1870).

Haarnoja, T., Zhou, A., Hartikainen, K., Tucker, G., Ha, S., Tan, J., Kumar, V., Zhu, H., Gupta, A., Abbeel, P., & Levine, S. (2018b). Soft actor-critic algorithms and applications. arXiv preprint arXiv:181205905

He, F., Liu, T., & Tao, D. (2019). Control batch size and learning rate to generalize well: Theoretical and empirical evidence. *Advances in Neural Information Processing Systems, 32*, 1143–1152.

He, K., Gkioxari, G., Dollár, P., & Girshick, R. (2017). Mask R-CNN. In *Proceedings of the IEEE International Conference on Computer Vision* (pp. 2961–2969).

Hochreiter, S., & Schmidhuber, J. (1997). Long short-term memory. *Neural Computation, 9*(8), 1735–1780.

Hornik, K. (1991). Approximation capabilities of multilayer feedforward networks. *Neural Networks, 4*(2), 251–257.

Ivakhnenko, A. G., & Lapa, V. G. (1965). *Cybernetic predicting devices.* CCM Information Corporation.

Kendall, A., Grimes, M., & Cipolla, R. (2015). PoseNet: A convolutional network for real-time 6-DOF camera relocalization. In *Proceedings of the IEEE International Conference on Computer Vision* (pp. 2938–2946).

Kingma, D. P.,& Ba, J. (2014). Adam: A method for stochastic optimization. arXiv preprint arXiv:14126980

Krizhevsky, A., Sutskever, I., & Hinton, G. E. (2012). ImageNet classification with deep convolutional neural networks. In *Advances in Neural Information Processing Systems* (Vol. 25, pp. 1097–1105).

Kumar, A., Fu, Z., Pathak, D., & Malik, J. (2021). RMA: Rapid motor adaptation for legged robots. arXiv preprint arXiv:210704034

Linnainmaa, S. (1970). *The representation of the cumulative rounding error of an algorithm as a Taylor expansion of the local rounding errors* (Master's Thesis), University of Helsinki, pp. 6–7 (in Finnish).

Long, X., Deng, K., Wang, G., Zhang, Y., Dang, Q., Gao, Y., Shen, H., Ren, J., Han, S., Ding, E., & Wen, S. (2020). PP-YOLO: An effective and efficient implementation of object detector. arXiv preprint arXiv:200712099

Mnih, V., Kavukcuoglu, K., Silver, D., Rusu, A. A., Veness, J., Bellemare, M. G., Graves, A., Riedmiller, M., Fidjeland, A. K., Ostrovski, G., Petersen, S., Beattie, C., Sadik, A., Antonoglou, I., King, H., Kumaran, D., Wierstra, D., Legg, S., & Hassabis, D. (2015). Human-level control through deep reinforcement learning. *Nature, 518*(7540), 529–533.

Moon, G., Chang, J. Y., & Lee, K. M. (2018). V2V-PoseNet: Voxel-to-voxel prediction network for accurate 3D hand and human pose estimation from a single depth map. In *Proceedings of the IEEE Conference on Computer Vision and Pattern Recognition* (pp. 5079–5088).

Nielsen, M. A. (2015). *Neural networks and deep learning* (Vol. 25). Determination Press.

Oord, A., Dieleman, S., Zen, H., Simonyan, K., Vinyals, O., Graves, A., Kalchbrenner, N., Senior, A., & Kavukcuoglu, K. (2016). WaveNet: A generative model for raw audio. arXiv preprint arXiv:160903499

Pham, H., Dai, Z., Xie, Q., & Le, Q. V. (2021). Meta pseudo labels. In *Proceedings of the IEEE/CVF Conference on Computer Vision and Pattern Recognition* (pp. 11557–11568).

Redmon, J., Divvala, S., Girshick, R., & Farhadi, A. (2016). You only look once: Unified, real-time object detection. In *Proceedings of the IEEE Conference on Computer Vision and Pattern Recognition* (pp. 779–788).

Schrittwieser, J., Antonoglou, I., Hubert, T., Simonyan, K., Sifre, L., Schmitt, S., Guez, A., Lockhart, E., Hassabis, D., Graepel, T., Lillicrap, T, & Silver, D. (2020). Mastering Atari, Go, chess and shogi by planning with a learned model. *Nature, 588*(7839), 604–609.

Schulman, J., Wolski, F., Dhariwal, P., Radford, A., & Klimov, O. (2017). Proximal policy optimization algorithms. arXiv preprint arXiv:170706347

Stigler, S. M. (1981). Gauss and the invention of least squares. *The Annals of Statistics,* 465–474.

Sutton, R. S., & Barto, A. G. (2018). *Reinforcement learning: An introduction.* MIT Press.

Tan, M., & Le, Q. (2019). EfficientNet: Rethinking model scaling for convolutional neural networks. In: *International Conference on Machine Learning, PMLR* (pp. 6105–6114).

Vaswani, A., Shazeer, N., Parmar, N., Uszkoreit, J., Jones, L., Gomez, A. N., Kaiser, Ł., & Polosukhin, I. (2017). Attention is all you need. In *Advances in neural information processing systems* (pp. 5998–6008).

Wang, M., & Deng, W. (2018). Deep visual domain adaptation: A survey. *Neurocomputing, 312,* 135–153.

Wang, Z., Chen, J., & Hoi, S. C. (2020). Deep learning for image super-resolution: A survey. *IEEE Transactions on Pattern Analysis and Machine Intelligence.*

Zhu, J. Y., Park, T., Isola, P., & Efros, A. A. (2017). Unpaired image-to-image translation using cycle-consistent adversarial networks. In *Proceedings of the IEEE International Conference on Computer Vision* (pp. 2223–2232).

Yann Bouteiller is an engineer from École des Mines de Saint-Étienne (France) working as a research associate at the Computer Science Department of Polytechnique Montreal (Canada). His research focuses on machine learning (ML) and more specifically on designing deep reinforcement learning algorithms for real-world applications. At the junction between theoretical and practical ML, he aims at facilitating the transfer of recent, simulation-based deep learning successes to the industry. His work includes advances in reinforcement learning theory as well as practical deep learning-based advances in neuroscience, autonomous driving, robot learning, video games, real-time embedded systems, etc.

Chapter 16
Robot Ethics: Ethical Design Considerations

Dylan Cawthorne

16.1 Learning Objectives

- Identify some of the ethical design considerations in the robotics field.
- Become familiar with the three main normative ethical theories.
- Understand why technology is not ethically neutral.
- Appreciate that technology exists in a context.
- Utilize the PPPP model to identify and design for a preferable future.
- Identify some human values relevant in robot design.
- Apply the value-sensitive design methodology.
- Identify impacted stakeholders.
- Utilize ethics checklists, standards, design principles, and frameworks.
- Identify specific examples of design features which support human values.
- Learn about the AIRR framework for responsible innovation.
- Be able to answer the questions: "should I build this robot? If so, why? If not, why not?"
- Apply theories and tools in the chapter to design your own ethically informed robot.

16.2 Introduction

In May of 2019, a twelve-kilogram medical delivery drone crashed in Switzerland, only 50 m from kindergarten children (Ackerman, 2019) (Fig. 16.1). No one was hurt

D. Cawthorne (✉)
Unmanned Aerial Systems Center, University of Southern Denmark, Odense, Denmark
e-mail: dyca@sdu.dk

© The Author(s) 2022
D. Herath and D. St-Onge (eds.), *Foundations of Robotics*,
https://doi.org/10.1007/978-981-19-1983-1_16

Fig. 16.1 Matternet (2021) drone which crashed in a wooded area of Switzerland near a kindergarten. The emergency parachute had been deployed, but the chord connecting it to the drone was cut allowing the 12 kg drone to freefall to the ground. Image used with permission, from SUST (2019)

or killed in the incident, but the failure of the drone and its parachute system caused the Swiss Post to immediately suspend operations of the large quadcopters made by the Silicon Valley company Matternet (2021). The system, which was designed to quickly transport up to 2 kg of urgently needed medical samples between hospitals and save lives, could have inadvertently caused the death of a small child. Should the drone still be used if its benefits outweigh the risks? Is it fair to subject people in the cities below the drone's flightpath to risk of injury or death? What about the sick people who need urgent medical care—isn't the drone helping them? And what responsibility do we as robot engineers and builders have for our creations? In this chapter, some of the ethical considerations in the field are presented, and theories and tools are offered for designing ethically informed robotic systems.

An Industry Perspective

Nolwenn Briquet, Control Team Lead, Kortex.

Kinova inc.

During my studies, I was always interested in Mechatronics courses. By combining the disciplines of automation, mechanics, and electronics, I felt that these subjects were the closest to cutting-edge systems. During all my studies and even after, I made all my choices little by little to work in this direction. In 2014, I started working as a robotics engineer in a robotics research laboratory on collaborative robotics applications before doing a Ph.D. in the field of human–robot interaction. What I particularly retain from this experience is that it allowed me not only to develop my expertise but also to ask myself questions about the place of the human being in automation through robotics and our responsibility in this evolution as engineers designing these systems. I finally joined Kinova in 2019 as a robotics control developer and I recently became the team lead of the control team.

When you work in robotics, the question of the societal impact of our actions is unavoidable. In some countries, robotization is seen as a way to overcome labor shortages, while in others it is seen as a real threat. When I was offered the opportunity to work in collaborative robotics, I was attracted by the concept of exploiting the complementarity of the human and the robot and thus keeping the human in the loop. The common discourse is that the human takes care of the tasks that require decision-making skills, critical thinking, and sometimes even dexterity, while the repetitive and non-value added tasks are left to the robot. However, in many situations, we can see that the real interest of these robots is that they are easy to integrate into a workspace as they do not need to be placed in a cage. In some cases, the only limitation to taking humans out of the loop is the technological limit, which I contribute to pushing back in my job. Therefore, to this day, I am faced with the following dilemma: Are these the real reasons for the success of collaborative robotics, or is it more a matter of politicization of public discourse and cobot-washing?

The field of ethics in robotics has been a hot topic in recent years. The question of responsibility can be found today at several levels: at the level of the researcher who shares his or her contributions in the public domain, at the level of the robot manufacturer who designs the functionalities of the robotic system that will be solved on the market, or at the level of the integrator who will be responsible for the security of the robotic application. At the level of the research community, there is an awareness among researchers on the integrity of the data and results that are shared. In the industrial field, this issue is covered by standards that have evolved in recent years to take into account safety concerns. However, further steps need to be taken to address the ethical issues related to the progress of artificial intelligence, whose results appear extremely promising but also less well-controlled and predictable.

16.3 Ethics

Ethics is the branch of philosophy that deals with questions about right and wrong, and how best to live and act in the world ("Ethics," 2021). But what does ethics have to do with robot design? According to philosopher of technology Peter-Paul Verbeek, "most scholars in the field agree that technologies actively help to shape culture and society, rather than being neutral means for realizing human ends" (Verbeek, 2008). This means that the robots we design and build won't just perform a task, their capabilities will allow some actions to be easier to perform and others more difficult—which has moral consequences.

For example, a healthcare drone may make it faster to deliver urgent medical samples between hospitals but make it more difficult to ensure the security of the samples during the trip. An industrial robot arm designed to weld a car's frame together could make assembly faster, but it might make it more difficult for factory workers to cultivate their welding skills. The complexity of the task becomes even more challenging—and crucial—as we consider the impacts if these technologies are scaled up. Will jobs as medical couriers and welders disappear completely? What will be the long-term impact on peoples' physical, psychological, and material welfare? Technologies always exist in a context, so we might ask where will this robot be implemented? Is it in a country where workers are likely to be retrained to build or collaborate with robots? Or will these people become redundant? Thus, we as responsible technologists need to be aware of the ethical considerations relevant in the domain, the context of use, and the potential long-term impacts—and make well-reasoned choices about the capabilities our new robots should have.

16.3.1 Normative Ethics

Luckily, ethics has been an area of study for thousands of years, and there are a lot of theories and tools we can apply when designing robots. Ethical questions can be approached at the level of normative ethics ("Ethics," 2021). Normative ethical theories can be useful to robot designers as they provide guidance on ways to view what is a morally good technology—or at least, technologies that support actions which are morally good. It is important to note that the way in which morally good consequences, actions, and behaviors are defined can vary depending on the context and across cultures.

16.3.2 Consequentialism

One type of normative ethics puts focus on the results or consequences of one's actions and is called consequentialism ("Consequentialism," 2003). This includes utilitarianism which states that we should act in a way that the consequences of our actions result in the most good for the most people ("Consequentialism," 2003). For example, heart disease is the leading cause of death globally at 16% (World_Health_Organization, 2020)—if we could design a mobile robot or drone which encouraged people to exercise we could potentially help a lot of people lead longer, healthier lives. However, the benefits of the drone would have to outweigh the negative outcomes such as injury, privacy violations, and environmental impact caused during production and at the end of the drone's useful life. And the context will matter too; in some countries, heart disease may be much less prevalent than others, and the way privacy is exercised could vary across cultures.

16.3.3 Deontology

Another normative ethics approach is deontology which puts focus on the rightness or wrongness of an action, rather than the outcome or consequence of that action ("Deontological Ethics," 2020). Deontology is a rule-based approach where actions that conform to moral norms—"the Right"—are allowed, and those that do not should not be undertaken. For example, if one were to follow the rule not to kill innocent people, then we should not design a weaponized robot that targets innocent people. And if one should save lives, then we ought to design a mobile robot or drone which encouraged people to exercise. With this last example, we can see that different normative ethical theories may suggest we perform the same actions in a given situation, but perhaps for different reasons.

However, there are sometimes important differences between normative theories. In consequentialism, it would be acceptable to perform a wrong action if it leads to an overall positive outcome for more people. This would not be accepted from a deontological standpoint, where "the Right is said to have priority over the Good" ("Deontological Ethics," 2020).

16.3.4 Virtue Ethics

A third type of normative theory is virtue ethics; here the focus is on the moral character of a person, and the theory aims to guide one in what type of person to be or become ("Virtue Ethics," 2016). Examples of virtues to strive for and cultivate over a long period of time include honesty, courage, care, and wisdom (Vallor, 2016). Designing an industrial robot arm that reduces workplace injuries would be a way of (indirectly) cultivating the virtue of care for other people. Developing a drone that provides rapid medical care in a context where this increased efficiency allows medical staff more time with patients would be another way to support the cultivation of care. Again, how different virtues are manifested could vary depending on context and across cultures. The three normative theories are summarized in Table 16.1.

All three normative theories share an emphasis on human values—values are what a person or group of people consider important in life (Friedman et al., 2013). Human values relevant in technology design include human welfare, privacy, freedom, calmness, and environmental sustainability (Friedman et al., 2013). Later in this chapter, we will see how human values can be utilized throughout the robot design process to enhance human flourishing.

16.4   The Non-Neutrality of Technology

Technology interacts with and impacts people and society in complex ways—but never in an ethically neutral way. Designing and building something alters the range of capabilities and possible actions available to people (Verbeek, 2008). Sometimes, technologies are described as "platforms"—this is the case with social media applications such as Facebook and YouTube (Gillespie, 2010), as well as drones (Cawthorne & Devos, 2020) and robots. These claims represent an older concept in the philosophy of technology called technological neutrality (Vermaas et al. 2007). Technological neutrality allows companies, governments, and engineers to distance themselves from responsibility for the uses of their products. For example, if someone uses a drone to transport illegal drugs the drone manufacturer could claim that it was

Table 16.1 Three main normative ethical theories

Consequentialism	The result/consequence of actions matters most
	Guides actions
	Includes utilitarianism—the most "good" for the most people/lifeforms
Deontology	Rule-based approach
	Guides actions
Virtue ethics	Focus on virtues/moral character
	Guides the type of person one should be/aim to become

the user that *mis*-used their product. However, the drone does clearly play a role in the crime, and as good robot designers, we should be aware of lots of different possible uses of our systems and design them so they prevent—or at least make it more difficult—to do unethical things with.

The concept of technological neutrality has since been replaced by a contextually situated and interactional model. This means that the context, the user, and the technology itself all play a role in the resulting mediation and possible resulting action. As stated before, this means that technology plays a key role in human actions, and since human actions are morally relevant then technology design is also morally relevant.

16.4.1 Dual-Use

Creating ethically informed robots may not be easy, especially given the nature of drone and robotic systems as dual-use technologies. Dual-use refers to a system's capability to be used within civilian contexts as well as in military contexts (Novitzky et al., 2018). Many normative approaches allow for the use of military technologies, especially to protect oneself or as part of a "just war" (as in a "justifiable" war) (Lin et al., 2008), but a person that develops a drone to map fields to help farmers may not have intended for their system to be used for military reconnaissance. Or someone developing a robot to carry injured soldiers out of harm's way may not expect the system to be used to deliver packages in crowded cities.

In practice, there is a lot of technology transfer that takes place both from civil contexts to military and from military to civil contexts. Drones were initially developed in university research labs, then proliferated in the military context, and now are seeing rapid growth in civil contexts (Choi-Fitzpatrick et al., 2016). As well, it may be difficult to determine if a technology is civil or military—take for example robots and drones that perform border patrol or those that are used in private security. Still, there are certain capabilities that are relevant in one context over another, and we can design for the intended context. For example, in a military context where the enemy will be trying to destroy the system the survivability of a drone will be a highly relevant capability, while this capability is much less relevant in a civil context (Van Wynsberghe & Nagenborg, 2016). How to design for relevant capabilities will be explained in more detail in the section on value sensitive design.

16.5 Technological Determinism and Multiple Futures

It is sometimes claimed that technology moves in certain ways, and that we are powerless to stop it. For example, that it is inevitable that in the future there will be more drones and robots. This idea is called technological determinism (Verbeek, 2008). In the philosophy of technology, this conception has mostly been superseded by the idea

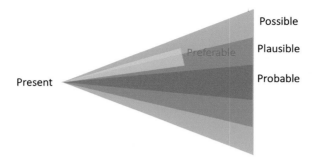

Fig. 16.2 The "PPPP"—or "probable, plausible, possible, preferable"—model shows the future as contingent on what we do in the present, and that we can choose to design our robotic systems for a preferable future rather than for the most likely (probable) future. Graphic by the author, based on Dunne and Raby (2013)

of multiple possible futures—such as that shown in the PPPP—or "probable, plausible, possible, preferable"- model in Fig. 16.2—with an emphasis on human agency and the role we play in shaping technological development. Clearly, if everyone for some reason decided to stop developing drones and robots, companies decided to stop producing them, governments outlawed them, and people stopped buying them then the future would not contain more drones and robots. There are lots of economic forces such as profitability and human forces such as curiosity which make it likely that robots will proliferate in the future, but this is not inevitable. Therefore, we as robot designers hold a lot of power when it comes to the trajectory of future technological developments and need to act in a responsible manner in doing so; this topic is explored in more detail in the section about **Responsibility.**

So in our case as designers, some critical first questions might be to ask ourselves "should we build this drone or robot at all?" "What are some of the possible opportunities, risks, and changes that it will support?" "Who will benefit the most, and who will be at the greatest risk?" "And if we should design the system, what capabilities and characteristics should it have—and which should it not have?" Later in the section on **Value Sensitive Design,** we will look at specific ways to address these questions.

16.6   Human Values in Design

Many human values that are relevant to technology design in general—and for us in robotics design—have been proposed and are shown in Table 16.2. Here, values refer to those things which humans find important and meaningful in life (Friedman et al., 2013). Values are different than preferences; preferences are opinions that individuals hold while values are more universal and are held by most people (Van de Poel, 2009). For example, I might like the color blue (it is my preference) while you might like the color green (your preference). But we both deeply value our own physical safety and the safety of others. The importance and universality of human values makes them critical to a flourishing life, and why they are so relevant

Table 16.2 Twelve human values that are considered relevant in technology design. Based on Table 4.1 in Friedman et al. (2013)

Human welfare	Includes physical, psychological, and material welfare Physical welfare deals with bodily well-being such as physical health Psychological welfare concerns mental health such as stress Material welfare refers to physical circumstances and is related to economics and employment
Ownership and property	The right to possess an object
Privacy	The ability to determine what information about one's self can be communicated to others
Freedom from bias	Without systematic unfairness toward individuals or groups
Universal usability	Technology that can be successfully used by all people
Trust	The expectation to experience goodwill from others
Human autonomy	The ability to decide, plan, and act in ways which allow one to achieve one's goals
Informed consent	Garnering voluntary agreement, such as in the use of information systems
Accountability	Ensuring that actions may be traced uniquely to the person, people, or institution responsible
Calmness	A peaceful and composed psychological state
Identity	The understanding of who one is over time
Environmental sustainability	Sustaining ecosystems such that they meet the needs of the present without compromising future generations

to designers and engineers since our technologies can support (or diminish) these values.

16.7 Value Sensitive Design

Value sensitive design, or VSD, is a way to systematically incorporate the ethical and social impacts of technologies early in the design process (Friedman et al., 2013). The process is shown in Fig. 16.3 and includes three phases: (1) conceptual, (2) empirical, and (3) technological.

16.7.1 Conceptual Phase

In the conceptual phase of VSD, the ethical considerations are identified, as well as the impacted stakeholders. If we consider the example of a cobot in a factory, this includes direct stakeholders such as those working alongside the robot as well as

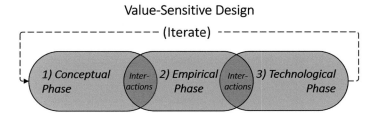

Fig. 16.3 The value sensitive design process consists of three phases: (1) conceptual, (2) empirical, and (3) technological. There are interactions between all phases, and the process itself is iterated many times throughout the design process as the technology is developed. Graphic by the author based on Friedman et al. (2013)

indirect stakeholders such as customers who buy products produced at the factory. Philosophers and technology ethicists are particularly well-suited to perform work on the ethical considerations in the conceptual phase, and social scientists can identify stakeholders and help to understand their values.

16.7.2 Empirical Phase

In the empirical phase of VSD, the interactions between the technology and people are investigated. Human–robot interaction (HRI) studies are a good example here. Continuing with the cobot example, how do workers expect a cobot to behave? Will they trust it and work in close proximity to it, or will they be afraid of it and stay away? (Read more in the previous chapter on social robots.) The empirical phase includes both the interaction of technology with individuals, but also society more broadly. Taking a drone example, will their more widespread use lead to a "chilling effect" where people assume that they do not have privacy anywhere because of surveillance from drone cameras? (Cawthorne & Cenci, 2019) Can we design drones so it is more obvious what their function is and who is controlling them? (Cawthorne & Frederiksen, 2020) Social scientists and HRI experts have a lot to offer in the empirical phase, and they can use a wide variety of quantitative and qualitative methods to better understand human–technology interactions such as by using surveys, interviews, and focus groups.

16.7.3 Technological Phase

In the technological phase of VSD, these inputs from the conceptual and empirical phases are used to design a technology—such as a cobot or drone—that supports the beneficial human values and positive social impacts identified earlier. "The technical phase is dedicated to understanding the artifact (i.e., technology, robot) in context and how it manifests values or fails to do so" (Van Wynsberghe & Nagenborg, 2016). Alternatively, a technology can be chosen first and then the social and ethical implications can be assessed, or a social phenomenon can provide inspiration for a new technology—the VSD process can be started at any phase (see the section "Practical suggestions for using value sensitive design" in Friedman et al., (2013)).

16.7.4   Contextual Design

VSD is an example of a contextual and embedded design approach—each individual technology is considered within the location of its eventual use and in relation to the people and systems that will be impacted by its uptake. And VSD is an inherently multidisciplinary design approach, since experts from fields such as philosophy and ethics of technology can contribute to the conceptual phase, social scientists to the empirical phase, and engineers and computer scientists to the technological phase. Therefore, it is useful for us as robot systems designers to at least be aware of some of the relevant issues with regard to ethical and social impacts, and collaborate with experts in these fields when developing technology responsibly. Of course, we cannot all become philosophers or social scientists overnight, but taking into consideration philosophical and human interaction issues is part of a holistic, contextually aware, and responsible design practice.

16.8   Ethics Tools

Although it is a developing field, there are already many tools available to make it easier to incorporate ethics into the design of robotic systems.

16.8.1   Checklists

Perhaps the easiest to use is an ethical checklist such as the one utilized in European Union Horizon 2020 projects (European_Union, 2019). The checklist asks yes or no questions about the project, and the questions should identify relevant ethical issues. For example, "does your research involve human participants?" and are they volunteers, vulnerable individuals, or children? "Does your research involve the processing of personal data?", and does this involve the processing of special categories of personal data such as genetics, sexual lifestyle, ethnicity, religion, etc.? A limitation of such checklists is that they are typically self-administered, and researchers with limited experience working with ethics may not see the potential risks of their technologies. In addition, most ethical issues do not easily resolve themselves to simple yes or no questions and involve complex reasoning and justification. And it is possible that the checklist may simply omit a relevant ethical issue.

16.8.2 Standards

Another source for ethics guidelines is industry standards. Within robotics, the Institute of Electrical and Electronics Engineers (IEEE) is the "world's largest technical professional organization for the advancement of technology" (IEEE, 2021). They have just released the 7000 series of standards to address ethical concerns during system design. The standard utilizes human values in design (see the earlier section on human values in design) and contains many elements of value sensitive design (see the earlier section on VSD). Industry standards help engineers design to similar requirements and promote an approach that allows companies to compare their technology to others'. However, standards can be expensive to buy, which can prevent individuals and small businesses from being able to access them.

16.8.3 Design Principles

Design principles and guidelines developed by researchers and organizations can also be helpful. For example, the "privacy by design" guidelines for drones were proposed in 2012 by the Canadian Information and Privacy commissioner (Cavoukian, 2012). These guidelines include proactively designing for privacy preservation (rather than reacting after privacy violations have occurred), privacy as the default setting, and visible and transparent operation—see Table 16.3. These design principles have since been utilized to improve privacy in drones compared to traditional approaches (Cawthorne & Devos, 2020).

The visible and transparent operation of robotic systems can be challenging—how does the robot work? What capabilities does it have? and who is controlling it? These considerations are called "explicability," and they describe to what extent a system is transparent in its operation, and if its actions can be attributed to a person

Table 16.3 Seven privacy by design guidelines. Any robotic system that uses a camera to sense the world will need to consider privacy issues. Table from Cawthorne and Devos (2020) based on Cavoukian (2012)

Taking a proactive rather than reactive approach
Privacy as the default setting
Embedding privacy in the design
Aiming for full functionality while maintaining privacy
Ensuring full life-cycle protection of sensitive data
Visible and transparent (i.e., explicable) operations
Taking a stakeholder-inclusive approach

or organization that is responsible for it. Design for explicability principles has been proposed within artificial intelligence (AI) (Floridi et al., 2018) and drone design (Cawthorne & Frederiksen, 2020) as both can appear from the outside as "black boxes." A series of questions to consider in designing drones for explicability have been developed, including "how can the drone be designed to convey the organization and person responsible for it?" and "how can the purpose of the drone (e.g., health care) be easily identified from a distance?" Another example of a design guidelines intended to limit the mis-use or risks drones is the five capability caution principles which ask the designer to consider aspects such as the context of use, the impact on jobs and human skills, and long-term impacts on society and the environment (Cawthorne & Devos, 2020).

Design principles and guidelines can be useful for designers since they pose open-ended questions or offer suggestions which allows room for creativity and context-specific solutions. However, they can be more difficult to apply than a checklist since they are more abstract and require ethically informed critical thinking.

16.8.4   Ethical Frameworks

A final category of tools at our disposal is ethical frameworks. Ethical frameworks are often high level which makes them useful for assessing the overall direction technologies should take and in determining what risks and opportunities may be ahead in the development of a new technology. One ethical framework concerns AI for the good of society (Floridi et al., 2018). It utilizes the four bioethics principles as its foundation—beneficence (do good), non-maleficence (do not do harm), human autonomy, justice, and a new enabling principle for AI—explicability. This framework has subsequently been translated into a drone context, producing an ethical framework for the development of drones in public healthcare (Cawthorne & Robbins-van Wynsberghe, 2020). The framework has been used to develop a prototype fixed-wing drone for rapid delivery of blood samples (Cawthorne & Robbins-van Wynsberghe, 2019); this case study is examined in the next section.

16.9   Case Study: VSD of a Danish Healthcare Drone

Can ethics and value sensitive design help us to design robotic systems that enhance human flourishing? Can the technology be designed so we can avoid some of the risks that we read about at the beginning of the chapter, such as the risk of injuring small children? In this section, we will look at a case study of a Danish healthcare drone developed using VSD (Cawthorne & Robbins-van Wynsberghe, 2019) and an ethical framework (Cawthorne & Robbins-van Wynsberghe, 2020) as a practical example of how these approaches can be used to enhance the design of real robotic systems. The prototype drone is shown in Fig. 16.4.

Fig. 16.4 Prototype Danish healthcare drone; it is the first known example of a drone developed using VSD methods. Image by the author, with permission granted by the subject in the image

Value sensitive design is holistic and contextual, so first it is important to understand the place the drone will operate and the process it could replace. The small, affluent country of Denmark consists of two large islands and a peninsula at the northern tip of Germany, along with many smaller islands. Healthcare services at these small communities may be limited since they are remote, and the small population makes it hard to justify very expensive testing equipment such as those used to analyze blood samples for certain ailments. In addition, the Danish Ministry of Health has been undergoing a process of centralizing healthcare services and has closed several regional clinics while upgrading hospitals in the larger cities into "superhospitals"—instead of 41 hospitals with 24 h care, there will soon only be 20 (Danish_Municipalities, 2015).

For this case study, we focus on the small island of Ærø, located about 25 km south of the central Danish island of Fyn. Currently, an average of 32 blood samples per day are generated at the regional hospital at Ærø and are transported twice a day on weekdays and once a day on weekends by a courier (Sand, 2019). The courier loads the samples into an insulated box and drives them to the port. There, the ferry is used to cross the 25 km of ocean to Fyn. Then, the courier drives a few kilometers to the larger hospital at Svendborg where the samples are analyzed. The infrequent deliveries and dependence on the ferry schedule mean that it can take several hours to get test results (Health_Drone, 2021) meaning some patients could be quarantined unnecessarily or go without proper treatment for some time. Several stakeholders are

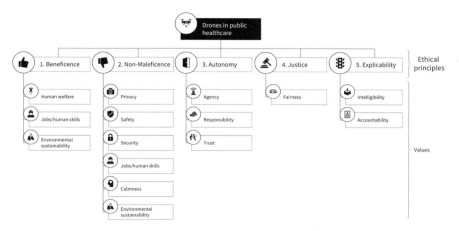

Fig. 16.5 Ethical framework for the design of drones in public health care based on the bioethics principles and AI ethics principles. The high-level ethical principles are made more specific in the second level of the framework which highlights relevant human values such as human welfare, jobs, safety, privacy, and fairness (Cawthorne & Robbins-van Wynsberghe, 2020)

relevant: citizens and sick people living on Ærø, healthcare workers, couriers, ferry operators, and Danish taxpayers among many others.

An ethical framework, shown in Fig. 16.5, was developed for drones used in public healthcare in collaboration with a robot ethicist (Cawthorne & Robbins-van Wynsberghe, 2020). The framework is designed to help the drone designer translate human values into design requirements (Van de Poel, 2013) and is based on bioethics principles since the drone could become part of the healthcare system. This values hierarchy includes four levels: at the top are ethical principles such as beneficence and non-maleficence, next are human values such as human welfare and privacy. The next lower level (not shown) is about contextual norms—specific considerations that pertain to the use-case in question. For example, we could compare the safety of the drone system to that of the current process of driving and taking the ferry. At the base of the hierarchy (not shown), we need to determine the design requirements that will support the ethical principles, human values, and norms—and enhance human flourishing. A detailed account of the development of the drone can be found in the references (Cawthorne, 2020; Cawthorne & Robbins-van Wynsberghe, 2019, 2020) along with its specifications and performance (i.e., fulfillment of the design requirements coming from the ethical framework.

The Danish health care drone is a fixed-wing aircraft which means it can be much smaller and lighter weight than a multirotor drone since flying on wings is more efficient than flying with powered rotors. The drone is so lightweight that it would not cause a fatality even if it were to hit a person on the ground—it is safe by design. The payload of the Danish drone is small, making it useful in urgent cases but not for routine transportation. The drone's cargo compartment includes a security system making it more difficult to carry unauthorized cargo. The drone is controlled manually by a pilot using a privacy-preserving camera system which cultivates drone piloting skills and makes responsibility more direct than with an automated system. And the drone is painted bright yellow with dark green checkers like a Danish ambulance, making it clearer what its purpose is and who is responsible for it. If we compare this drone to the one in the opening paragraph of the chapter, we see key differences. The contexts of use are not the same so we should not compare them directly, but the Danish drone exhibits a high level of safety, privacy, security, responsibility, and explicability which could provide health benefits—while protecting those on the ground.

As you can see from this example, developing robots in a holistic and value sensitive way is complex, and there are many impacted stakeholders—some who will benefit from the technology, and some who may be harmed. What is our responsibility as robot developers in this complex process? We will explore the topic in the next section on **Responsible research and innovation.**

16.10   Responsible Research and Innovation

In the previous sections, we saw how technology is not ethically neutral (**the non-neutrality of technology**), which means we need to consider ethics when we design robotic systems. This is the first step in responsible research and innovation—accepting that the things we design have ethical importance. Then, the question turns to how we actually design using ethics as a design input. Here, we could utilize normative ethical theories in Table 16.1, consider human values listed in Table 16.2, and utilize **value sensitive design, checklists, standards, design principles, and ethical frameworks.** These theories and tools can help us to combat moral de-skilling—the process where we become less adept at making ethically informed decisions (Vallor, 2015). Ideally, our moral progress should keep pace with our technological progress.

16.10.1   AIRR Framework

An often-cited framework for responsible research and innovation is called AIRR: anticipation, inclusion, reflexivity, and responsiveness (Stilgoe et al., 2013). "Anticipation prompts researchers and organizations to ask 'what if…?' questions, to consider contingency, what is known, what is likely, what is plausible, and what is possible" in the future (Stilgoe et al., 2013)—as we saw earlier in the PPPP model in Fig. 16.2. Inclusion means considering not just powerful stakeholders, but all those that will be impacted—directly or indirectly—by our robots (**the conceptual phase of VSD**). Reflexivity "means holding a mirror up to one's own activities, commitments, and assumptions, being aware of the limits of knowledge and being mindful that a particular framing of an issue may not be universally held"… "reflexivity means rethinking prevailing conceptions about the moral division of labor within science and innovation" (Stilgoe et al., 2013)—as exemplified in VSD interdisciplinary approach. Responsiveness "requires a capacity to change shape or direction in response to stakeholder and public values and changing circumstances" (Stilgoe et al., 2013). Responsiveness can be seen in the iterative nature of the VSD process (Fig. 16.3)—as circumstances change, we must adapt our robots to the new situation.

16.11 Chapter Summary

In summary, interdisciplinary collaboration, a holistic perspective, and the ethically informed design of robotic systems give us the best chance to perform responsible research and innovation—and ultimately enhance human flourishing.

16.12 Revision Questions

- What are three normative ethical theories, and what do they say?
- List some human values that are relevant in robot design.
- What is dual-use and what are some implications to your robot design?
- List the three phases of value sensitive design; what activities take place in each phase? Which research areas are most relevant in each phase?
- What are some benefits and limitations of ethical checklists? What about ethical frameworks?
- Identify industry standards related to ethics in technology design?
- Which design principles would be useful in the design of your robot?
- What does AIRR stand for? How could the four phases of the framework be applied to your robot?
- Consider an existing or proposed robot or drone system:
  - Who are the direct and indirect stakeholders?
  - What is the context of use?
  - What might be some social impacts of the system?
  - Should this robot be built? If so, why? If not, why not?
  - How will this robot enhance human flourishing?

References

Ackerman, E. (2019). *Swiss post suspends drone delivery service after second crash: An emergency parachute failure raises questions about the safety of urban delivery drones.* https://spectrum. ieee.org/swiss-post-suspends-drone-delivery-service-after-second-crash

Cavoukian, A. (2012). *Privacy and drones: Unmanned aerial vehicles.* Information and Privacy Commissioner of Ontario, Canada Ontario.

Cawthorne, D. (2020). *Value sensitive design of unmanned aerial systems.* (PhD Thesis). University of Southern Denmark.

Cawthorne, D., & Cenci, A. (2019). *Value sensitive design of a humanitarian cargo drone.* Paper presented at the 2019 International conference on unmanned aircraft systems (ICUAS).

Cawthorne, D., & Devos, A. (2020). *Capability caution in UAV design.* Paper presented at the 2020 International Conference on Unmanned Aircraft Systems (ICUAS).

Cawthorne, D., & Frederiksen, M. H. (2020). *Using the public perception of drones to design for explicability.* Paper presented at the International Conference on Robot Ethics and Standards (ICRES), Taipei, Taiwan.

Cawthorne, D., & Robbins-van Wynsberghe, A. (2019). *From HealthDrone to FrugalDrone: Value-sensitive design of a blood sample transportation drone.* Paper presented at the 2019 IEEE International Symposium on Technology and Society (ISTAS).

Cawthorne, D., & Robbins-van Wynsberghe, A. (2020). An ethical framework for the design, development, implementation, and assessment of drones used in public healthcare. *Science and Engineering Ethics, 26*(5), 2867–2891.

Choi-Fitzpatrick, A., Chavarria, D., Cychosz, E., Dingens, J. P., Duffey, M., Koebel, K., Siriphanh, S., Yurika Tulen, M., Watanabe, H., Holland, J., & Juskauskas, T. (2016). *Up in the air: A global estimate of non-violent drone use 2009–2015.*

Consequentialism. (2003). *Stanford encyclopedia of philosophy.* https://plato.stanford.edu/entries/consequentialism/

Danish_Municipalities. (2015). *Here is where your superhospital is located.* http://www.danskekommuner.dk/Global/Artikelbilleder/2015/DK-3/DK-3-side-26-27.pdf

Deontological Ethics. (2020). *Stanford encyclopedia of philosophy.* https://plato.stanford.edu/entries/ethics-deontological/

Dunne, A., & Raby, F. (2013). *Speculative everything: Design, fiction, and social dreaming.* MIT press.

Ethics. (2021). *Wikipedia.* https://en.wikipedia.org/wiki/Ethics

European_Union. (2019). *How to complete your ethics self-assessment.* https://ec.europa.eu/research/participants/data/ref/h2020/grants_manual/hi/ethics/h2020_hi_ethics-self-assess_en.pdf

Floridi, L., Cowls, J., Beltrametti, M., Chatila, R., Chazerand, P., Dignum, V., Luetge, C., Madelin, R., Pagallo, U., Rossi, F., & Schafer, B. (2018). AI4People—an ethical framework for a good AI society: Opportunities, risks, principles, and recommendations. *Minds and Machines, 28*(4), 689–707.

Friedman, B., Kahn, P. H., Borning, A., & Huldtgren, A. (2013). Value sensitive design and information systems. In *Early engagement and new technologies: Opening up the laboratory* (pp. 55–95). Springer.

Gillespie, T. (2010). The politics of 'platforms.' *New Media Society, 12*(3), 347–364.

Health_Drone. (2021). Health Drone: The project. https://sundhedsdroner.dk/index.php?page=the-project

IEEE. (2021). *7000-2021 standard addressing ethical concerns during systems design.*

Lin, P., Bekey, G., & Abney, K. (2008). *Autonomous military robotics: Risk, ethics, and design.* https://apps.dtic.mil/sti/pdfs/ADA534697.pdf

Matternet. (2021). *Matternet website.* https://mttr.net/

Novitzky, P., Kokkeler, B., & Verbeek, P.-P. (2018). The dual-use of drones. *Tijdschrift Voor Veiligheid, 17*(1–2), 79–95.

Sand, P. S. (2019). Director after flight in medical drone: I felt completely safe. *Fyns.* https://www.fyens.dk/erhverv/Direktoer-efterflyvetur-i-laege-drone-Jeg-foelte-mig-helt-tryg/artikel/3318057

Stilgoe, J., Owen, R., & MacNaghten, P. (2013). Developing a framework for responsible innovation. *Research Policy, 42*(9), 1568–1580.

SUST. (2019). *Interim report of the Swiss Safety investigation agency SUST about the accident involving the M2 V9 drone.* https://www.sust.admin.ch/inhalte/AV-berichte/ZB_SUI-9903.pdf

Vallor, S. (2015). Moral deskilling and upskilling in a new machine age: Reflections on the ambiguous future of character. *Philosophy and Technology,28*(1), 107–124.

Vallor, S. (2016). *Technology and the virtues: A philosophical guide to a future worth wanting.* Oxford University Press.

Van de Poel, I. (2009). Values in engineering design. In *Philosophy of technology and engineering sciences* (pp. 973–1006). Elsevier.

Van de Poel, I. (2013). Translating values into design requirements. In *Philosophy and engineering: Reflections on practice, principles and process* (pp. 253–266). Springer.

Van Wynsberghe, A., & Nagenborg, M. (2016). Civilizing drones by design. In *Drones and responsibility* (pp. 148–165). Routledge.

Verbeek, P.-P. (2008). Morality in design: Design ethics and the morality of technological artifacts. In *Philosophy and design* (pp. 91–103). Springer.

Vermaas, P. E., Kroes, P., Light, A., & Moore, S. (2007). *Philosophy and design: From engineering to architecture.* Springer.

Virtue Ethics. (2016). *Stanford encyclopedia of philosophy.* https://plato.stanford.edu/entries/ethics-virtue/

World_Health_Organization. (2020). *The top 10 causes of death.* https://www.who.int/news-room/fact-sheets/detail/the-top-10-causes-of-death

Dylan Cawthorne is an Associate Professor at the Unmanned Aerial Systems Center at the University of Southern Denmark. His aim is to support human flourishing through the development of ethically informed technologies. His main area of research is using value sensitive design methods and ethical principles to develop and build prototype drones for humanitarian, public health care, and search and rescue operations. He works across disciplines, utilizing art, craft, and creativity to enhance engineering practice. A common theme in his work is the use of holistic and contextually situated technological development. He considers himself an activist engineer, and in his free time, he teaches and performs repairs with the local nonprofit organization Repair Café Odense, and volunteers with Engineers Without Borders Denmark to develop low-cost cloth masks and mapping drones for use in Africa.

Part IV
Projects

Chapter 17
Robot Hexapod Build Labs

David Hinwood and Damith Herath

17.1 Introduction

Robotics is a practical field of study. As we discussed earlier in the book, it is essential to actively construct your own knowledge of the subject through experience. The series of projects presented in this chapter builds on the theoretical foundation developed throughout the book and flags the need to explore further when there are gaps in your current knowledge based on the material covered. These projects expand on the embodied design and prototyping concepts covered in Chap. 12 about the hexapod robot. We provide you with a hands-on guide to build a robot from scratch using "first principles". Once completed, you will gain experience in implementing mathematical concepts in a practical application and communicating with hardware. Please note that this chapter references online resources associated with this book regularly (https://foundations-of-robotics.org). You will;

- Learn about programming techniques for deriving kinematic equations from the leg component of the robot hexapod. These equations include the direct kinematic (DK) homogenous transformation and inverse kinematic (IK) equations of the end-effector (EE) frame.
- Create and manipulate the geometric Jacobian with programming techniques to analyse the behaviour of the robotic leg.
- Understand and implement serial-based communication, a standard method by which we can communicate with robotic systems.

D. Hinwood (✉) · D. Herath
University of Canberra, Bruce ACT 2617, Australia
e-mail: David.Hinwood@canberra.edu.au

D. Herath
e-mail: Damith.Herath@Canberra.edu.au

We will discuss the development of a single leg from the hexapod and perform a kinematic analysis across the first three projects within this chapter. We emphasise intuitive descriptions of mathematical concepts within robotics with Python code examples. The final project will explore serial communication using C++ code, discussing bit-wise operations to efficiently send numerical data to a microcontroller. A provided docker configuration from the associated online resources ensures examples introduced in this chapter work out of the box. Please use Python 3.6 or newer to operate the code segments presented in this chapter.

17.2   Project One: Defining the Robot System

17.2.1   Project Objectives

- Define a mechanical system from a conceptual design and plan modelling techniques.
- Research and identify suitable actuators for the task.

17.2.2   Project Description

Let us begin by summarising the requirements of our hexapod system. As described in Chap. 12, the hexapod has six legs, each leg containing four actuators. The hexapod moved with the tripod gait, as discussed in Chap. 8. First, we actuate the base of the leg in a lateral motion. Next, three rotational link pairs follow this first actuator, making up the rest of the leg component, as shown in Fig. 17.1.

The actuators used in the original case study described in Chap. 12 were the rotational Dynamixel MX28r (Dynamixel, 2021) actuators. Each MX28r contains a

Fig. 17.1 A rendering of a single leg from the proposed hexapod platform. Bryce Cronin/CC BY-NC-ND 4.0. www.cronin.cloud/hexapod

microcontroller running a lower-level PID loop with an encoder. These lower-level controllers enabled the original developers of the hexapod case study to command the actuator's desired angle and velocity and be confident that the behaviour would execute correctly.

17.2.3 Project Tasks

17.2.3.1 Task One: Basic Questions

- Take the robot leg shown in Fig. 17.1 and indicate the type of mechanical device of the leg, i.e. a parallel structure or a serial-link robot.
- Referring to the definitions in Chap. 10, list suitable techniques for modelling the following relationships.

  - The position relationships, how would the position of the tip/foot position of the robot leg relate to the actuator positions?
  - The velocity relationships, how would the speed of the tip/foot position of the robot leg relate to the actuator velocities?
  - The dynamics relationship, how would the desired motion, forces and torques of the foot/tip of our robot relate to the torques exerted by our actuators?

17.2.3.2 Task Two: Research Components

- The original design of the hexapod called for the MX28r actuator, as described earlier. What design considerations do you think were made in selecting these actuators? You may make assumptions in answering this point, think about;

  - The weight of the hexapod and its distribution over the legs
  - Components such as the battery and electronics
  - Actuator weights and torque capacity

- You can find a list of suitable actuators for this project listed on the associated website of this book. Go through the list of actuators present on this page and analyse each actuator's datasheet while taking note of the various benefits. Then, identify the optimal solution for the hexapod model. Consider parameters such as tracking the position/velocity, handling collisions while walking and actuator strength.

Table 17.1 DH parameters of the robot leg

i	1	2	3	4
a_i	0.071	0.104	0.048	0.041
d_i	0	0	0	0
α_i	$\frac{\pi}{2}$	0	0	$-\frac{\pi}{2}$
$qlim$	$-\frac{2\pi}{9}, \frac{2\pi}{9}$	$-\frac{\pi}{2}, \frac{\pi}{2}$	$-\frac{\pi}{2}, \frac{\pi}{2}$	$-\frac{\pi}{2}, \frac{\pi}{2}$

17.3 Project Two: Modelling the Position Kinematics

17.3.1 Project Objectives

- Implement the DH parameters of the hexapod leg using Python.
- Programmatically calculate the direct kinematics homogenous transformation matrix.
- Create an inverse kinematic solution of the robotic leg.
- Learn how we can validate positional relationships.

17.3.2 Project Description

As our robot leg follows a standard joint-link pair structure, we can define the system with the DH parameters. Table 17.1 shows the parameters assuming the world coordinate frame equals the leg base, i.e. the connection point to the hexapod body. These parameters also include the additional *qlim* parameter (which holds the position limits of each actuator). Finally, we visualise the parameters with a kinematic diagram illustrated in Fig. 17.2.

17.3.3 Project Tasks

17.3.3.1 Task One: Codifying the DH Parameters

Note: Coding segments referenced in these questions are in Sect. 3.4

- Take our presented DH parameters from Table 17.1 and Fig. 17.2. Using the numerical values (Table 17.1), create a simulated version of the robot leg using the robotics toolbox in Python.
- When creating the robot with the toolbox, make sure you include the joint limit (*qlim*) variables in Table 17.1. See Coding Segment 2 for a starting point on this task.

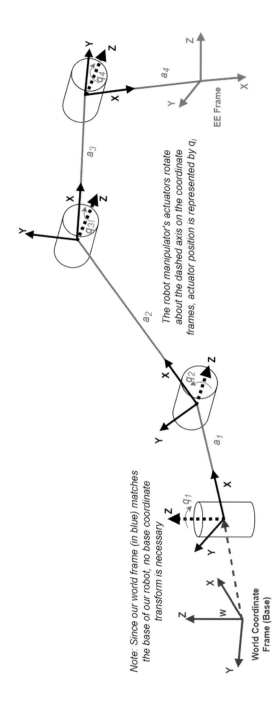

Fig. 17.2 A visualisation of the presented DH parameters from Table 17.1. We highlight the EE position in a red coordinate frame, with the world coordinate frame displayed in blue. Note that the shown configuration is with the four joints at arbitrary angles

17.3.3.2   Task Two: Kinematics

- While using the spatial math library and numerical values from Table 17.1, calculate the DK homogenous transformation of the robotic leg. See Coding Segments 1 and 3 for help with this problem.
- Confirm the DK homogenous transform calculation by comparing the output to the "fkine" function from the robot created in Task One. Hint, Look at Coding Segments 2 and 3 for help with this problem. This task requires you to investigate the robotics toolbox function "fkine".

17.3.3.3   Task Three: Advanced Kinematics

- Modify Coding Segment 3 to find the leg EE's x-, y- and z-positions using symbolic variables. Remember the structure of the homogenous transformation matrix for this step.
- Validate your equations by manually deriving the position equations with diagrams. For a guide on this step, see Sect. 3.4. Do the equations derived match the symbolic expression previously calculated?

Note: The following two items are an extension exercise for the readers without explicit instructions in this chapter. However, answers are available with the associated online resources.

- Derive an inverse kinematics algorithm for our robotic leg using Python. We leave this question as an open exercise to readers who can use their knowledge of kinematics and previous information presented throughout the book.
- Validate your inverse kinematics algorithm. Validation can occur by running through the workspace of the leg and comparing outputs from your direct and inverse kinematic solutions.

17.3.4   Case Study Example

17.3.4.1   Representing the DH Parameters with Code

We wanted to go back briefly to the mathematical representation of the DH parameters presented in Chap. 10. In this chapter, we may use some slightly different notations. The first significant change concerns the values of θ. Henceforth, we represent the values of θ with q, which indicates the angle of an actuator. The method by which we represent homogenous transformations is also slightly different.

A homogenous transformation, A_{i-1}^{i}, would represent the ith link-joint pair of the DH parameters. We utilise the spatial math Python library to hold various transformations and rotations of the DH parameters. In code, this may look slightly different to the previously presented notation. Assuming that we're utilising the standard

DH convention, we can represent the transformation of A_{i-1}^i with the expression $A_{i-1}^i = R_z(q_i)T_z(d_i)T_x(a_i)R_x(\alpha_i)$. In this case, R and T represent functions that produce a homogenous transformation matrix from a single rotation or translation. The subscript denotes the axis the transformation takes place along. In the case of our robot leg, remember that for the ith joint-link pair, the variable q_i represents the position of our actuator while the other parameters are constant. In Python code, the calculation of A_{i-1}^i is illustrated in Code Segment 1.

Code Segment 1. An example of a single joint-link pair represented with four sequential transformations

```python
#import requirements - Note we are using python 3.6
import spatialmath as sm
#A joint link pair of the DH parameters, assuming the variables
of the variables of q_i, d_i, a_i and alpha_i already exist
A_i = sm.SE3.Rz(q_i) * sm.SE3.Tz(d_i) *
sm.SE3.Tx(a_i) * sm.SE3.Rx(alpha_i)
```

Additionally, in this chapter, we utilise the robotics toolbox, a Python library presented by Corke and Haviland (Corke & Haviland, 2021) that can model serial-link structures like the proposed leg. We show an example of the robotics toolbox below, creating a simple two-link manipulator.

Code Segment 2. An example of the robotics toolbox creating a simple two-link simulated robot.

```python
#import requirements - Note we are using python 3.6
import roboticstoolbox as rtb
import math as rwm
import spatialmath as sm
import numpy as np
#We delcare link lengths for the a variables in this robot
a0 = 0.5
a1 = 0.5
#We also include a base transform variable
base_transform=sm.SE3(np.identity(4))
#base_transform= sm.SE3.Rx(rwm.pi/2)
#base_transform=sm.SE3.Ry(rwm.pi/2)*sm.SE3.Tz(0.5)
#Create the robot with the toolbox, note how theta is not
present as the joint position is not a constant
linkjoint_0 = rtb.RevoluteDH(d=0, alpha=0, a=a0, offset=0,
qlim=None)
linkjoint_1 = rtb.RevoluteDH(d=0, alpha=0, a=a1, offset=0,
qlim=None)
example_robot = rtb.DHRobot([linkjoint_0,
linkjoint_1], base = base_transform, name =
```

```
'Simple_Example')
example_robot.teach()
```

Note how in Code Segment 2, there are several additional parameters, including the *base* parameter in the function *DHRobot*. Also, in each joint-link pair, defined by *RevoluteDH*, there are two parameters of *offset* and *qlim*. These parameters operate as follows:

- Offset ($offset_i =$)—Adds a constant offset to the position of our *i*th actuator. Take Code Segment 2 and modify it. Change the *offset* parameter for **linkjoint_1**, currently *0*, and change it to $\pi/2$ (in code `rwm.pi/2`). Observe how this change impacts the simulation generated.
- Joint Limits ($qlim_i =$)—Set our *i*th actuator's upper and lower position limits. Change the *qlim* variable in Code Segment 2 for either **linkjoint_0** or **linkjoint_1**. Currently, this variable is *None*. Try changing this variable to `[-0.3, 0.3]`. Once again, observe how changing this variable impacts the generated simulation.
- Base Position (*base =*)—The pose of your manipulator's initial position relative to the world coordinate frame. It is essential that as you move forward in this chapter, you define your world coordinate and where your robot is relatively located when developing your system. In Code Segment 2, several alternative definitions for the variable *base_transform* are in commented lines of code. Uncomment different variations of *base_transform* and observe how that impacts the simulation.

While not part of the traditional mathematical definition of the DH convention, these parameters can help define a robot to a desired zero configuration or implement multiple serial-link systems in a single world.

17.3.4.2 Deriving the Forward Kinematics

The next step in modelling a robot is defining the forward kinematics, estimating the EE's position and rotation relative to the robot's base. We highlight this problem in Fig. 17.2. We want to establish the location and orientation of the red coordinate frame (EE), previously referred to as the tip/foot, relative to the blue frame (base) based on actuator positions. We split the problem into the rotation and position components.

Let us first discuss the orientation problem. Broadly, we want to estimate the rotation matrix using the DH parameters and the actuator positions. We can do this using Python with the spatial math library. Thus, we present Coding Segment 3, which retrieves the rotation matrix of an end-effector from the base of our robot leg.

Code Segment 3. The Python code for calculating the rotation matrix of our robot leg.

```
#import requirements
import spatialmath as sm
```

```
import spatialmath.base as base
import numpy as np
#Creates a set of symbolic variables
a_1, a_2, a_3, a_4 = base.sym.symbol('a_1, a_2, a_3, a_4')
d_1, d_2, d_3, d_4 = base.sym.symbol('d_1, d_2, d_3, d_4')
alpha_1, alpha_2, alpha_3, alpha_4 = base.sym.symbol('alpha_1,
alpha_2, alpha_3, alpha_4')
q_1, q_2, q_3, q_4 = base.sym.symbol('q_1, q_2, q_3, q_4')
#Base transform, equivalent to an identity matrix since no
base transform exists on this leg
base_transform=sm.SE3(np.identity(4))
#A joint link pair of the DH parameters, remembering our
leg contains no base transform
leg_linkjoint_1   =   sm.SE3.Rz(q_1)   *   sm.SE3.Tz(d_1)   *
sm.SE3.Tx(a_1) * sm.SE3.Rx(alpha_1)
leg_linkjoint_2   =   sm.SE3.Rz(q_2)   *   sm.SE3.Tz(d_2)   *
sm.SE3.Tx(a_2) * sm.SE3.Rx(alpha_2)
leg_linkjoint_3   =   sm.SE3.Rz(q_3)   *   sm.SE3.Tz(d_3)   *
sm.SE3.Tx(a_3) * sm.SE3.Rx(alpha_3)
leg_linkjoint_4   =   sm.SE3.Rz(q_4)   *   sm.SE3.Tz(d_4)   *
sm.SE3.Tx(a_4) * sm.SE3.Rx(alpha_4)
#A joint link pair of the DH parameters
DK_transform    =    base_transform    *    leg_linkjoint_1    *
leg_linkjoint_2 * leg_linkjoint_3 * leg_linkjoint_4
#Extract and print the rotation matrix component
RotationMatrix = DK_transform.R
print(RotationMatrix)
```

Code Segment 3 presents a valuable tool for deriving the rotation matrix using the spatial math library. But we also need to know the x, y and z displacements from the base frame to the EE frame. It is possible to gather these variables using a similar technique to Code Segment 3. However, let us extrapolate these equations manually. There are several reasons for extracting these equations manually, but the primary motivation is validation. While the symbolic calculations and the robotic toolbox are helpful, using them in conjecture with our own manually extrapolated equations reinforces that we understand our system.

So we begin with two images of our robotic system as shown in Figs. 3 and 4. These are the two images we use to calculate our forward kinematic transformation. Figure 17.3 shows a top-down view perpendicular to the Z-axis, which observes the X- and Y-axes and the manipulator's EE position within those axes. Alternatively, in Fig. 17.4, we illustrate a side view of a robot leg highlighting the position of the final three actuators. This figure also highlights the Z-position of the EE. In both images, the EE position is the red node at the end of the leg. These images also contain a variable, mx, that highlights the extended length of the leg. Please note that both figures' axis are from the base/world coordinate frame of Fig. 17.2.

We start by estimating the values of mx and z from Fig. 17.4. One method of thinking about these values is that $mx = a_1 + w_2 + w_3 + w_4$ and $z = h_2 + h_3$

Fig. 17.3 Position of our
robot leg from a top-down
view highlights the *x*- and
y-positions of our EE

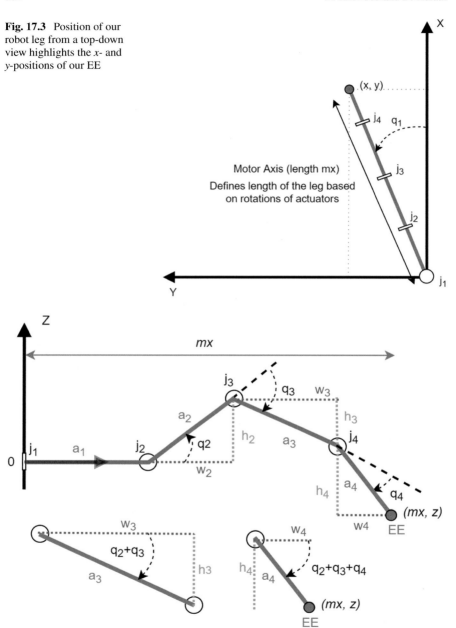

Fig. 17.4 By observing our manipulator from a side view, we display the positions of the final
three actuators along with the length of mx and the *z* displacement

$+ h_4$. The equations below can explicitly express these variables from standard trigonometric functions.

$$mx = a_1 + a_2 \cos(q_2) + a_3 \cos(q_2 + q_3) + a_4 \cos(q_2 + q_3 + q_4)$$

$$z = a_2 sin(q_2) + a_3 sin(q_2 + q_3) + a_4 sin(q_2 + q_3 + q_4)$$

Having established these variables, we can now use the length of mx to estimate the x- and y-positions in Fig. 17.3.

$$x = mx \cos(q_1)$$

$$y = mx \sin(q_1)$$

17.4 Project Three: Modelling the Velocity Kinematics with Python

17.4.1 Project Objectives

- Using programming techniques, derive the geometric Jacobian of the serial-link leg relating the actuators and the EE coordinate frame's velocity.
- Using programming techniques, manipulate the geometric Jacobian and find properties such as the determinant, inverse and transpose.
- Learn how the Jacobian operates by relating joint speed and EE velocity parameters.

17.4.2 Project Description

Modelling the velocity relationship between the leg EE coordinate frame and the actuators is crucial for building a robotic system, as described in Chap. 10. To briefly reintroduce the Jacobian matrix for serial-link robots, it is generally a $6 \times N$ sized matrix where N equals the number of actuators within our system. We split our Jacobian into the position and orientation components to calculate this matrix. First, let's discuss the position component, henceforth called $J_P(q)$. Next, we take our direct kinematics and partially derive each equation by every actuator variable in the structure below. As can be observed, the equation is simply a matrix of partial derivatives.

$$J_p(q) = \begin{bmatrix} \frac{\partial x}{\partial q_1} & \frac{\partial x}{\partial q_2} & \frac{\partial x}{\partial q_3} & \frac{\partial x}{\partial q_4} \\ \frac{\partial y}{\partial q_1} & \frac{\partial y}{\partial q_2} & \frac{\partial y}{\partial q_3} & \frac{\partial y}{\partial q_4} \\ \frac{\partial z}{\partial q_1} & \frac{\partial z}{\partial q_2} & \frac{\partial z}{\partial q_3} & \frac{\partial z}{\partial q_4} \end{bmatrix}$$

Once we know the entire matrix of $J_P(q)$, we can utilise it in the equation below, where multiplying the Jacobian by the $\dot{q}$ vector (speed of the actuators) produces the velocities of the EE in the x-, y- and z-directions.

$$\begin{bmatrix} \dot{x} \\ \dot{y} \\ \dot{z} \end{bmatrix} = J_P(q) \begin{bmatrix} \dot{q}_1 \\ \dot{q}_2 \\ \dot{q}_3 \\ \dot{q}_4 \end{bmatrix}$$

The orientation Jacobian, $J_R(q)$, requires a slightly different calculation to the previously presented translation components. A rule of thumb to consider is that, $J_R(q)$ and $J_P(q)$ should be the same size. Since we have a rotation matrix and no explicit equations to differentiate, we need to calculate the partial derivative of a rotation matrix. The broad rule to consider when building your orientation Jacobian is that the ith column, J_{Ri}, is equal to the rotation matrix of the previous joint-link pair multiplied by the vector $[0, 0, 1]$. If we consider R_{i-1} as the rotation matrix before $R_z(q_i)$, the expression below represents ith column, J_{Ri}. Please note that this methodology is specific to the constraints on the hexapod leg as it uses the standard DH convention and only contains rotational actuators.

$$J_{Ri} = R_{i-1} \begin{bmatrix} 0 \\ 0 \\ 1 \end{bmatrix}$$

When we put together this matrix, we express our orientation Jacobian in the form below (where R_0 is the rotation matrix of our base transform);

$$J_R(q) = \begin{bmatrix} R_0 \begin{bmatrix} 0 \\ 0 \\ 1 \end{bmatrix} & R_1 \begin{bmatrix} 0 \\ 0 \\ 1 \end{bmatrix} & R_2 \begin{bmatrix} 0 \\ 0 \\ 1 \end{bmatrix} & R_3 \begin{bmatrix} 0 \\ 0 \\ 1 \end{bmatrix} \end{bmatrix}$$

Once calculated, $J_R(q)$ can perform the matrix multiplication operation displayed below, in which ω_x, ω_y and ω_z are angular velocity components about the three-axis.

$$\begin{bmatrix} \omega_x \\ \omega_y \\ \omega_z \end{bmatrix} = J_R(q) \begin{bmatrix} \dot{q}_0 \\ \dot{q}_1 \\ \dot{q}_2 \\ \dot{q}_3 \end{bmatrix}$$

Once the calculations for $J_P(q)$ and $J_R(q)$ are complete, we can create what is known as the Geometric Jacobian $J(q)$. The calculation for the geometric Jacobian is the vertical concatenation of $J_P(q)$ and $J_R(q)$. It can calculate the velocities of EE DOF as shown in the expression below.

$$\begin{bmatrix} \dot{x} \\ \dot{y} \\ \dot{z} \\ \omega_x \\ \omega_y \\ \omega_z \end{bmatrix} = \begin{bmatrix} J_P(q) \\ J_R(q) \end{bmatrix} \begin{bmatrix} \dot{q}_0 \\ \dot{q}_1 \\ \dot{q}_2 \\ \dot{q}_3 \end{bmatrix} = J(q)\dot{q}$$

17.4.3 Project Tasks

17.4.3.1 Task One: Calculating and Using the Geometric Jacobian Components

- Using the DH parameters and direct kinematic equations, calculate the matrix $J_P(q)$. See Coding Segment 4 in Sect. 4.4 for assistance.
- Assuming our joint speed vector $\dot{q} = [0.1, 0.5, 0.2, -0.3]$, what is the speed of the EE in the x-, y- and z-directions?
- Using the DH parameters, calculate the matrix $J_R(q)$. See Coding Segment 5 in Sect. 4.4 for assistance.
- Assuming our joint speed vector $\dot{q} = [0.1, 0.5, 0.2, -0.3]$, what is the angular velocity of the EE for the values of ω_x, ω_y and ω_z?

17.4.3.2 Task Two: Completing and Manipulating the Geometric Jacobian

Note: The case study presented in Sect. 4.4 only deals with the Jacobian calculation. Please peruse the online resources for a more in-depth guide for the steps below.

- Put together the complete geometric Jacobian $J(q)$.
- Make $J(q)$ a square matrix by eliminating rows. Present a summary of how this operation impacts your Jacobian matrix and why we need to perform this step.
- Establish the following features of our now square Jacobian matrix;

  - Using Python code, calculate the determinant. As previously mentioned in Chap. 10, the determinant can find singularities. Using the determinant, can you find any possible kinematic singularities?
  - Using Python code, calculate the inverse of the square matrix. What can this matrix do?

- Using Python code, calculate the transpose of the square matrix. What can this matrix do?

17.4.4 Case Study Example

We again use Python and the sympy library to find the Jacobian of our system. We require the forward kinematics we have previously calculated in Project Two to proceed with the case study. The Python sympy library calculates $J_P(q)$ with in-built differentiation. In Code Segment 4, we provide an example by calculating the first element of $J_P(q)$, the term of $\frac{\partial x}{\partial q_1}$.

Code Segment 4. An example of calculating a single element of position Jacobian $J_P(q)$

```python
#import requirements - Note we are using python 3.6
from sympy import *
import spatialmath.base as base
#Create symbolic variables to use
a1, a2, a3, a4, q1, q2, q3, q4 = base.sym.symbol('a0,a1,a2,a3,
q0,q1,q2,q3')
#Rememeber our DK expression for x? If not check the
forward kinematics derivation of the robot leg
x = (a1 + a2 * cos(q2) + a3 * cos(q2 + q3) + a4 * cos(q2 + q3 +
q4)) * cos(q1)
dx_dq1 = diff(x, q1)
print(dx_dq1)
```

Coding Segment 5 shows how we would calculate the first two columns of $J_R(q)$. Remember from our previous description of $J_R(q)$ how each column corresponds to a rotation matrix before an actuator motion.

Code Segment 5. Python code which calculates the first two columns of orientation Jacobian $J_R(q)$

```python
#import requirements - Note we are using python 3.6
from sympy import *
import spatialmath.base as base
import spatialmath as sm
import numpy as np
a1,    a2,    a3,    a4,    q1,    q2,    q3,    q4,    Pi    =
base.sym.symbol('a0,a1,a2,a3,q0,q1,q2,q3,Pi')
```

```
#The base transformation of the DH parameters, see how
this is transformation before Rz(q_0)
baseTransformation = sm.SE3(np.identity(4))
#The first joint link pair, see how this is transformation
before Rz(q_1) of the DH parameters)
linkjoint_1 = sm.SE3.Rz(q1) * sm.SE3.Tz(0) * sm.SE3.Tx(a1) *
sm.SE3.Rx(Pi / 2)
#The second row of our DH parameters i.e. A1
linkjoint_2 = linkjoint_1 * sm.SE3.Rz(q2) * sm.SE3.Tz(0) *
sm.SE3.Tx(a2) * sm.SE3.Rx(0)
#Calculate the first column of the orientation jacobian
C1Jac = Matrix(baseTransformation.R) * Matrix([0, 0, 1])
#Calculate the second column of the orientation jacobian.
See how we combine the first link-joint pair followed by
the second
C2Jac = Matrix(baseTransformation.R) *
Matrix(linkjoint_1.R) * Matrix([0, 0, 1])
```

17.5   Project Four: Building Communication Protocols

17.5.1   Project Objectives

- Learn about bytes and different types of integer variables.
- Learn about basic serial (TTL) communication through implementing C++ code.
- Implement a ROS package to communicate with an Arduino microcontroller.

17.5.2   Project Description

This final project discusses communication protocols between a robot and a host PC. This section describes how the original case study contributors controlled the robot leg using ROS. The techniques presented will be helpful in many other robotics projects. When developing robots, we rarely send radian commands or decimal values directly to an actuator when commanding the system to move to a position. Instead, many actuators simply take a tick or step value input, usually an unsigned 8-bit or 16-bit integer. The MX28r actuators used originally received encoder values from 0 to 4095 to indicate the desired position of our actuator and other feedback or command values.

The method by which we implemented a solution was by designing a custom message protocol that utilised TTL, or serial communication. We write custom 8-bit integer arrays to the serial port. If you are unfamiliar with the terms unsigned or 8-bit integer, let's very briefly go over what these mean. These terms relate to the binary numeral system. Binary numbers are sequences of 1 s or 0 s that can represent integers. An 8-bit number is a binary number with eight 1 s and 0 s. A 16-bit number would have 16 values, and so on.

So how do sequences of 1 s and 0 s represent other numbers? First, we treat each value as 2^i with i determined by its place in the sequence by reading from right to left. We then sum up all the values where a 1 is in the sequence. For example, we would treat the number 1011 in binary as the sum of $2^3 + 2^1 + 2^0$. Notice how we omit 2^2 since the third number in our sequence is 0 (remembering we are reading from right to left). So, in binary, the number 1011 is equivalent to 11.

This example is also what we would refer to as unsigned. An unsigned integer simply means that we do not consider negative values and sum up all the observed values in sequence. Therefore, it stands to reason that the maximum value of an unsigned 8-bit integer (also known as a byte) is 255, i.e. 11,111,111. So this isn't particularly useful for a robot system, particularly our system where the target and feedback values range from 0–4095.

So, for each value in our 8-bit integer array, we're limited to values of 255 and cannot utilise negative numbers. Let's first address negative numbers. We can use signed integers which utilise a mathematical operation to represent negative numbers with binary sequences. We refer those curious to the "two's complement" method for in-depth information about this operation. Using "two's complement", a signed 8-bit binary sequence can now represent a value between −128 and 127. We can apply this technique to any binary sequence, including 16- and 32-bit numbers.

However, a constraint of serial communication is that we are sending bytes (unsigned 8-bit integers) across. Thus, we are still left with the problem of how we can communicate with our microcontroller with only unsigned 8-bit values. Especially when considering we want to send both unsigned integers and larger values.

Thus we utilise bit-shifting. For example, let's say we have a *signed 16-bit* integer of 1058. In binary, this would be the value of 00,000,100 00,100,010 using "two's complement". Such a value would be inconvenient to send in an 8-bit unsigned integer array. Essentially, we would write a function that would split our 16-bit integer in half and represent it as two unsigned 8-bit numbers. We highlight this process in the image below. As we can observe, we input a signed 16-bit integer and return two 8-bit unsigned values. We can then recombine these two numbers if required.

Let's now take a look at bit-shifting in code. Note that we now use C++ rather than Python. Code Segment 6 demonstrates the operation illustrated in Fig. 17.5, in

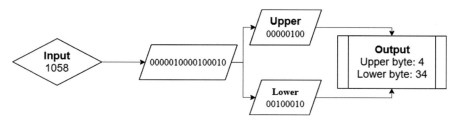

Fig. 17.5 Splitting a signed 16-bit integer into two unsigned 8-bit integers

which we split a single signed 16-bit number into two unsigned 8-bit numbers. It also performs the step of rejoining the two unsigned 8-bit integers.

Code Segment 6. A simple C++ example that shows bit-shifting operations in a coding context

```
#include <iostream>
#include <string>
using namespace std;
//Our functions for bit shifting and altering data
#define UPPER_BYTE(b) (b >> 8)
#define LOWER_BYTE(b) (b & 0xff)
#define INT_JOIN_BYTE(u, l) (u << 8) | l
int main()
{
int16_t exampleNumber = 1058;
 //Manipulates our 16-bit integer in a variety of methods
uint8_t upper = UPPER_BYTE(exampleNumber);//will be 4
uint8_t lower = LOWER_BYTE(exampleNumber);//will be 34
int16_t rejoined = INT_JOIN_BYTE(upper, lower);//will be
1058
 //print the newly calculated variables
cout << "input = " << exampleNumber << endl;
cout << "lower = " << int(lower) << endl;
cout << "upper = " << int(upper) << endl;
cout << "rejoined = " << rejoined << endl;
return 0;
}
```

17.5.3 Project Tasks

17.5.3.1 Task One: Basic Bit-Shifting

- We performed some basic bit-shifting in Code Segment 6. By researching C++ and bit-wise operations, discuss what the functions UPPER_BYTE, LOWER_BYTE and INT_JOIN_BYTE are doing to the inputs of these functions.
- Take a random number represented by a signed 16-bit integer and calculate the three outputs from the above functions (Hint: See the process illustrated in Fig. 17.5 for inspiration).

17.5.3.2   Task Two: A ROS Example of Serial Communication

- For serial communication, the online resources present a ROS1 package with an Arduino Uno microcontroller [Arduino, 2021]. Examine both of these repositories and implement the ROS package by following the instructions available in the README.md file of the ROS package.
- Draw a diagram of how the data passes through this package's serial communication and ROS network. Include screenshots demonstrating how you input the data and view any published information within the ROS network.
- The ROS and Arduino example are limited to sending 16-bit values within the serial communication module. Modify both the ROS package and Arduino code to receive signed 32-bit integers instead.
- Describe your methodology and what variables you had to modify. Include screenshots demonstrating that you can successfully send these larger integers to the Arduino and the ROS network.

17.6   Some Final Thoughts

This chapter discussed the practical implementation of hardware and software related to the Hexapod case study in Chap. 12. By presenting the underlying concepts and applied techniques, we hope that readers can take these skills and apply them to their robotic systems. Please be aware that there are many different software packages and libraries to develop robots. However, understanding the core concepts will maximise your impact and use of these various tools.

References

Arduino UNO & Genuino UNO. (2021). Retrieved December 19, 2021, from https://www.arduino.cc/en/pmwiki.php?n=Main/arduinoBoardUno

Corke, P., & Haviland, J. (2021). Not your grandmother's toolbox—the Robotics Toolbox reinvented for Python. In *IEEE International Conference on Robotics and Automation*.

Robotis e-Manual—MX-28AR, MX-28AT. (2021). Retrieved December 19, 2021, from https://emanual.robotis.com/docs/en/dxl/mx/mx-28/

David Hinwood is a Ph.D. candidate studying robot manipulators at the University of Canberra, Australia. Initially completing a Bachelor of Software Engineering, he then moved into robotics under the supervision of Dr Damith Herath. His research interests include the field of HRI and the design of human-like robotic systems. Previously, he has assisted HRI studies in embodied interaction, working on the Articulated Head Exhibit. He has also worked on creating human–robot interactions that play tic-tac-toe and sketch images with members of the public—resulting in several awards from the International Conference on Social Robotics. His current research focuses on developing and learning dexterous skills for the robust autonomous manipulation of fabric.

Damith Herath is an Associate Professor in Robotics and Art at the University of Canberra. He is a multi-award winning entrepreneur and a roboticist with extensive experience leading multidisciplinary research teams on complex robotic integration, industrial and research projects for over two decades. He founded Australia's first collaborative robotics startup in 2011 and was named one of the most innovative young tech companies in Australia in 2014. Teams he led in 2015 and 2016 consecutively became finalists and, in 2016, a top-ten category winner in the coveted Amazon Robotics Challenge—an industry-focused competition amongst the robotics research elite. In addition, he has chaired several international workshops on Robots and Art and is the lead editor of the book "Robots and Art: Exploring an Unlikely Symbiosis"—the first significant work to feature leading roboticists and artists together in the field of Robotic Art.

Chapter 18
Deployment of Advanced Robotic Solutions: The ROS Mobile Manipulator Laboratories

David St-Onge, Corentin Boucher, and Bruno Belzile

18.1 Introduction

Throughout the book, we introduced the various disciplines and topics involved in the development of robotic systems. Most of them are individually covered in dedicated books showing the extent of the knowledge required in robotics. Fortunately, we learned in Chap. 5 that open-source community-based software ecosystems, such as the Robotic Operating System (ROS), can support several of the integrations and ease the deployment process. ROS provides users with access to the latest research algorithms and software deployment tools for code maintenance, visualization, simulation, and more. Thanks to ROS, we expect all readers of this book to be able to complete the challenging advanced mobile manipulator tasks of this chapter.

For that purpose, we designed a robotic platform and we built a custom dedicated ROS workspace to support our academic teaching laboratories in robotics, from Gazebo simulations to the physical deployment (see Fig. 18.1). While undergoing a single-semester introduction course in robotics, the students are not expected to install and deploy the ROS workspace on their personal computer, but rather use a server infrastructure available at the university. Nevertheless, we provide the complete ROS workspace with detailed installation instruction,[1] including a Docker container to ease the deployment. The laboratories were designed at École de technologie supérieure, in Montréal, Canada, where we host the physical infrastructure to which these tools are tailored. We have eight robotic platforms for the students

---

[1] https://github.com/Foundations-of-Robotics/mobile_manip_ws

D. St-Onge (✉) · C. Boucher · B. Belzile
Department of Mechanical Engineering, École de Technologie Supérieure, Montréal, Canada
e-mail: david.st-onge@etsmtl.ca

B. Belzile
e-mail: bruno.belzile.1@ens.etsmtl.ca

© The Author(s) 2022
D. Herath and D. St-Onge (eds.), *Foundations of Robotics*,
https://doi.org/10.1007/978-981-19-1983-1_18

Fig. 18.1 A group of students testing the robots outside of the laboratory

to test their algorithms and several preconfigured desktop stations. The stations are accessible remotely to run the full simulation stack including remote visual rendering of the simulation on a web browser. We tested up to three teams running their Gazebo simulations in parallel on a single station without degrading performances.[2] Each station is accessible to all teams and so team's accounts are set on each station. Since most of the students' work is done in Jupyter notebooks, we deploy on each station (the littlest) JupyterHub.[3] Details on users' management and stations configuration, including Gazeboweb (gzweb) and JupyterHub are also available online.[4]

18.1.1 Dingo and Gen3 Lite

The motivation behind this set of assessments is to cover several topics relevant to both mobile robots and manipulator robots with a single integrated robotic platform. Several engineering courses are available with hands-on laboratory on industrial manipulators and custom-made mobile robot prototypes (a great project in this last category is covered in Chap. 17 of this book). What we found was lacking is teaching material leveraging a commercial robotic platform fit (safe) for proximity with the users. Our suggested platform, to which we tailored all this chapter's content, brings

---

[2] Tested on Dell ThinkStation P340 Tiny i7, 16 Gb RAM with NVidia Quadro1000 4 Gb.

[3] https://tljh.jupyter.org/en/latest/

[4] https://github.com/Foundations-of-Robotics/mobile_manip_ws/tree/master/doc.

Fig. 18.2 CAD model of the robotic platform with its custom turret

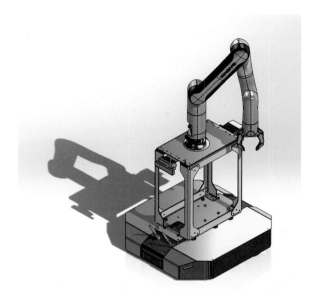

together a Clearpath Dingo[5] differential drive mobile base with a Kinova Gen3 lite[6] six-degree-of-freedom manipulator.

This recent version (2020) of the now-famous Kinova arms has been designed specifically for teaching. The same goes for Clearpath's mobile platform, the Dingo, which was released just a couple of months after the Gen3 lite. Clearpath (Toronto) and Kinova (Montreal) are considered good potential employers for students passioned by robotics and their hardware can be found in thousands of companies and universities around the world. Kinova's arms are deployed in disability assistance centers and hospitals, not to mention human–robot collaborative research laboratories, and the company is currently working to increase its presence in the industrial sector. Clearpath provides products for NASA and several emergency response agencies around the world and in recent years has become a major player in warehouse automation. Among everything, both companies are contributing actively to the ROS community with nodes for their robots and several additional ROS tools to ease robotics deployment (Fig. 18.2).

We put together the arm and the base using a custom-made turret that ensures the arm can easily reach door handles and objects on tables at human height. The turret also hosts a set of Realsense cameras: one for position tracking, the other used for mapping, obstacle avoidance and user teleoperation. To increase the platform safety, we added an emergency stop button on the back of the turret, which can be locked to prevent any motion of the robots. Anyhow, both the arm and the base have

---

[5] https://clearpathrobotics.com/dingo-indoor-mobile-robot/

[6] https://www.kinovarobotics.com/product/gen3-lite-robots.

their own remote controller for manual control and to take over when autonomous control is not behaving properly.

18.1.2 Recommended Tools and Base Skill Set Required

The reader undergoing the projects of this chapter must have a good understanding of the content of most of the chapters in this book. However, the essentials are the ROS environment (Chap. 5), Python basic programming (Chap. 4), differential drive kinematics (Chap. 8), homogeneous transformation (Chap. 6), Kalman state estimation (Chap. 8), serial manipulator kinematics (Chap. 10), navigation (Chap. 9) and designing a user study (Chap. 13). In the following, the first three projects are well framed in order to help the student's focus on the theoretical content and their learning of the tools, namely ROS and Python. The last two projects are less structured as they are meant to allow the students to explore more advanced topics and integrate the knowledge they acquired in the previous projects and throughout the book. All projects notebooks, such as the one shown in Fig. 18.3, are available online.[7] The following set of instructions take for granted that you have local access to a ROS-configured station and access the robots' onboard computer (also preconfigured) through a local (wireless) network. The ROS workspace repository also contains instructions for the Dingo setup and to launch a simulation on a remote station.[8]

18.2 Project 1: Discovering ROS and the Dingo

18.2.1 Project Objectives

- Get familiar with the ROS environment (basics);
- Get familiar with the Gazebo simulator (visualization only);
- Get familiar with Python programming (from notebooks);
- Control a mobile robot manually;
- Control a robot with simple Python instructions;
- Program a robot's differential drive kinematics.

---

[7] https://github.com/Foundations-of-Robotics/mobile_manip_notebooks.

[8] https://github.com/Foundations-of-Robotics/mobile_manip_ws/tree/master/doc.

Fig. 18.3 Notebook example from Project 1 shown on the GitHub repository

18.2.2   Project Description

This project aims at comparing the performance of a mobile robot (Clearpath Dingo) in simulation and in reality, through the extraction of the resulting trajectories' noise. In order to compare the two, an autonomous open-loop trajectory must be programed and to do so the Dingo's differential drive inverse kinematics is required.

18.2.3   First Task: Manual Control in Simulation

- Open a terminal and use this command to start the simulation:
  roslaunch mobile_manip dingo_arenasim.launch
- Open a second terminal and use the command **rostopic** (**list, info** and **echo**) to find the topics' names for the IMU (sensor_msg/imu), the encoders (nav_msgs/odometry), and the velocity command (geometry_msgs/Twist).
- Use these topics to record a rosbag with this command:
  rosbag record /topic1 /topic2 /topic3
- Open a third terminal and run the teleoperation node (Fig. 18.4):
  rosrun   teleop_twist_keyboard   teleop_twist_keyboard.py cmd_vel:=mobile_manip/cmd_vel

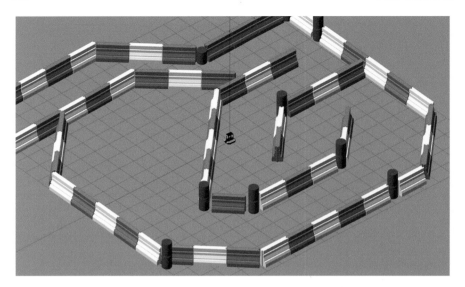

Fig. 18.4 Dingo robot in a simulated maze

Keeping a focus in the last terminal window, you can use the keyboard (u, i, o, j, k, l, m, <) to manually control the robot toward the end of the maze. When you are out of the maze, stop the rosbag (control-C), and then the simulation.

18.2.4 First Task: Manual Control in Reality

- Start the robot by pressing the power button and wait for the front lights to turn white.
- Power on the controller. You can now control the robot with the controller (Fig. 18.5).

18.2.5 Second Task: Inverse Kinematics

The controller and the state estimator need information from the robot's sensors to work properly. The provided Python notebook **Project1–2** for this task contains missing ROS topics names and some geometry information of the robot that you need to fill out.

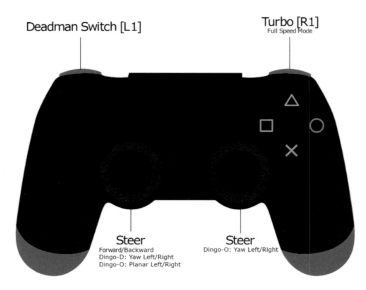

Fig. 18.5 Clearpath remote controller, from the Dingo manual

You can now program your differential drive robot's inverse kinematics. Look for the function "move_robot" in the notebook, it's the one that needs to be modified. This function uses two arguments as inputs:

- **vlin** is the desired linear velocity in m/s in the robot's frame.
- **vang** is the desired angular velocity in rad/s in the robot's frame, positive counterclockwise (Z is up).

Your work consists in using these variables and the robot's geometry to calculate the velocity commands sent to each wheel. The variables **vel_left** and **vel_right** will contain the results of your derivation. The function **move_robot** will then automatically send the commands to the robot whenever its called.

You can test your code by moving the robot in an empty world simulation with:
roslaunch mobile_manip gen3_lite_dingo_emptysim.launch
using the variables **vlin** and **vang** in the last cell of the Python notebook. When you are ready, record a rosbag of the topic **/tf** only (it includes the transforms of each rigid body in the simulation) for the following movements:

- Moving forward in a straight line.
- Moving backward in a straight line.
- Rotate on itself.
- Turn on a circle with a diameter of 1 m.

18.2.6 Third Task: Simulation Versus Reality

Use the **project1–3** notebook and enter the topics' names in the publishers and subscribers where it is needed. Edit the values for the velocity of each wheel to make the robot turn on a circle of 1 m in diameter. This notebook will be used to control the robot both in simulation and in reality. You will need to record a rosbag for each and then compare the results.

To record the **simulation**:

- Open a terminal and run the simulation with
  roslaunch mobile_manip gen3_lite_dingo_emptysim.launch
- With **rostopic** command, find the topics' names for the IMU (sensor_msg/imu), the encoders (nav_msgs/odometry), and the motor command (jackal_msgs/cmd_drive) and edit **project1–3** notebook accordingly.
- The last cell of the notebook records a rosbag with the needed topics while the robot is moving. Make sure to close the rosbag once the circle is done at least once.

To record the **real** robot:

- Turn on the robot using the power button on the back (ensure the emergency stop button is disabled) and wait until the light indicator is on below the Wi-Fi symbol.
- Open a new terminal on your computer and connect to the robot over SSH (where X is your Dingo's number):
  ssh mecbot@cpr-ets05-0X (user mecbot and the password given by your professor)
- Then when you are connected to it, launch all the custom nodes required for the laboratories' task onboard:
  roslaunch mobile_manip gen3_lite_dingo_real.launch

Finally, to compare the results, open the notebook **Analyse1**, enter the name of your rosbags and edit the code to calculate the variance on the circle trajectory. Using at most one page, answer this question:

- What quantifiable difference(s) can you observe between the circle trajectory in simulations and on the real robot? What can you say about the source(s) of these differences?

18.3 Project 2: Kalman for Differential Drive

18.3.1 Project Objectives

- Design a Kalman filter for position estimation (sensor fusion);
- Estimate sensors' noise from real data.

18.3.2   Project Description

This project aims at developing a state estimator for the mobile base pose from the available onboard sensors' measurements. You will use a Kalman filter to fuse the different measurements and get the best out of each sensor. Since such a piece of code requires quite a bit of debugging, use only the simulation until you get a working behavior:

- To launch a simulation, use this launch file:
  roslaunch mobile_manip gen3_lite_dingo_emptysim.launch

18.3.3   First Task: Extract Encoders Information (Notebook Project2_1)

Your first task is to use the encoders' values to estimate the robot's movements (x, y, Vx, Vy, and the heading θ). The function **encoders_callback** will be called each time a new update is received from the wheels' encoders. Modify its content to estimate progressively the robot's position and velocity from the encoders' readings (given in radians from 0 to infinity).

18.3.4   Second Task: Estimate the Sensor's Noise (CSV_Analyse)

To include sensors in a Kalman filter, you first need to know the noise in the measurements. Previous experimental data has been saved in .csv files and are provided to you alongside a template notebook to help with loading and processing the data (**CSV_Analyse**). You need to find the variance of every measurement you will use in your Kalman filter in order to build your measurement covariance matrix.

18.3.5   Third Task: Design a Kalman Filter (Project2_2)

Now that you extracted meaningful information from the encoders and that you estimated the encoders and IMU noise, you need to fuse these measurements into a more robust state estimator. Indeed, since the wheels can slip, the IMU measurements can improve the pose estimation. This fusion will be made with a Kalman Filter. The given Python notebook already imports the library **filterpy.kalman** which is responsible for most of the Kalman filter implementation, but you need to configure the filter (states, covariance matrices, transition matrices, etc.). Follow the instructions in the notebook and refer to the library documentation for more details.[9]

There is not a unique solution to this task. The filter can be configured in different ways, for example using (or not) command inputs (u). If you choose a configuration without commands, consider the system's variance to be $\sigma_S = 10$.

18.3.6　Fourth Task: Design Justification and Validation

The last cell of the **project2_2** notebook sends commands to move the robot. This cell also creates a rosbag recording the trajectory estimated by your Kalman filter and the ground truth trajectory from gazebo. You can then test your Kalman filter performance with the **KF-Analyse** notebook.

In a page, describe and justify the design of your Kalman filter: Why did you choose these states? Why do you think this model (configuration matrices) is a better fit to the problem? What other possibilities were available?

18.4　Project 3: 3-DoF Kinematics

18.4.1　Project Objectives

- Compute direct and inverse kinematics;
- Calculate the Jacobian and the points of singularity;
- Validate the results with the real robot.

18.4.2　Project Description

This project aims at the application of the direct and inverse kinematics to the use case of a simplified manipulator. The robot used for this laboratory is the Kinova Gen3 Lite manipulator (see Fig. 18.6). In order to simplify this project (and avoid the need for a symbolic computation software), the number of axes to be controlled is reduced from 6 to 3.

As shown in Fig. 18.7, axes 1, 2, and 3 are the ones that can be controlled here. Assume all other joints are fixed in their initial position (0°), as depicted in Fig. 18.6. A 90° angle is applied to the third joint on the left-hand side of the figure for the sake of comparison. The parameters of each of the robot segments are given in Fig. 18.7 (in mm).

---

[9] https://filterpy.readthedocs.io/en/latest/kalman/KalmanFilter.html.

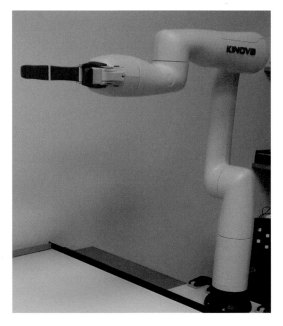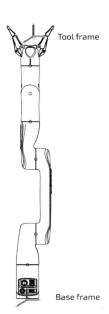

Fig. 18.6 Gen3 lite manipulator: photograph on the left, base, and tool frame locations on the right

18.4.3   First Task: Denavit–Hartenberg Table

You must complete the DH table (again, for the 3-DoF robot) with the parameters given (their numerical values, not only the variables). Add comments if necessary to clarify the meaning of the values used.

18.4.4   Second Task: Transformation Matrices

Compute all sequential homogeneous transformation matrices (Q) from the DH table, then use the resulting matrices to obtain the final concatenated transformation matrix, or direct kinematics.

18.4.5   Third Task: Inverse Kinematics

Derive the equations to compute each joint coordinate explicitly as a function of a Cartesian position (not orientation) to be reached within the workspace with the end-effector. A drawing might help you in your work.

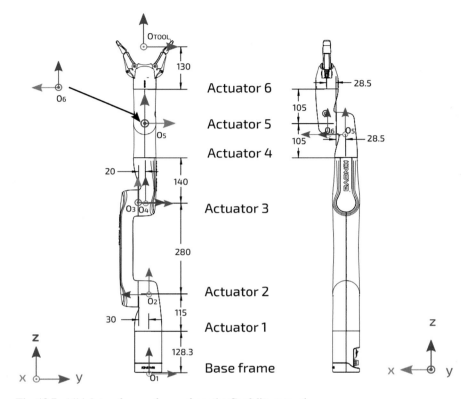

Fig. 18.7 All joints reference frames from the Gen3 lite manual

Suggested procedure:

1. Extract the three components (x, y, z) defining the position from the homogeneous transformation matrix computed in the previous task.
2. Eliminate unknown variables to obtain an equation with only one unknown left. You can do this similarly to the procedure used to solve the inverse position problem of a wrist-partitioned serial manipulator, in Chap. 10.

   (a) Use the equations corresponding to the coordinates x and y, then compute the sum of their squares (to eliminate θ_1) and isolate $\cos(\theta_2)$. Then, after isolating $\sin(\theta_2)$ in the equation of the z-coordinate, use the identity $\sin2(\theta_2) + \cos2(\theta_2) = 1$ to make one unknown joint coordinate disappear, namely θ_2.

   (b) Use the Weierstrass substitution with $(\theta_2 - \theta_3)$ in the equation obtained in the previous step and find the roots of the obtained quadratic equation. The possible values of $(\theta_2 - \theta_3)$ can then be computed.

   (c) Going back to the expressions of $\sin\theta_2$ and $\cos\theta_2$ obtained earlier, you can now compute θ_2 with the arctan2 function.

(d) θ_2 and $(\theta_2 - \theta_3)$ known, it is trivial to obtain θ_3.

(e)   Finally, return to the equations corresponding to the coordinates x and y, then cast them in matrix form $Ax = b$, where $x = [\sin\theta_1 \ \cos\theta_1]^T$ since θ_1 is the only remaining unknown. Then, after solving for x, use arctan2 to find the solution for θ_1.

18.4.6   Fourth Task: Validation

In order to validate your solution, you must now apply the results obtained in the previous step to the robot. To do this, the robot must follow a path passing through the following Cartesian positions:

[mm]	x	y	z
Pose 1	−66.9	−50.2	965.8
Pose 2	223.3	366.8	873.0
Pose 3	10.0	678.0	384.3

Start the robot simulation using the following command:
roslaunch mobile_manip gen3_lite_sim.launch
Use the ***Kinova_3DoF_Joint_Control*** notebook: this file contains a function named ***dof3control(j1, j2, j3)***. This function sends angle commands (in degrees) to the 3 joints of the simulated robot in Gazebo. You need to add your code implementing the solution to the inverse kinematics problem (previous task).

You are encouraged to validate the angle values returned by your inverse kinematics using the ***DirectKinematics*** notebook. Finally, you can confirm that the position reached by the robot is good with the call at the bottom of the ***Kinova_3DoF_Joint_Control*** notebook.

When your code has proven to work properly in Gazebo, you can validate on the real robot by changing the ROS_MASTER_URI in the first cell to the IP of your robot. Then launch the nodes on the real robot with:

- Turn on the robot using the power button on the back (ensure the emergency stop button is disabled) and wait until the light indicator is on below the Wi-Fi symbol.
- When the front lights turn white, turn on the Gen3 Lite (button on the back on the base of the arm)
- Open a new terminal on your computer and connect to the robot over SSH (where X is your Dingo's number):
  ssh mecbot@cpr-ets05-0X (user `mecbot` and password given by your professor)
- Then when you are connected to it, launch all the custom nodes required for the laboratories' task onboard:
  roslaunch mobile_manip gen3_lite_dingo_real.launch

If it launches correctly, you should see the arm reach its home position.

In less than a page, answer the following question: To your knowledge, what criteria could be used to select one solution to the inverse kinematics problem over another? Justify, it is not necessary to have an exhaustive list of criteria.

18.5 Project 4: Let's Bring It Back Together!

18.5.1 Project Objectives

- navigate with a mobile manipulator;
- manipulate the environment;
- identify phenomena potentially dangerous involving mobile manipulators and analyze the risk.

18.5.2 Project Description

This project consists of deploying a mobile manipulator in a real application scenario: the handling of machined parts in a factory. More specifically, your objective is to recover a part that has just been machined (in the corridor) and transfer it to a surface treatment tank (in the laboratory). Two robots are available for this task. The two robots are initially in the laboratory and must therefore open the door themselves to reach the part. All tasks can either be run in simulation or with the real robot using the same Python notebook.

To test any part of your code in a safe environment, do it in simulation. Open a Linux terminal and start the simulation with the command:

roslaunch mobile_manip gen3_lite_dingo_labsim.launch

After you validate your code in the simulation, you can launch it with the real robot:

- Turn on the robot using the power button on the back (ensure the emergency stop button is disabled) and wait until the light indicator is on below the Wi-Fi symbol.
- When the front lights turn white, turn on the Gen3 Lite remote (button on the back on the base of the arm).
- Open a new terminal on your computer and connect to the robot over SSH (where X is your Dingo's number):

  ssh mecbot@cpr-ets05-0X (user `mecbot` and password given by your professor)
- Then when you are connected to it, launch all the custom nodes required for the laboratories' task onboard (Fig. 18.8):

  roslaunch mobile_manip gen3_lite_dingo_real.launch

Fig. 18.8 Mobile
manipulator platform in its
parking space

If it launches correctly, you should see the arm reach its home position. Keep the remote control in your hands at all times: the deadman switch allows you to take back manual control of the robot if your code reacts badly. *Warning! The remote control only interrupts the commands sent to the Dingo not the ones sent to the arm.*

18.5.3   First Task: Teleoperation

Use the notebooks prepared for this task in order to pilot the robot in the room and the corridor in front. You need to test:

- manual piloting with the remote control, following the robot (not available in the simulation);
- remote manual control with the notebook and using visual feedback from the front color camera (**manual-control.ipynb**—remember to test in simulations first!);
- autopilot using the notebook, but immediately regaining control if a collision is imminent. (**autonomous.ipynb**—remember to test in simulations first!).

18.5.4   Second Task: Hit a Marker!

Use the **TagTouch** notebook to detect markers in front of the robot using the Realsense T265 fisheye camera. The coordinates obtained are in the reference frame of the camera, you must transpose them into the frame of reference of the manipulator base. The position of the camera in the reference frame of the base of the manipulators is illustrated in Fig. 18.9:

- in the simulation, $[x, y, z] = [0.16, 0.04, 0.463]$ m;
- on the real robot, $[x, y, z] = [0.0, 0.05, -0.137]$ m.

Remember that the frame of reference of an image (camera) always has the z-axis coming out of the camera. Then use the inverse kinematics (Cartesian control) provided by the Gen3 lite controller to touch the marker on the wall. Test first in the simulation, then in the laboratory. You can move the robot manually using one of the strategies from step 1 to position the robot (camera) in front of the marker.

You cannot use Kinova's web service for this and subsequent steps.

Fig. 18.9 Mobile manipulator model in Gazebo with reference frames for the arm base, the mobile base, and the camera. *X*-axis is shown in red, *Y-axis* in green, and *Z-axis* in blue

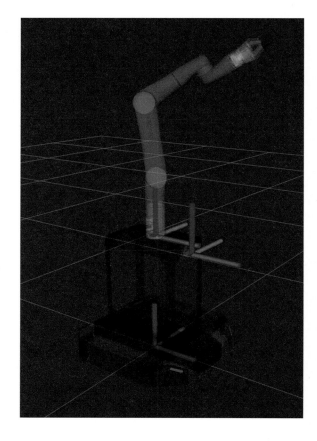

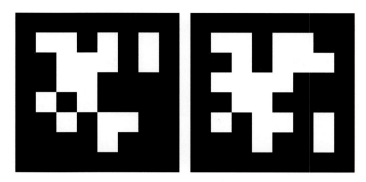

Fig. 18.10 Two Apriltags were used for this task, both in the simulation and in the real deployment

18.5.5 Third Task: Grasping

The door handle and the object to be picked up are both at predetermined poses relative to their fiduciary marker, as shown in Fig. 18.10. To the transformation from the previous step, you must now add the transformation required to change frame:

- of the handle (given in the referential of the marker), $[x, y, z] = [0.05, 0.3, 0.1]$ m;
- of the object (given in the referential of the marker), $[x, y, z] = [0.05, -0.45, 0.2]$ m.

When you reach your goal, you must then tighten the gripper on it. Write down the movements required to orient the gripper and apply the required force. You can then use the Dingo and Gen3 lite together to pull the door and bring the object back. **Note:** To send an opening or closing command to the gripper, you must use the ROS service for this purpose (look for *reach_gripper* in the notebooks).

18.5.6 Fourth Task: Risk Assessment

Based on what your learnings from Chap. 14, you have to take care of the risk management for the mobile manipulator in this scenario. You are the integrator, you must therefore deliver an initial analysis of the risks as well as the relevant means of risk mitigation. Keep in mind that the initial analysis is done without considering the existing protection mechanisms on the robotic system. Then, propose risk mitigation measures and clearly state which ones are already included in the robotic system and those which should be added (to the system, to the environment or by the management).

To help you in your assignment, the operator of the mobile manipulator system has shared the following information with you:

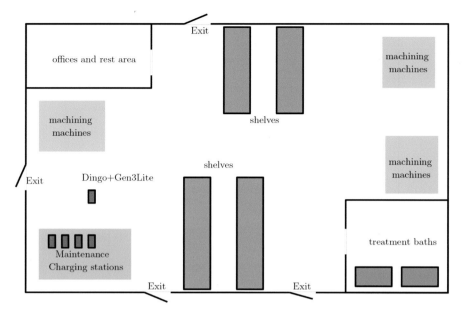

Fig. 18.11 Fictitious plan of a factory cell with mobile manipulators

- the robotic system to be deployed is a Clearpath Dingo equipped with a Kinova Gen3 lite arm;
- the task of the industrial robotic system (Dingo and Gen3 lite) is to pick the parts from various machining machines to place them in surface treatment baths (*machine tending*);
- the machines manufacturing the parts are spread over the entire factory floor, with workers circulating in between regularly;
- treatment baths are in a closed room to meet ventilation standards for the chemicals used;
- production runs 24/7.

  To help you with this task, a plan layout is illustrated in Fig. 18.11.

18.6 Project 5: Save the Day!

18.6.1 Project Objectives

- Program autonomous navigation for the Dingo;
- Program obstacle avoidance;
- Design a remote command interface;
- Plan, conduct, and analyze the ergonomy of a command interface.

18.6.2 Project Description

This project consists in studying the ease of use of a teleoperation system you will be designing. You will first need to design a controller and a user interface before conducting a user study. The mission of the participants will be to explore the floor of a building to find some objects. The building is evacuated, only the mobile manipulator is left inside (no human to avoid).

To test any part of your code in a safe environment, do it in simulation. Open a Linux terminal and start the simulation with the command:

roslaunch mobile_manip gen3_lite_dingo_labsim.launch

After you validate your code in the simulation, you can launch it with the real robot:

- Turn on the robot using the power button on the back (ensure the emergency stop button is disabled) and wait until the light indicator is on below the Wi-Fi symbol.
- When the front lights turn white, turn on the Gen3 Lite remote (button on the back on the base of the arm).
- Open a new terminal on your computer and connect to the robot over SSH (where X is your Dingo's number):
  ssh mecbot@cpr-ets05-0X (user `mecbot` and password given by your professor)
- Then when you are connected to it, launch all the custom nodes required for the laboratories' task onboard:
  roslaunch mobile_manip gen3_lite_dingo_real.launch

If it launches correctly, you should see the arm reach its home position. Keep the remote control in your hands at all times: the deadman switch allows you to take back manual control of the robot if your code reacts badly. *Warning! The remote control only interrupts the commands sent to the Dingo not the ones sent to the arm.*

18.6.3 First Task: Autonomous Navigation

Teleoperation requires a good deal of autonomy from the robotic system so it can deal with complex maneuvers and leave the operator to attend more sensitive tasks. You need to design your navigation solution with various levels of autonomy for 1. Path planning, 2. Collision avoidance, and 3. Objects (Apriltags) detection. A set of notebooks is provided with A* and RRT path planners, including an occupancy grid of the environment as well as visualization tools for laser scans (extracted from the depth camera) and video feeds (including tags detection in the camera reference frame). Many solutions are possible for each aspect of the navigation. For the Apriltags detection, remember that the coordinates obtained are in the reference frame of the camera, you must transpose them into the frame of reference of the arm base for manipulation. The position of the camera in the reference frame of the base of the manipulator is:

- in the simulation, $[x, y, z] = [0.16, 0.04, 0.463]$ m;
- on the real robot, $[x, y, z] = [0.0, 0.05, -0.137]$ m.

18.6.4 Second Task: User Interface

Based on the navigation solution you designed in the previous step, code a tele-operation interface for an operator to complete the mission remotely (i.e., without following the real robot and looking at the simulator window). This interface must include visualization of relevant sensor information and input modalities to send commands to the robot. A minimal interface for the manual control is given as an example (see Fig. 18.12).

This task is a creative step where you should try to imagine what would help the user the most. You can then do some research and find what is possible to do in a Python notebook to make the integration in your interface.

Fig. 18.12 Jupyter notebook minimal interface provided for the teleoperation of a mobile manipulator (camera view from simulation)

18.6.5 Third Task: Ergonomy Study

You can now design a user study (see Chap. 13) to assess the potential of your teleoperation solution, namely the impact of your solution on the operator's performance and his appreciation. You can either do an explorative study or a confirmative (comparative) one. To conduct a comparative study, you need an interface that the users can test in two different conditions (i.e., manual and assisted). In both cases, you need to define and justify the selected statistical tools (see Chap. 6). At the end of the term, you will conduct your study on some students of the course group.

You need to complete a protocol for your user study including the questionnaires and the metrics you will analyze to answer your research question. For instance, a common questionnaire used to measure the cognitive task load is the NASATLX:

NASATLX questionnaire example

Item	Endpoints	Description
Mental demand	1–10 Low/High	How much mental and perceptual activity was required (e.g., thinking, deciding, calculating, remembering, looking, searching, etc.)? Was the task easy or demanding, simple or complex, exacting or forgiving?
Physical demand	1–10 Low/High	How much physical activity was required (e.g., pushing, pulling, turning, controlling, activating, etc.)? Was the task easy or demanding, slow or brisk, slack or strenuous, restful or laborious?
Temporal demand	1–10 Low/High	How much time pressure did you feel due to the rate or pace at which the tasks occurred? Was the pace slow and leisurely or rapid and frantic?
Performance	1–10 Low/High	How successful do you think you were in accomplishing the goals of the task set by the experimenter (or yourself)? How satisfied were you with your performance in accomplishing these goals?
Effort	1–10 Low/High	How hard did you have to work (mentally and physically) to accomplish your level of performance?
Frustration level	1–10 Low/High	How insecure, discouraged, irritated, stressed and annoyed versus secure, gratified, content, relaxed and complacent did you feel during the task?

Your ergonomy study report is expected to follow this structure:

1. INTRODUCTION—Introduce the content of this report, the main characteristics of your navigation and user interface and the results of the study.
2. USER STUDY DESIGN—Describe the elements of the study with the research question, causal or correlated effect sought, impact of the robotic system type, place of the study and the selection of the candidates. You must also list and justify the measures used.

3. STUDY PROTOCOL—Describe how you will proceed in the study sessions.
4. DATA ANALYSIS—Use statistical tools to demonstrate the results distribution.
5. DISCUSSION—Discuss the most important observations you made in the data analysis section. Mention the limitations of the study and add recommendations.

David St-Onge (Ph.D., Mech. Eng.) is an Associate Professor in the Mechanical Engineering Department at the École de technologie supérieure and director of the INIT Robots Lab (initrobots.ca). David's research focuses on human-swarm collaboration more specifically with respect to operators' cognitive load and motion-based interactions. He has over 10 years' experience in the field of interactive media (structure, automatization and sensing) as workshop production director and as R&D engineer. He is an active member of national clusters centered on human-robot interaction (REPARTI) and art-science collaborations (Hexagram). He participates in national training programs for highly qualified personnel for drone services (UTILI), as well as for the deployment of industrial cobots (CoRoM). He led the team effort to present the first large-scale symbiotic integration of robotic art at the IEEE International Conference on Robotics and Automation (ICRA 2019).

Corentin Boucher is a research student at the École de Technologie Supérieure (ÉTS). The interest in robotics that he developed during his studies pushed him to continue his journey and to carry out research in the field.

Bruno Belzile is a postdoctoral fellow at the INIT Robots Lab. of ÉTS Montréal in Canada. He holds a B.Eng. degree and Ph.D. in mechanical engineering from Polytechnique Montréal. His thesis focused on underactuated robotic grippers and proprioceptive tactile sensing. He then worked at the Center for Intelligent Machines at McGill University, where his main areas of research were kinematics, dynamics, and control of parallel robots. At ÉTS Montréal, he aims at creating spherical mobile robots for planetary exploration, from the conceptual design to the prototype.

Index

A
A* algorithm, 230
Abstraction, 92
Accuracy, 181
Ackerman drive, 213
Action, 114
Activation function, 450
Active, 180
Aerostabiles, 27
Algorithms, 189
Alternative hypothesis, 157
Ament, 109
ANSI, 416, 417
ANSI/ITSDF B56.5, 420
ANSI/RIA R15.06, 424
ANSI/RIA R15.08, 420, 427, 431
Anticipation, Inclusion, Reflexivity, and
        Responsiveness (AIRR), 486
A quadruped gait, 218
Architecture, 267
Archival research, 387
Arm type robots, 191
Association for Advancing Automation,
        417
Assuming correlation equals causation, 407
Auction, 305
Augmented Reality (AR), 351
Automata, 7
Automated guided vehicle, 416, 418, 431
Automatic methods to swarm design, 310
Automatophones, 17
Autonomous boat, 20
Autonomous mobile robot, 416, 431

B
Ballet Robotique, 28

Bandwidth, 181
Basic parametric elements, 341
Battery, 432
Bayes' theorem, 146, 249
Beacons, 244
Behaviourism, 53
Bellman equation, 463
Benefits, 399
Berger, 107
Between-subjects design, 389
Bicycle, 211
Bill Vorn, 27
Blinding, 395
Body of knowledge, 44
Boursier-Mougenot, 30
Bug algorithm, 234
Buzz, 315

C
CAD assembly files, 340
CAD drawing files, 344
CAD file transfer, 351
CAD file types, 338
CAD Modelling Guidelines for Assembly
        Files, 339
CAD model validation, 343
CAD systems, 336
Callback, 91
Capek, 5
Case studies, 387
Catkin, 109
Centralized muti-robot systems, 304
Central limit theorem, 147
Circular references, 339
Class, 92
Classifier, 442

Printed in the United States
by Baker & Taylor Publisher Services